T0140107

Advances in Intelligent Systems and Computing

Volume 981

The series "Advances in Intelligent Systems and Computing" contains publications on theory, applications, and design methods of Intelligent Systems and Intelligent Computing. Virtually all disciplines such as engineering, natural sciences, computer and information science, ICT, economics, business, e-commerce, environment, healthcare, life science are covered. The list of topics spans all the areas of modern intelligent systems and computing such as: computational intelligence, soft computing including neural networks, fuzzy systems, evolutionary computing and the fusion of these paradigms, social intelligence, ambient intelligence, computational neuroscience, artificial life, virtual worlds and society, cognitive science and systems, Perception and Vision, DNA and immune based systems, self-organizing and adaptive systems, e-Learning and teaching, human-centered and human-centric computing, recommender systems, intelligent control, robotics and mechatronics including human-machine teaming, knowledge-based paradigms, learning paradigms, machine ethics, intelligent data analysis, knowledge management, intelligent agents, intelligent decision making and support, intelligent network security, trust management, interactive entertainment, Web intelligence and multimedia.

The publications within "Advances in Intelligent Systems and Computing" are primarily proceedings of important conferences, symposia and congresses. They cover significant recent developments in the field, both of a foundational and applicable character. An important characteristic feature of the series is the short publication time and world-wide distribution. This permits a rapid and broad dissemination of research results.

** Indexing: The books of this series are submitted to ISI Proceedings, EI-Compendex, DBLP, SCOPUS, Google Scholar and Springerlink **

More information about this series at http://www.springer.com/series/11156

Radomír Halaš · Marek Gagolewski ·
Radko Mesiar

Editors

New Trends in Aggregation Theory

 Springer

Editors
Radomír Halaš
Faculty of Science
Palacký University
Olomouc, Czech Republic

Radko Mesiar
Slovak University of Technology
Bratislava, Slovakia

Marek Gagolewski
Systems Research Institute
Polish Academy of Science
Warsaw, Poland

Faculty of Mathematics and Information Science
Warsaw University of Technology
Warsaw, Poland

ISSN 2194-5357   ISSN 2194-5365 (electronic)
Advances in Intelligent Systems and Computing
ISBN 978-3-030-19493-2   ISBN 978-3-030-19494-9 (eBook)
https://doi.org/10.1007/978-3-030-19494-9

This Springer imprint is published by the registered company Springer Nature Switzerland AG
The registered company address is: Gewerbestrasse 11, 6330 Cham, Switzerland

Preface

Aggregation functions have numerous applications, including, but not limited to, data fusion, statistics, image processing, and decision making. They are usually defined as those functions that are monotone with respect to each input and that satisfy some natural boundary conditions. In particular settings, these conditions might be relaxed or otherwise customized according to a user's needs. Noteworthy classes of aggregation functions include means, t-norms and t-conorms, uninorms and nullnorms, copulas and fuzzy integrals (e.g., the Choquet and Sugeno integrals).

Besides the aggregation of real inputs, the aggregation functions on general/particular lattices or other spaces are also considered. This volume collects the finally revised manuscripts of 26 accepted contributions submitted to the International Summer School on Aggregation Functions—AGOP—that took place in Olomouc, Czech Republic, on July 1–4, 2019. The AGOP conferences are organized biannually by the AGOP working group of EUSFLAT—European Society for Fuzzy Logic and Technology. It is the tenth in a series of summer schools, including AGOP 2001 (Oviedo, Spain), AGOP 2003 (Alcalá de Henares, Spain), AGOP 2005 (Lugano, Switzerland), AGOP 2007 (Gent, Belgium), AGOP 2009 (Palma de Mallorca, Spain), AGOP 2011 (Benevento, Italy), AGOP 2013 (Pamplona, Spain), AGOP 2015 (Katowice, Poland), and AGOP 2017 (Skövde, Sweden). The volume also includes five papers related to the invited talks given during the course of the school. All the included contributions were reviewed by the Program Committee members and several external reviewers, and they include works ranging from the theory and foundations of the aggregation functions to their rich applications. Together, they provide a good overview of recent trends in research on the aggregation functions.

March 2019

Radomír Halaš
Marek Gagolewski
Radko Mesiar

Contents

Invited Speakers

Qualitative Integrals and Cointegrals: A Survey

Agnès Rico[✉]

ERIC & Univ. Claude Bernard Lyon 1, 43 bld du 11 novembre,
69100 Villeurbanne, France
agnes.rico@univ-lyon1.fr

Abstract. Qualitatives integrals and cointegrals are aggregation functions defined on qualitative scales where only minimum, maximum and order reversing map are allowed. Their definition uses a conjunction or an implication linked by property of semi-duality. They generalise Sugeno integrals which are defined using the Kleene-Dienes implication or conjunction. This survey remembers the definitions, the properties and some characterisation theorems for these qualitative (co)integrals.

1 Introduction

In the setting of Artificial Intelligence (recommender systems, cognitive robotics, ...) the use of numerical aggregation functions is not always natural. For instance by lack of time (we can't ask too many questions to users) or lack of precision the collected data are imprecise. In such cases it seems more natural to use qualitative aggregation functions instead of quantitative ones.

This survey deals with qualitative integrals defined on totally ordered bounded chain. In such a context aggregation operations classicaly use only minimum, maximum and an order-reversering map. The main qualitative aggregation functions are the Sugeno integrals [5,20]. They are used in decision under uncertainty or in multiple criteria decision making [21,22]. Their definition is based on capacities or fuzzy measures representing the likelihood of events or the importance of sets of criteria. More precisely the weight of the subsets of criteria acts on the local ratings and then they are aggregated in a global evaluation. For Sugeno integrals this action of the weight on the ratings is calculate using the Kleene-Dienes implication or conjunction. In this case the capacity is used as a bound that restricts the global evaluation from below and above.

The qualitative integrals and cointegrals are defined using other conjunctions and implications [1,11]. We assume that they are linked by semi duality. For example we will use Gödel and contrapositive Gödel conjunctions and implications. In these cases, capacities are used as thresholds on utility values: either the local rating remains as it stands or it is modified. Technically speaking we can study this approach in a more general context where the conjunctions and the implications are not necessarily commutative.

R. Halaš et al. (Eds.): AGOP 2019, AISC 981, pp. 3–14, 2019.
https://doi.org/10.1007/978-3-030-19494-9_1

The survey is structured as follows. Section 2 presents the weighted qualitative aggregation functions and it presents the different ways the weights can act on the local ratings. Section 3 deals with the definitions of the q-integrals and q-cointegrals. Section 4 recalls their properties and presents the cases where both q-integrals and q-cointegrals are equal. Section 5 presents the characterization theorems using comonotonic functions. Before concluding Sect. 6 proposes to see qualitative integrals as upper or lower possibility qualitative (co)integrals.

2 Motivations

We adopt the terminology and notations used in multi-criteria decision making, where some alternatives are evaluated according to a set $C = \{1, \dots, n\} = [n]$ of criteria. A common evaluation scale L is assumed to provide ratings according to the criteria: each alternative is thus identified with a function $f \in L^C$ which maps every criterion i of C to the local rating f_i of the alternative with regard to this criterion.

We assume that L is a finite totally ordered set with 1 and 0 as top and bottom, respectively (L may be a subset of the real unit interval $[0, 1]$ for instance). For any $a \in L$, we denote by $\mathbf{a}_C$ the constant alternative whose ratings equal a for all criteria. L is equipped with an involutive negation denoted by $1-$. Such a scale $(L, \geq, 1-, 0, 1)$ is called *qualitative* in this paper. The minimum operator is denoted by $\wedge$ and the maximum operator is denoted by $\vee$.

In such a context, there are two elementary qualitative aggregation schemes:

- The pessimistic one $\bigwedge_{i=1}^{n} f_i$ which is very demanding since in order to obtain a good evaluation an object needs to satisfy all the criteria.
- The optimistic one, $\bigvee_{i=1}^{n} f_i$ which is very loose since one fulfilled criterion is enough to obtain a good evaluation.

These two aggregation schemes can be slightly generalised by means of importance levels or priorities $\pi_i \in L$, on the criteria $i \in [n]$, thus yielding weighted minimum and maximum [3]:

$$MIN_\pi(f) = \bigwedge_{i=1}^{n} \left((1 - \pi_i) \vee f_i\right); \quad MAX_\pi(f) = \bigvee_{i=1}^{n} (\pi_i \wedge f_i). \tag{1}$$

MIN_π (resp. MAX_π) is a conjunctive (resp. disjunctive) aggregation in which the action of the criterion weight on the local rating is given by $(1 - \pi_i) \vee f_i$ (resp. $\pi_i \wedge f_i$). In this case low (resp. high) ratings of unimportant criteria are upgraded (resp. downgraded) and a fully important criterion ($\pi_i = 1$) can alone bring the global score to 0 (resp. 1).

In the ordinal setting, the construction of the qualitative integrals is based on the exploration of various alternatives for the action of the weight on the local rating. This action depends on the scheme of aggregation used: conjunctive scheme or disjunctive scheme.

Conjunctive Aggregation Scheme

In a conjunctive aggregation an alternative should not be rejected due to a poor rating on an unimportant criterion, especially the global rating should not be affected by useless criteria, whatever the corresponding rating. So the local rating on a useless criterion should always be turned into 1, since $a \wedge 1 = a$. Moreover, if the criterion has maximal importance, a very poor local rating on this criterion should be enough to bring the global evaluation down to 0. Besides it is natural that the better the local rating f_i, the greater the modified local rating; and, the less important the criterion, the more tolerant should be the modified local rating. Hence, the rating modification operator should be an implication [2]:

Definition 1. *A* fuzzy implication *is an operation* $\rightarrow: L \times L \rightarrow L$ *which is decreasing in first argument, increasing in second argument such that* $0 \rightarrow 1 = 1$, $1 \rightarrow 0 = 0$, $1 \rightarrow 1 = 1$ *and* $0 \rightarrow 0 = 1$.

We are going to calculate the implication between π_i and f_i. As we use a finite scale with an involutive negation, we shall consider the special case when the set of possible values for $\pi_i \rightarrow f_i$ is restricted to $\{0, 1 - \pi_i, 1 - f_i, \pi_i, f_i, 1\}$ according to the relative position of f_i with respect to π_i or $1 - \pi_i$. Some well-known implications used in [4] or [1] are retrieved (see [11] for more details):

– Rescher-Gaines implication $a \rightarrow_{RG} b = \begin{cases} 1 \text{ if } a \leq b, \\ 0 \text{ otherwise;} \end{cases}$

– Gödel implication $a \rightarrow_G b = \begin{cases} 1 \text{ if } a \leq b, \\ b \text{ otherwise;} \end{cases}$

– The contrapositive symmetric of Gödel implication:

$a \rightarrow_{GC} b = \begin{cases} 1 \text{ if } a \leq b, \\ 1 - a \text{ otherwise;} \end{cases}$

– Kleene-Dienes implication: $a \rightarrow_{KD} b = (1 - a) \vee b$.

– the Łukasiewicz implication $a \rightarrow_L b = (1 - a + b) \vee 0$.

– The residuum of the nilpotent minimum: $a \Rightarrow b = \begin{cases} 1 \text{ if } a \geq b \\ (1 - a) \vee b \text{ otherwise.} \end{cases}$

Disjunctive Aggregation Scheme

In disjunctive aggregation an alternative should be rejected only due to a poor rating on all important criteria. Similarly to the previous case, the global rating should not be affected by useless criteria, whatever the corresponding rating. So the local rating on a useless criterion should always be turned into 0, since $a \vee 0 = a$. Moreover, if the criterion has maximal importance, a very good local rating on this criterion should be enough to bring the global evaluation up to 1. It is natural that the better the local rating f_i, the better the modified local rating; and a given rating on an important criterion should have a more positive influence on the global evaluation than the same rating on a little important criterion. Hence, the rating modification operation should be a conjunction [2]:

Definition 2. *A fuzzy conjunction is an operation* $\otimes : L \times L \to L$ *increasing in both arguments such that* $0 \otimes 1 = 0$, $1 \otimes 0 = 0$, $1 \otimes 1 = 1$ *and* $0 \otimes 0 = 1$.

In order to generate interesting conjunctions $\pi_i \otimes f_i$, we extend the De Morgan relationship existing between conjunctive and disjunctive aggregations, namely $\vee_{i=1}^n f_i = 1 - \wedge_{i=1}^n (1 - f_i)$. One obtains $\pi_i \otimes f_i = 1 - \pi_i \to (1 - f_i)$. This semi duality relation is denoted by $\mathcal{S}(\otimes) = \to$ which is equivalent to $\mathcal{S}(\to) = \otimes$. Hence, the aggregation schemes are generalised as follows:

$$MIN_\pi^\to(f) = \bigwedge_{i=1}^n \pi_i \to f_i; \quad MAX_\pi^\otimes(f) = \bigvee_{i=1}^n \pi_i \otimes f_i. \tag{2}$$

Another way to define an implication with a conjunction is by mean of the residuation operator, $Res(\cdot)$, defined as follows: $b Res(\otimes) c = \vee \{a : a \otimes b \leq c\}$. In [26] it was proved that the generation process of conjunctions modeled by the minimum $\wedge$ is closed as represented on Fig. 1:

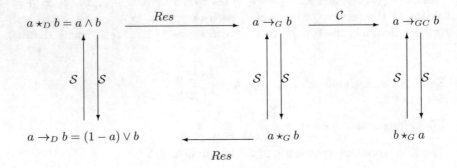

Fig. 1. Conjunctions and implications on a finite chain

This diagram can be generalised as follows where $\star$ is a conjunction and $\Rightarrow$ is an implication (Fig. 2):

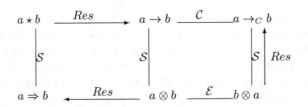

Fig. 2. Residuation and semiduality on a finite chain

Fodor [2] has proved that this diagram commutes if $\star$ is commutative.

3   Qualitative Integrals and Cointegrals

In this section we are going to generalise the aggregation schemes (2) assigning relative weights to subsets of criteria via a capacity.

A capacity is an order-preserving map $\gamma \colon 2^{\mathcal{C}} \to L$ that satisfies $\gamma(\varnothing) = 0$ and $\gamma(\mathcal{C}) = 1$. Particular cases of capacities are: the necessity measures satisfying $N(A \cap B) = N(A) \wedge N(B)$ and the possibility measures satisfying $\Pi(A \cup B) = \Pi(A) \vee \Pi(B)$ for all A, B subsets of criteria. For each criterion i let us denote by π_i the value $\Pi(\{i\})$. $\{\pi_i\}_{i=0,\dots,n}$ is the possibility distribution of Π. We have $\Pi(A) = \vee_{i \in A} \pi_i$ and $N(A) = \wedge_{i \notin A}(1 - \pi_i)$ for all subsets of criteria A.

The Sugeno integral [5] of an alternative f can be defined by means of several equivalent expressions:

$$\int_{\gamma} f = \bigvee_{A \subseteq \mathcal{C}} \left(\gamma(A) \wedge \bigwedge_{i \in A} f_i \right) = \bigwedge_{A \subseteq \mathcal{C}} \left(\gamma(\overline{A}) \vee \bigvee_{i \in A} f_i \right), \tag{3}$$

where $\overline{A}$ is the complement of A. These expressions can be simplified as follows:

$$\int_{\gamma} f = \bigvee_{a \in L} \gamma(\{i : f_i \geq a\}) \wedge a = \bigwedge_{a \in L} \gamma(\{i : f_i > a\}) \vee a. \tag{4}$$

Note that for the necessity measure N associated with a possibility distribution π, we have $\int_N f = MIN_{\pi}(f)$; and for the possibility measure Π associated with π, we have $\int_{\Pi} f = MAX_{\pi}(f)$.

The conjugate capacity γ^c of γ is defined by $\gamma^c(A) = 1 - \gamma(\overline{A})$ for every $A \subseteq \mathcal{C}$. This duality relation extends to Sugeno integral with respect to the conjugate capacity [6]:

$$\int_{\gamma} f = 1 - \int_{\gamma^c} (1 - f). \tag{5}$$

The equality (3) can be expressed using the conjugate capacity, the Kleene-Dienes implication $\to_{KD}$ and its semi dual conjunction $\otimes_{KD}$ which is $\wedge$:

$$\int_{\gamma} f = \bigvee_{A \subseteq \mathcal{C}} \left(\gamma(A) \otimes_{KD} \bigwedge_{i \in A} f_i \right) = \bigwedge_{A \subseteq \mathcal{C}} \left(\gamma^c(A) \to_{KD} \bigvee_{i \in A} f_i \right). \tag{6}$$

Sugeno integral can be generalised using other conjunctions or implications.

Definition 3. *Let $\gamma \colon 2^{\mathcal{C}} \to L$ be a capacity.*
A q-integral is defined by

$$\int_{\gamma}^{\otimes} f = \bigvee_{A \subseteq \mathcal{C}} \left(\gamma(A) \otimes \bigwedge_{i \in A} f_i \right), \text{ for all } f \in L^{\mathcal{C}}.$$

A q-cointegral is defined by

$$\int_{\gamma}^{\to} f = \bigwedge_{A \subseteq \mathcal{C}} \left(\gamma^c(A) \to \bigvee_{i \in A} f_i \right), \text{ for all } f \in L^{\mathcal{C}}.$$

When $\otimes$ is the product and $L = [0,1]$ this is Shilkret integral presented in [10] in 1971 (and later Kaufmann [9] in 1979 under the name of "admissibility"). Borzová-Molnárová et al. [7] study this type of integrals in the continuous case as well when $\otimes$ is a semi-copula (1 is the identity on both sides). This kind of definition is also proposed by Dvořák and Holčapek [8] assuming $(L, \otimes, 1)$ is a commutative monoid and considering the complete residuated lattice generated by this monoidal operation. The q-integral is also the smallest universal integral proposed in [18] if $\otimes$ is a pseudo-multiplication.

The q-integrals and the q-cointegrals act differently on the local ratings according to the conjunction and implication used [1]:

Saturation levels. The Kleene-Dienes implication and conjunction reduce the scale od the evaluations all the more as criteria are less important.

Softening threshold. With Gödel implication and conjunction, the weight makes local evaluation all the less demanding as the criterion is less important.

Drastic threshold. With contrapositive Gödel implication and conjunction, if the criterion is violated, the rating is all the more severely decreased if the criterion is important.

4 Elementary Properties

According to the equations (6), Sugeno integrals coincide but, as already seen using Gödel implication in [4], this is not generally the case. So we present the properties of both q-integrals and q-cointegrals separately.

4.1 Comparison Between q-Cointegrals and q-Integrals

This section presents more details concerning the equality between q-integrals and q-cointegrals.

The inequality: $\int_\gamma^\otimes f \geq \int_\gamma^\rightarrow f$ holds for $(\otimes, \rightarrow) \in \{(\otimes_G, \rightarrow_G), (\otimes_{GC}, \rightarrow_{GC})\}$ [4], but it cannot even be generalised to other conjunctions. For example, consider the nilpotent minimum $\overline{\wedge}$ and its De Morgan dual the nilpotent maximum

$$a\underline{\vee}b = \begin{cases} a \vee b \text{ if } a \leq 1 - b \\ 1 \text{ otherwise.} \end{cases}$$

The relations $a\overline{\wedge}b \leq a \wedge b$ and $a \vee b \leq a\underline{\vee}b$ imply the opposite inequality $\int_\gamma^{\overline{\wedge}} f \leq \int_\gamma^{\overrightarrow{}} f$, where the implication $\overrightarrow{\Rightarrow} = \mathcal{S}(\overline{\wedge})$.

Example 1. We consider the case of a profile $f = aAb$ such that $f_i = a > b$ if $i \in A$ and $f_i = b$ otherwise. The q-integral and q-cointegral of aAb reduce to

$$\int_\gamma^\otimes (aAb) = (1 \otimes b) \vee (\gamma(A) \otimes a); \qquad \int_\gamma^\rightarrow (aAb) = (\gamma^c(\overline{A}) \rightarrow b) \wedge (1 \rightarrow a).$$

Suppose for simplicity that $\otimes$ is a right conjunction and $\rightarrow$ is its semi-dual. Then these expressions simplify as follows:

$$\int_\gamma^\otimes (aAb) = b \vee (\gamma(A) \otimes a); \qquad \int_\gamma^\rightarrow (aAb) = a \wedge (\gamma(A) \oplus b)$$

where $\oplus$ is the De Morgan dual of $\otimes$.

These two expressions suggest the following remarks

- Since when $\otimes = \wedge$ the q-integral and q-cointegrals coincide, we notice that if $\otimes$ is a triangular norm, or a copula, then $\int_\gamma^{\to}(aAb) > \int_\gamma^{\otimes}(aAb)$ in general.
- However, if $\otimes$ is greater than the minimum (which is the case with the semidual of the contrapositive symmetric of Gödel implication), then $\int_\gamma^{\to}(aAb) < \int_\gamma^{\otimes}(aAb)$ in general.

In [19] the pair (conjunction $\otimes$, implication $\to$) such that the associated q-integrals and q-cointegrals coincide are characterised. Let us recall these results. $\mathcal{G}$ denotes the set of all nondecreasing functions $\varphi : L \to L$ such that $\varphi(0) = 0$ and $\varphi(1) = 1$. We have the following result:

Theorem 1. *If the q-integral and q-cointegral coincide for some $\otimes$, $\to$ then there exist $\varphi, \psi \in \mathcal{G}$ such that $a \to b = \varphi(1 - a) \vee \psi(b)$, $a \otimes b = \varphi(a) \wedge \psi(b)$ and $(\varphi(a), \psi(b)) = ((1 - a) \to 0, 1 \otimes b))$ for all $a, b \in L$.*

Moreover if $\varphi, \psi \in \mathcal{G}$, $a \to b = \varphi(1 - a) \vee \psi(b)$ and $a \otimes b = \varphi(a) \wedge \psi(b)$, for all $a, b \in L$, then the q-integral and the q-cointegral coincide.

In [19] the following example in which the q-integrals and q-cointegrals coincide with other conjunction and implication is presented.

Example 2. We suppose that $\to$ and $\otimes$ are linked by semi duality. We can consider $\varphi(a) = \begin{cases} \rho(a) & \text{if } a \in [0, 1/2] \\ 1 - \rho(1 - a) & \text{if } a \in (1/2, 1/2] \end{cases}$ where $\rho : [0, 1/2] \to [0, 1/2]$ is a nondecreasing function such that $\rho(0) = 0$ and $\rho(1/2) = 1/2$. ψ has the same expression than φ where ρ is now a nondecreasing function such that $\rho(0) = 0$ and $\rho(1/2) = 1/2$.

Note that one obtains the Kleene conjunction and implication where for both applications g and h we consider that f is the identity.

4.2   Elementary Properties

First note that if $\otimes$ and $\to$ are linked by semi-duality, then $\int_\gamma^{\to} f = 1 - \int_{\gamma^c}^{\otimes}(1 - f)$. The q-integrals and q-cointegrals are monotonic (Borzová-Molnárová et al. [7] also prove this result when $\otimes$ is a semi-copula) and have equivalent expressions:

$$\int_\gamma^{\otimes} f = \bigvee_{i=1}^{n} \gamma(F_{(i)}) \otimes f_{(i)} = \bigvee_{a \in L} \gamma(\{f \geq a\}) \otimes a \qquad (7)$$

$$\int_\gamma^{\to} f = \bigwedge_{i=1}^{n} \gamma^c(\overline{F_{(i+1)}}) \to f_{(i)} = \bigwedge_{a \in L} \gamma^c(\{f \leq a\}) \to a \qquad (8)$$

where $(\cdot)$ denotes a permutation on the set of criteria such that $f_{(1)} \leq \cdots \leq f_{(n)}$ and we let $F_{(i)} = \{(i), \cdots, (n)\}$ with the convention $F_{(n+1)} = \varnothing$.

If γ is a possibility measure Π, we retrieve the $\otimes$-weighted maximum using the integral (7): $\int_{\Pi}^{\otimes} f = MAX_{\pi}^{\otimes}(f)$. However, if γ is a necessity measure N, $\int_{N}^{\otimes} f \neq \wedge_{i=1}^{n}((1-\pi_i) \otimes f_i)$ except when $\otimes = \wedge$.

Similarly, if γ is a necessity measure N based on possibility measure π, the q-cointegral (8) reduces to the $\rightarrow$-weighted minimum: $\int_{N}^{\rightarrow} f = MIN_{\pi}^{\rightarrow}(f)$. However the q-cointegral with respect to a possibility measure $\int_{\Pi}^{\rightarrow} f$ does not reduce to a weighted maximum.

4.3   Inequalities for q-Integrals

This section presents the Chebyshev inequality for the q-integrals and q-cointegrals presented in [17].

Theorem 2. *Let $\otimes$ be a conjunction and $\star : L^2 \rightarrow L$ be nondecreasing and $1 \star 0 = 0 \star 1 = 0$. Then the following statements are equivalent:*

1. *For any $a, b, c \in L$ $a \otimes (b \star c) \geq ((a \otimes b) \star (1 \otimes c)) \vee ((1 \otimes b) \star (a \otimes c))$*
2. $\int_{\gamma}^{\otimes} f \star g \geq \int_{\gamma}^{\otimes} f \star \int_{\gamma}^{\otimes} g.$

Theorem 3. *Let $\rightarrow$ be an implication with $1 \rightarrow x \geq x$ and $\star : L^2 \rightarrow L$ be nondecreasing and $1 \star 0 = 0 \star 1 = 0$. Then the following statements are equivalent:*

1. *For any $a, b, c \in L$ $a \rightarrow (b \star c) \leq ((a \rightarrow b) \star (1 \rightarrow c)) \wedge ((1 \rightarrow b) \star (a \rightarrow c))$*
2. $\int_{\gamma}^{\rightarrow} f \star g \leq \int_{\gamma}^{\rightarrow} f \star \int_{\gamma}^{\rightarrow} g.$

Note that such inequalities for Sugeno integrals have been studied by many authors [12–14, 16].

5   Characterization Results

This part presents the characterisation theorems. We will see that comonotonic maxitivity is specific to q-integrals and comonotonic minitivity is specific to q-cointegrals. Moreover q-integrals and q-cointegrals do not satisfy the same homogeneity conditions, the former being homogeneous with respect to $\wedge$ and the latter with respect to $\vee$. It is also important to remember that q-integrals simplify if computed wrt a possibility measure and q-cointegral simplify wrt a necessity measure but not conversely.

Let us recall that f and g are comonotonic functions if for i, j if $f_i > f_j$ then $g_i \geq g_j$.

5.1   Characterisation Results for q-Integrals

Theorem 4. *Let $I: L^C \rightarrow L$ be a mapping. There are a capacity γ and a fuzzy conjunction $\otimes$ such that $I(f) = \int_{\gamma}^{\otimes} f$ for every $f \in L^C$ if and only if*

1. $I(f \vee g) = I(f) \vee I(g)$, *for any comonotone $f, g \in L^C$.*

2. *There are a capacity* $\lambda \colon 2^{\mathcal{C}} \to L$ *and a binary operation* $\star$ *on* L *such that*
   $I(a \wedge \mathbf{1}_A) = \lambda(A) \star a$ *for every* $a \in L$ *and every* $A \subseteq \mathcal{C}$.
3. $I(\mathbf{1}_{\mathcal{C}}) = 1$ *and* $I(\mathbf{0}_{\mathcal{C}}) = 0$.

In that case, we have $\gamma = \lambda$ *and* $\otimes = \star$.

Note that this theorem is not a generalization of the characterisation of Sugeno integral because generally $\gamma(A) \otimes 1$ is a capacity not equal to γ. However it is the case if $\otimes$ is a right conjunction i.e., if $a \otimes 1 = 1$ for all $a \in L$.

Corollary 1. *If the fuzzy conjunction* $\otimes$ *is a right conjunction, then under the assumptions of Theorem 4, the functional* I *is of the form* $I(f) = \int_{\gamma}^{\otimes} f$ *where* $\gamma(A) = I(\mathbf{1}_A)$.

For the representation theorem when the capacity is a possibility measure, we need the following definition: A binary operation $\odot$ is said to be *x-right-cancellative* if and only if for every $a, b \in L$, if $a \odot x = b \odot x$ then $a = b$.

Theorem 5. *Let* $I \colon L^{\mathcal{C}} \to L$ *be a mapping. There are a possibility measure* Π *and a fuzzy conjunction* $\otimes$ *such that* $I(f) = \int_{\Pi}^{\otimes} f$ *for every* $f \in L^{\mathcal{C}}$ *if and only if* I *satisfies the following properties:*

1. $I(f \vee g) = I(f) \vee I(g)$, *for any* $f, g \in L^{\mathcal{C}}$.
2. *There are a capacity* $\lambda \colon 2^{\mathcal{C}} \to L$ *and a binary operation* $\star$ *on* L *such that*
   $I(a \wedge \mathbf{1}_A) = \lambda(A) \star a$ *for every* $a \in L$ *and every* $A \subseteq \mathcal{C}$.
3. $I(\mathbf{1}_{\mathcal{C}}) = 1$ *and* $I(\mathbf{0}_{\mathcal{C}}) = 0$.

In that case, we have $\star = \otimes$ *and, if* $\star$ *is 1-right-cancellative, we have* $\Pi = \lambda$.

Since the fuzzy conjunction $\otimes$ is not supposed to be commutative, there is a companion q-integral defined as follows.

Definition 4. *Let* $\gamma \colon 2^{\mathcal{C}} \to L$ *be a capacity. The q-integral,* $\int_{\gamma}^{\mathcal{E}(\otimes)}$ *is the mapping* $\int_{\gamma}^{\mathcal{E}(\otimes)} \colon L^{\mathcal{C}} \to L$ *defined by*

$$\int_{\gamma}^{\mathcal{E}(\otimes)} f = \bigvee_{A \subseteq \mathcal{C}} \left(\bigwedge_{i \in A} f_i \otimes \gamma(A) \right).$$

This q-integral generally differs from $\int_{\gamma}^{\otimes} f$ [4] but it satisfies the same property since we only use the increasingness property of the fuzzy conjunction in both places, and the limit conditions $0 \otimes 1 = 1 \otimes 0 = 0$ and $1 \otimes 1 = 1$.

5.2 Characterisation Theorems for q-cointegrals

Theorem 6. *Let* $I \colon L^{\mathcal{C}} \to L$ *be a mapping. There are a capacity* γ *and a fuzzy implication* $\to$ *such that* $I(f) = \int_{\gamma}^{\to} f$ *for every* $f \in L^{\mathcal{C}}$ *if and only if the following properties are satisfied.*

1. $I(f \wedge g) = I(f) \wedge I(g)$, *for any comonotone* $f, g \in L^{\mathcal{C}}$.
2. *There are a capacity* $\rho \colon 2^{\mathcal{C}} \to L$ *and a binary operation* $\triangleright$ *such that*
   $I(a \vee \mathbf{1}_A) = \rho^c(\overline{A}) \triangleright a, \forall a \in L$.
3. $I(\mathbf{1}_{\mathcal{C}}) = 1$ *and* $I(\mathbf{0}_{\mathcal{C}}) = 0$.

In that case $\rho = \gamma$, *and* $\triangleright = \to$.

In the case when the functional I is fully minitive, we prove the following:

Theorem 7. *Let* $I \colon L^{\mathcal{C}} \to L$ *be a mapping. There are a necessity measure* N *and a fuzzy implication* $\to$ *such that* $I(f) = \int_N^{\to} f$ *for every* $f \in L^{\mathcal{C}}$ *if and only if* I *satisfies the following properties:*

1. $I(f \wedge g) = I(f) \wedge I(g)$, *for any* $f, g \in L^{\mathcal{C}}$.
2. *There are a capacity* $\rho \colon 2^{\mathcal{C}} \to L$ *and a binary operation* $\triangleright$ *such that*
   $I(a \vee \mathbf{1}_A) = \rho^c(\overline{A}) \triangleright a$, *for every* $a \in L$ *and every* $A \subseteq \mathcal{C}$.
3. $I(\mathbf{1}_{\mathcal{C}}) = 1$ *and* $I(\mathbf{0}_{\mathcal{C}}) = 0$.

In that case, we have that $\triangleright = \to$ *is a fuzzy implication and, if the fuzzy implication* $\to$ *is 0-right-cancellative,* $\rho = N$ *is a necessity measure.*

The companion q-cointegral defined from a q-cointegral by contrapositive symmetry.

Definition 5. *Let* $\gamma \colon 2^{\mathcal{C}} \to L$ *be a capacity. The contrapositive q-cointegral,* $\int_{\gamma}^{\mathcal{C}(\to)}$ *is the mapping* $\int_{\gamma}^{\mathcal{C}(\to)} \colon L^{\mathcal{C}} \to L$ *defined by*

$$\int_{\gamma}^{\mathcal{C}(\to)} f = \bigwedge_{A \subseteq \mathcal{C}} ((1 - \bigvee_{i \in A} f_i) \to (1 - \gamma^c(A))).$$

This q-integral generally differs from $\int_{\gamma}^{\to} f$ [4] but it satisfies the same properties since $\mathcal{C}(\to)$ is also a fuzzy implication.

6   Upper and Lower Qualitative Possibility Integrals

The possibilistic core of a capacity γ, defined by $S(\gamma) = \{\pi : \Pi(A) \geq \gamma(A), \forall A \subseteq S\}$, is never empty since it always contains the vacuous possibility measure, based on the distribution $\pi^?$ expressing ignorance ($\forall A \neq \emptyset \subset C, \Pi^?(A) = 1 \geq \gamma(A), \forall$ capacity γ, and $\Pi^?(\emptyset) = \gamma(\emptyset) = 0$).

Some possibility distributions in the core can be generated by permutations of elements. Let σ be a permutation of the $n = |C|$ elements in C. The ith element of the permutation is denoted by $\sigma(i)$. Hence the sets $C_{\sigma}^i = \{\sigma(i), \ldots, \sigma(n)\}$ define the possibility distribution $\pi_{\sigma}^{\gamma} \colon \forall i = 1 \ldots, n, \pi_{\sigma}^{\gamma}(\sigma(i)) = \gamma(C_{\sigma}^i)$. We have the following equality [24]: $\forall A \subseteq C, \gamma(A) = \bigwedge_{\sigma} \Pi_{\sigma}^{\gamma}(A)$.

Moreover, $S(\gamma)$ is an upper semi-lattice and we denote $\mathcal{S}_*(\gamma) = \min S(\gamma)$ the set of minimal elements in $S(\gamma)$. Any capacity can be viewed either as a lower possibility measure or as an upper necessity measure, defined on the minimal possibility distributions in the core [23,25]:

$$\gamma(A) = \bigwedge_{\pi \in \mathcal{S}_*(\gamma)} \Pi(A) = \bigvee_{\pi \in \mathcal{S}_*(\gamma^c)} N(A). \tag{9}$$

As a consequence of this result, it was proved in [23] that Sugeno integral is a lower prioritized maximum, as well as an upper prioritized minimum:

Proposition 1. $\displaystyle\int_\gamma f = \bigwedge_{\pi \in \mathcal{S}_*(\gamma)} \int_\Pi f \; and \; \int_\gamma f = \bigvee_{\pi \in \mathcal{S}_*(\gamma^c)} \int_N f.$

We obtain the same results for the q-integrals and q-cointegrals:

Proposition 2. $\displaystyle\int_\gamma^{\rightarrow} f = \bigvee_{\pi \in \mathcal{S}_*(\gamma^c)} \int_N^{\rightarrow} f \; and \; \int_\gamma^{\otimes} f = \bigwedge_{\pi \in \mathcal{S}_*(\gamma)} \int_\Pi^{\otimes} f.$

7 Conclusion

In this paper we have proposed a survey of q-integrals and q-cointegrals, their construction and their properties. Note that one future work will be to study an acts-based characterisation of these integrals.

There exists also variants of Sugeno integrals, named desintegrals, dealing with negative values scale [4]. These aggregation functions merge negative ratings into a global positive one. The desintegrals can be expressed in terms of qualitative integrals which allows us to establish their properties which are similar to those satisfies by the qualitative integrals. In the context of bipolar evaluations, an alternative f is represented by a vector of positive and negative components. Hence an alternative could be evaluate with a pair (integral, desintegral), the integral would be used for the positive ratings and desintegral for the negative ratings.

References

1. Dubois, D., Rico, A., Prade, H., Teheux, B.: Characterizing variants of qualitative Sugeno integrals in a totally ordered Heyting algebra. In: Proceedings of 9th Conference of the European Society for Fuzzy Logic and Technology (Eusflat), Gijon, pp. 862–872 (2015)
2. Fodor, J.: On fuzzy implication operators. Fuzzy Sets Syst. **42**, 293–300 (1991)
3. Dubois, D., Prade, H.: Weighted minimum and maximum operations. Inf. Sci. **39**, 205–210 (1986)
4. Dubois, D., Prade, H., Rico, A.: Residuated variants of Sugeno integrals. Inf. Sci. **329**, 765–781 (2016)
5. Sugeno, M.: Theory of Fuzzy Integrals and its Applications. Ph.D thesis, Tokyo Institute of Technology (1974)
6. Grabisch, M., Murofushi, T., Sugeno, M.: Fuzzy measure of fuzzy events defined by fuzzy integrals. Fuzzy Sets Syst. **50**(3), 293–313 (1992)

7. Borzová-Molnárová, J., Halčinová, L., Hutník, O.: The smallest semicopula-based universal integrals I: properties and characterizations. Fuzzy Sets Syst. **271**, 1–15 (2015)
8. Dvořák, A., Holčapek, M.: Fuzzy measures and integrals defined on algebras of fuzzy subsets over complete residuated lattices. Inf. Sci. **185**, 205–229 (2012)
9. Kaufman, A., Le calcul des admissibilités: Une idée nouvelle à partir de la théorie des sous-ensembles flous. In: Proceedings of Colloque International sur la Théorie et les Applications des Sous-Ensembles Flous, vol. I, p. 14, Marseilles (1978)
10. Shilkret, N.: Maxitive measure and integration. Indag. Math. **33**, 109–116 (1971)
11. Dubois, D., Prade, H., Rico, A., Teheux, B.: Generalized qualitative Sugeno integrals. Inf. Sci. **415**, 429–445 (2017)
12. Agahi, H., Mesiar, R., Ouyang, Y.: New general extensions of Chebyshev type inequalities for Sugeno integrals. Int. J. Approx. Reason. **51**, 135–140 (2009)
13. Agahi, H., Mesiar, R., Ouyang, Y.: General Minkowski type inequalities for Sugeno integrals. Fuzzy Sets Syst. **161**, 708–715 (2010)
14. Ouyang, Y., Mesiar, R., Agahi, H.: An inequality related to Minkowski type for Sugeno integrals. Inf. Sci. **180**, 2793–2801 (2010)
15. Yan, T., Ouyang, Y.: Chebyshev inequality for q-integrals. Int. J. Approx. Reason. **106**, 146–154 (2019)
16. Flores-Franulič, A., Román-Flores, H.: A Chebyshev type inequality for fuzzy integrals. Appl. Math. Comput. **190**, 1178–1184 (2007)
17. Mesiar, R., Ouyang, Y.: General Chebyshev type inequalities for Sugeno integrals. Fuzzy Sets Syst. **160**, 58–64 (2009)
18. Klement, E.P., Mesiar, R., Pap, E.: A universal integral as common frame for Choquet and Sugeno integral. IEEE Trans. Fuzzy Syst. **18**, 178–187 (2010)
19. Boczek, M., Kaluszka, M.: On conditions under which some generalized Sugeno integrals coincide: a solution to Dubois' problem. Fuzzy Sets Syst. **326**, 81–88 (2017)
20. Sugeno, M.: Fuzzy measures and fuzzy integrals: a survey. In: Gupta, M.M., et al. (eds.) Fuzzy Automata and Decision Processes, North-Holland, pp. 89–102 (1977)
21. Dubois, D., Prade, H., Sabbadin, R.: Qualitative decision theory with Sugeno integrals. In: Grabisch, M., et al. (eds.) Fuzzy Measures and Integrals Theory and Applications, pp. 314–322. Physica Vrlg, Heidelberg (2000)
22. Grabisch, M., Labreuche, C.: A decade of application of the Choquet and Sugeno integrals in multi-criteria decision aid. Ann. Oper. Res. **175**, 247–286 (2010)
23. Dubois, D.: Fuzzy measures on finite scales as families of possibility measures. In: Proceedings of EUSFLAT-LFA Conference, Aix-Les-Bains, France pp. 822–829 (2011)
24. Banon, G.: Constructive decomposition of fuzzy measures in terms of possibility and necessity measures. In: Proceedings of VIth IFSA World Congress, São Paulo, Brazil, vol. I, p. 217–220 (1995)
25. Dubois, D., Prade, H., Rico, A.: Qualitative capacities as imprecise possibilities. In: Van der Gaag, L. (ed.) Symbolic and Quantitative Approaches to Reasoning with Uncertainty, (ECSQARU 2013). Lecture Notes in Computer Science, vol. 7958, pp. 169–180. Springer, Heidelberg (2013)
26. Dubois, D., Prade, H.: A theorem on implication functions defined from triangular norms. Stochastica **8**, 267–279 (1984)

The Concave and Decomposition Integrals: A Review and Future Directions

AGOP Conference 2019 in Olomouc

Ehud Lehrer[1,2(✉)]

[1] School of Mathematical Sciences, Tel Aviv University, Tel Aviv 69978, Israel
lehrer@post.tau.ac.il
[2] INSEAD, Bd. de Constance, 77305 Fontainebleau Cedex, France

Abstract. The concave and decomposition-integrals for capacities defined over finite spaces are introduced. The decomposition integral is based on allowable decompositions of a random variable and it generalizes the Choquet, concave, Shilkret and Pan integrals. Further research directions are given at the end of the paper.

1 Introduction

A rational decision maker is often described in decision theory under uncertainty, as an expected utility maximizer. Different experiments, among which the Ellsberg's paradox [2], show that expected utility theory is often violated.

Schmeidler [8] incorporate Choquet [1] expected utility into the theory of decision making. The belief of the decision maker (DM) is represented by a non-additive probability (henceforth referred to as *capacity*). It reflects the idea that the DM has an incomplete or imprecise information about the uncertain aspects of the decision problem under consideration. Moreover, the expected value of a random variable is the respective integral, using the Choquet integral [1].

In this review we introduce first an alternative scheme of evaluating random variables: the concave integral: the concave integral (Lehrer [5]). We then introduce a generalization: the decomposition-integral (Even and Lehrer [4]). The latter generalizes Choquet integral, the concave integral, Pan integral (see Wang and Klir [12]) and Shilkret integral [10].

2 A Motivating Example

Three workers work on a joint project. However, each worker is willing to put a different amount of time on the project, and moreover, the workers' output depends on the team working together. For instance, if workers 1 and 2 are working one month together, they complete, 0.9 of the project. We say then that $v(12) = 0.9$. The following figures provide a full information about the teams' productivity rates per month. $v(1) = v(2) = v(3) = 0.2$, $v(23) = 0.5$,

© Springer Nature Switzerland AG 2019
R. Halaš et al. (Eds.): AGOP 2019, AISC 981, pp. 15–25, 2019.
https://doi.org/10.1007/978-3-030-19494-9_2

$v(13) = 0.8$ and $v(123) = 1$. We denote by X_i the time (in months fractions) that worker i is willing to invest on the project. Let $X_1 = 1, X_2 = 0.4, X_3 = 0.6$. This means, for instance, that worker 1 is willing to invest one month on the project. The question is what is the maximal product that can be obtained, given the workers' willingness to invest (henceforth, time endowment) and the teams' productivity rates.

Suppose that team $\{1, 2\}$ is working 0.4 of a month together and team $\{1, 3\}$ is working 0.6 of a month together. This way all workers exhaust their time endowment and the total product is $v(1, 2) \cdot 0.4 + v(1, 3) \cdot 0.6 = 0.9 \cdot 0.4 + 0.8 \cdot 0.6 = 0.84$. With this team structure, the output is 84% of the project. It turns out that this is the maximum that can be produced. In other words, any other team structure would result in a smaller product. This method is akin to what is later referred to as the concave integral (Lehrer [5]).

Suppose however, that the players are not free to choose the teams they are working with the way they want. Rather, the entire group should start working together and then workers gradually leave without returning to work again on the project. Under these constraints, the maximum that the workers could produce is attained when $\{1, 2, 3\}$ work 0.4 of a month together, 2 leaves and let $\{1, 3\}$ work 0.2 of a month together, and finally $\{1\}$ works 0.4 of a month alone. The output is then, $1 \cdot 0.4 + 0.8 \cdot 0.2 + 0.2 \cdot 0.4 = 0.64$. That is, due to the constraint on teams formation the output reduces to 64%. This method is the one induced by the Choquet integral (Choquet [1]).

While the method related to the concave integral seems to be more suitable to measuring the productivity of a group, the method defined by the Choquet integral is extensively used in the theory of decision making under uncertainty. The question arises as to what makes one method more suitable than the other in one context and less so in another. Furthermore, these two methods suggest that there might exist other methods, possibly more suitable for applications in some other contexts.

3   Capacity and Decompositions of a Random Variable

Let N be a finite set ($|N| = n$). A *collection* D is a set of subsets of N. That is, $D \subseteq 2^N$. A *capacity* v over N is a function $v : 2^N \to [0, \infty]$ satisfying: (i) $v(\varphi) = 0$; and (ii) $S \subseteq T \subseteq N$ implies $v(S) \leq v(T)$.

A *random variable* (r.v. or simply, a variable) X over N is a function $X : N \to \mathbb{R}$. A subset of N will be called an *event*. For any event $A \subseteq N$, $\mathbf{1}_A$ denotes the indicator of A, which is the random variable that takes the value 1 over A and the value 0 otherwise.

We deal here with non-negative random variables and therefore, when we say a random variable, we refer to a non-negative one.

Definition 1. *Let X be a random variable.*

1. A sub-decomposition *of X is a finite summation $\sum_{i=1}^{k} \alpha_i \mathbf{1}_{A_i}$ such that*

(i) $\sum_{i=1}^{k} \alpha_i \mathbf{1}_{A_i} \leq X$; and

(ii) $\alpha_i \geq 0$ and $A_i \subseteq N$ for every $i = 1, ..., k$.

2. Let D be a collection. $\sum_{i=1}^{k} \alpha_i \mathbf{1}_{A_i}$ is a D-sub-decomposition of X if it is a sub-decomposition of X and $A_i \in D$ for every $i = 1, ..., k$.

We say that $\sum_{i=1}^{k} \alpha_i \mathbf{1}_{A_i}$ is a *decomposition* of X if equality replaces inequality in (i). That is, $\sum_{i=1}^{k} \alpha_i \mathbf{1}_{A_i}$ is a decomposition of X if it is a sub-decomposition of X, and $\sum_{i=1}^{k} \alpha_i \mathbf{1}_{A_i} = X$. A similar definition applies to D-decomposition of X.

Suppose, for instance, that $D = 2^N$ and $X = \mathbf{1}_N$. Then, $X = \sum_{i=1}^{n} \mathbf{1}_{\{i\}}$, and at the same time, $X = \mathbf{1}_N$. Both decompositions use subsets in D.

3.1 The Concave and Choquet Integrals

Using the terminology of D-decompositions we can reiterate the definition of the concave integral with respect to (henceforth, w.r.t.) the capacity v (see Lehrer [5]):

$$\int^{cav} X dv = \max \left\{ \sum_{i=1}^{k} \alpha_i v(A_i); \ \sum_{i=1}^{k} \alpha_i \mathbf{1}_{A_i} \text{ is } 2^N\text{-sub-decomposition of } X \right\}.$$

(1)

Note that since v is monotonic w.r.t. inclusion, one can be replaced sub-decomposition in Eq. (1) by decomposition. That is,

$$\int^{cav} X dv = \max \left\{ \sum_{i=1}^{k} \alpha_i v(A_i); \ \sum_{i=1}^{k} \alpha_i \mathbf{1}_{A_i} \text{ is } 2^N\text{-decomposition of } X \right\}. \quad (2)$$

In words, $\int^{cav} X dv$ is the maximum of the values $\sum_{i=1}^{k} \alpha_i v(A_i)$ among all possible decompositions of X. The concave integral imposes no restriction over the decompositions being used: all possible decompositions are taken into account when considering the maximum.

We show that Choquet integral can also be expressed in terms of decompositions. However, unlike the concave integral, Choquet integral does impose restrictions. We recall first the traditional definition of the Choquet integral. Let σ be a permutation on N, such that $X_{\sigma(1)} \leq ... \leq X_{\sigma(n)}$. The Choquet integral of a r.v. X, denoted $\int^{Ch} X dv$, is defined by the following summation: $\sum_{i=1}^{n} (X_{\sigma(i)} - X_{\sigma(i-1)}) v(A_i(X))$, where $X_{\sigma(0)} = 0$ and $A_i(X) = \{\sigma(i), ..., \sigma(n)\}$, $i = 1, ..., n$. We say that two subsets A and B of N are *nested* if either $A \subseteq B$ or $B \subseteq A$. A collection D is called a *chain* if any two events $A, B \in D$ are nested. Denote by $\mathcal{F}^{Ch}$ the set of all chains.

The following proposition states that the Choquet integral is the maximum of $\sum_{i=1}^{k} \alpha_i v(A_i)$, among all decompositions in which every A_i and A_j are nested.

Proposition 1.

$$\int^{Ch} X dv = \max \left\{ \sum_{i=1}^{k} \alpha_i v(A_i); \; \sum_{i=1}^{k} \alpha_i \mathbf{1}_{A_i} \text{ is } \mathcal{F}^{Ch}\text{-sub-decomposition of } X \right\}$$

$$= \max \left\{ \sum_{i=1}^{k} \alpha_i v(A_i); \; \sum_{i=1}^{k} \alpha_i \mathbf{1}_{A_i} \text{ is } \mathcal{F}^{Ch}\text{-decomposition of } X \right\}. \quad (3)$$

Stated differently, Choquet integral is the maximum of $\sum_{i=1}^{k} \alpha_i v(A_i)$, over all decompositions in which every A_i and A_j are nested. The proof is deferred to the Appendix.

Since any chain is a subset of 2^N, it is evident from Eqs. (2) and (3) that

$$\int^{Ch} \cdot \, dv \leq \int^{cav} \cdot \, dv.$$

Example 1. Let $N = \{1, 2, 3\}$, $v(N) = 1$, $v(12) = v(13) = 1/2$, $v(23) = 11/12$ and $v(1) = v(2) = v(3) = 1/3$. Define $X = (3, 5, 2)$ to be a variable over N. The decomposition $X = 3 \mathbf{1}_{12} + 2 \mathbf{1}_{23}$ is the one at which the maximum of the right hand side of (1) is obtained. Therefore, the concave integral of X is

$$\int^{cav} X dv = 3 \cdot (1/2) + 2 \cdot (11/12) = \frac{10}{3}.$$

On the other hand, Choquet integral of X is obtained at the chain $\{(2), (12), (123)\}$, where the decomposition of X is $2 \mathbf{1}_2 + 1 \mathbf{1}_{12} + 2 \mathbf{1}_N$ and

$$\int^{Ch} X dv = 2 \cdot (1/3) + 1 \cdot (1/2) + 2 \cdot 1 = \frac{19}{6}.$$

3.2   The Decomposition-Integral

In this part we show that the method of sub-decomposition enables us to unify many well-known and useful methods of integration under one general method. Suppose that $\mathcal{F}$ is a set of collections. A sub-decomposition of X is $\mathcal{F}$-*allowable* if it is a D−sub-decomposition of X, with the restriction that $D \in \mathcal{F}$. In other words, it has the form $\sum_{A_i \in D} \alpha_i \mathbf{1}_{A_i}$, where $D \in \mathcal{F}$. Thus, in the sub-decomposition of X only events from the same collection D in $\mathcal{F}$ are allowed to be used. The key concept of this paper is introduced in the following definition.

Definition 2. *The* decomposition-integral w.r.t. *the set of collections* $\mathcal{F}$ *is defined as follows.*

$$\int_{\mathcal{F}} X dv = \max \left\{ \sum_{A \in D} \alpha_A v(A); \; \sum_{A \in D} \alpha_A \mathbf{1}_A \text{ is } \mathcal{F} - \text{allowable sub} - \text{decomposition of } X \right\}.$$

$$(4)$$

The integral $\int_{\mathcal{F}} \cdot \, dv$ is the maximum over all sub-decompositions that use only A_i's from the same collection $D \in \mathcal{F}$. The sub-decomposition attaining the maximum in Eq. (4) is called the v-*optimal* sub-decomposition (or decomposition) of X w.r.t. $\mathcal{F}$. When no ambiguity arises, we just call it an optimal sub-decomposition (or decomposition) of X.

Remark 1. The decomposition-integral is defined as the maximum over the set on the RHS of Eq. (4). Considering the maximum rather than the supremum is justified because for any collection $D \in \mathcal{F}$, the set of vectors

$$\left\{ (\alpha_A)_{A \in D} ; \ \alpha_A \geq 0 \text{ for every } A \in D \text{ and } \sum_{A \in D} \alpha_A \mathbf{1}_A \text{ is } D - \text{sub} - \text{decomposition of } X \right\}$$

is compact and the function $\sum_{A \in D} \alpha_A v(A)$ defined over this set is continuous. Therefore, for any collection $D \in \mathcal{F}$,

$$\max \left\{ \sum_{A \in D} \alpha_A v(A); \ \sum_{A \in D} \alpha_A \mathbf{1}_A \text{ is } D - \text{sub} - \text{decomposition of } X \right\}$$

exists. Since there are finitely many collections D in $\mathcal{F}$, writing the maximum in Eq. (4) is justified.

The following example illustrates the reason why in Definition 2 we allow for sub-decompositions and do not insist on decompositions.

Example 2. [Example 1 continued] Consider $\mathcal{F}$ defined as follows.

$$\mathcal{F} = \{\{(1),(23)\},\{(12)\},\{(2),(13)\}\}.$$

Here $\mathcal{F}$ consists of three collections. It turns out that a sub-decomposition, rather than a decomposition, attains the maximum in Eq. (4). The optimal sub-decomposition of X is $3\mathbf{1}_{(1)} + 2\mathbf{1}_{(23)}$ obtained at the collection $\{(1),(23)\}$, and $\int_{\mathcal{F}} X dv = 3 \cdot (1/3) + 2 \cdot (11/12) = \frac{34}{12}$.

Denote by $\mathcal{F}^{cav}$ the set of collections consisting of merely the collection 2^N. Then $\int_{\mathcal{F}^{cav}} \cdot \, dv = \int^{cav} \cdot \, dv$. Proposition 1 states that[1] $\int_{\mathcal{F}^{Ch}} \cdot \, dv = \int^{Ch} \cdot \, dv$. Hence, the concave and Choquet integral differ from each other in the decompositions that the respective sets of collections allow. While the concave integral allows for all possible decompositions, the Choquet integral allows for chain decompositions (or Choquet decompositions) only. Since the set of collections $\mathcal{F}^{cav}$ allows for all decompositions, the following statement (given without a proof) is obtained.

Proposition 2. *Suppose that $\mathcal{F}$ is a set of collections. Then,*

$$\int_{\mathcal{F}} \cdot \, dv \leq \int^{cav} \cdot \, dv$$

for every v.

---

[1] Coincidentally, the notation $\mathcal{F}^{Ch}$, derived from the word chain, resonates with the notation $\int^{Ch}$ that derives from Choquet.

In other words, of all the decomposition-integrals, the concave integral is the highest.

3.3 Pan Integral and Shilkret Integral

It turns out that other integration schemes also conform to the decomposition method. A partition of N is a collection $D = \{A_1, A_2, ..., A_k\}$ consisting of pairwise disjoint events whose union is N itself. Denote by $\mathcal{F}^{part}$ the set of all partitions of N. The integral $\int_{\mathcal{F}^{part}} \cdot dv$ is Pan integral (see Wang and Klir [12]).

Consider now the set $\mathcal{F}^{sing} = \{\{A\}; A \subseteq N\}$. This set of collections consists of all the singletons whose members are events. The maximum in Eq. (4) is obtained at the event that maximizes $\alpha v(A)$, subject to the constraint that $\alpha \mathbf{1}_A \leq X$. Formally,

$$\int_{\mathcal{F}^{sing}} X dv = \max\left\{\sum_i \alpha_i v(A_i); \ \sum_i \alpha_i \mathbf{1}_{A_i} \text{ is } \mathcal{F}^{sing}\text{-allowable sub-decomposition of } X\right\}$$

$$= \max\left\{\alpha v(A); \ \alpha \mathbf{1}_A \leq X, \ A \subseteq N, \ \alpha \geq 0\right\} = \max\left\{\alpha \cdot v(X \geq \alpha); \ \alpha \geq 0\right\}.$$

The right hand side is the scheme known as Shilkret integral of X w.r.t. v (Shilkret [10]).

Another natural set of collections is the one consisting of a single member: an algebra of sets. We say that D is an *algebra* of sets if it is closed under unions and complement, That is, if $A, B \in D$ implies that $A \cup B$ and $N \setminus A$ are also in D. It might occur that a decision maker is forced or would like to rely only on events in an algebra D. This might happen, for instance, when the decision maker suspects that the information embedded in the capacity about events out of the algebra is unreliable. In this case, employing the integral $\int_{\{D\}} X dv$ to evaluate the random variable X seems to be a natural choice.

Finally, consider the set of collections $\mathcal{F}$ that consists of $\{N\}$ alone. Then,

$$\int_{\mathcal{F}} X \ dv = \min X \cdot v(N).$$

4   Properties of the Decomposition-Integral

This section examines the family of decomposition-integrals with respect to four natural properties.

- **Positive homogeneity of degree one**

  The decomposition-integral is *positive homogeneous* for any set of collections $\mathcal{F}$. This means that for every $\lambda > 0$, $\int_{\mathcal{F}} \lambda X dv = \lambda \int_{\mathcal{F}} X dv$ for every X, v and $\mathcal{F}$.

- **The decomposition-integral and additive capacities**

  The integral w.r.t. a general capacity is meant to generalize the notion of expectation in case the capacity is probability. Riemann, Choquet and the concave integrals indeed coincide with the expectation whenever v is a probability, while Shilkret integral does not. The objective of this chapter is to find conditions on the set of collections which guarantee that the decomposition-integral coincides with the expectation in case the capacity is a probability distribution. Denote by $\mathbb{E}_P(X)$ the expectation of X w.r.t. probability P.

Proposition 3. $\mathbb{E}_P(X) = \int_{\mathcal{F}} X dP$ *for every r.v. X and every probability P, if and only if every X has a D-decomposition with $D \in \mathcal{F}$.*

• **Monotonicity**

The first observation regarding monotonicity refers to fixed sets of collections and capacity. Fix v and $\mathcal{F}$, and suppose that $X \leq Y$. Then, $\int_{\mathcal{F}} X dv \leq \int_{\mathcal{F}} Y dv$.

The second observation refers to comparison between two capacities. Fix a set of collections $\mathcal{F}$. If for every $D \in \mathcal{F}$ and every $A \in D$, $v(A) \geq u(A)$, then for every r.v. X, $\int_{\mathcal{F}} X du \leq \int_{\mathcal{F}} X dv$.

The third observation refers to the comparison between two sets of collections. Any set of collections $\mathcal{F}$ induces a decomposition-integral. The question arises as whether any two different sets of collections induce different integrals. The answer to this question is negative. The following proposition characterizes the circumstances in which the decomposition-integral w.r.t. $\mathcal{F}$ is always smaller than, or equal to, that w.r.t. $\mathcal{F}'$. For this purpose we need the following definition and lemma.

Definition 3. *Fix a collection $C \subseteq 2^N$ of subsets of N. We say that C is an* independent collection *if the variables $\mathbf{1}_A$, $A \in C$, are linearly independent.*

In other words, C is an independent collection if for every variable X there are no two different C-decompositions of X. The $C = \{(12), (1)\}$ is an independent collection, while $C = \{(12), (1), (2)\}$ is not because $\mathbf{1}_{(1)}, \mathbf{1}_{(2)}$ and $\mathbf{1}_{(12)}$ are linearly dependent. This is demonstrated also by the fact that $\mathbf{1}_{(1)} + \mathbf{1}_{(2)}$ and $\mathbf{1}_{(12)}$ are two different decompositions of the same variable, which employ indicators of events from C.

Lemma 1. *Fix v, $\mathcal{F}$ and X. Suppose that an optimal $\mathcal{F}$-allowable sub-decomposition of X is obtained by a D-sub-decomposition of X, where $D \in \mathcal{F}$. Then, there is an independent collection $C \subseteq D$ and a C-sub-decomposition which is an optimal $\mathcal{F}$-allowable sub-decomposition of X.*

Proposition 4. *Suppose that $\mathcal{F}$ and $\mathcal{F}'$ are two sets of collections. Then, $\int_{\mathcal{F}} \cdot dv \leq \int_{\mathcal{F}'} \cdot dv$ for every v, if and only if for every $D \in \mathcal{F}$ and every independent collection $C \subseteq D$, there is $D' \in \mathcal{F}'$ such that $C \subseteq D'$.*

• **Additivity**

A well known property of Choquet integral is comonotonic additivity. Two variables X and Y are *comonotone* if for every $i, j \in N$, $(X(i) - X(j))(Y(i) - Y(j)) \geq 0$. It turns out that this property can be expressed in terms of sets of collections and optimal decompositions. Consider the set of collections $\mathcal{F}^{Ch}$ (recall, it consists of all chains). Then, X and Y are comonotone if and only if the optimal decompositions of X and Y use the same D in $\mathcal{F}^{Ch}$. Comonotonic additivity means that if X and Y use the same D for their optimal decomposition, then $\int_{\mathcal{F}^{Ch}} X\,dv + \int_{\mathcal{F}^{Ch}} Y\,dv = \int_{\mathcal{F}^{Ch}} (X+Y)\,dv$. A natural question arises as to whether this is a general property of the decomposition-integral. That is, whether for any set of collections $\mathcal{F}$, if X and Y use the same $D \in \mathcal{F}$ for their optimal sub-decomposition, then $\int_{\mathcal{F}} X\,dv + \int_{\mathcal{F}} Y\,dv = \int_{\mathcal{F}} (X+Y)\,dv$.

The answer to this question proves to be negative. Indeed, consider the set of collections $\mathcal{F}^{part}$ (recall, the one consisting of all partitions of N), and a capacity v, defined on $N = \{1,2\}$ as follows: $v\,(1) = v\,(2) = 1/3$ and $v\,(12) = 1$. Define $X = (\varepsilon, 1), Y = (1, \varepsilon)$, $\varepsilon > 0$. Assume that ε is small enough, so that the optimal decomposition of both X and Y use $D = \{(1), (2)\}$. In this case, $\int_{\mathcal{F}} X\,dv = \int_{\mathcal{F}} Y\,dv = (1/3)\,(1+\varepsilon)$. As for the sum $X+Y$, taking $D' = \{(12)\}$ yields $\int_{\mathcal{F}} (X+Y)\,dv = 1+\varepsilon$, which is strictly greater than $\int_{\mathcal{F}} X\,dv + \int_{\mathcal{F}} Y\,dv = (2/3)\,(1+\varepsilon)$.

The following proposition refers to additivity in case two integrands use the same $D \in \mathcal{F}$ for their optimal decomposition w.r.t. to $\mathcal{F}$ and a specific v.

Fix a set of collections $\mathcal{F}$ and a capacity v. We say that the variable Y is *leaner than* the variable X if there exist (i) an optimal decomposition of Y: $\sum_{A \in C'} \beta_A \mathbf{1}_A$ with $\beta_A > 0$, $A \in C'$; and (ii) an optimal decomposition of X : $\sum_{A \in C} \alpha_A \mathbf{1}_A$ with $\alpha_A > 0, A \in C$, such that $C' \subseteq C$. In words, Y is leaner than X, if there are optimal decompositions in which X employs every indicator that Y employs.

Proposition 5 [Co-decomposition additivity]. *Fix a set of collections $\mathcal{F}$ such that every X has an optimal decomposition (not sub-decomposition) w.r.t. $\mathcal{F}$ for every capacity. Suppose that for every $D, D' \in \mathcal{F}$, whenever there are two different decompositions of the same variable, $\sum_{A \in D} \delta_A \mathbf{1}_A = \sum_{B \in D'} \gamma_B \mathbf{1}_B$, there is $D'' \in \mathcal{F}$ that contains all the A's with $\delta_A > 0$ and all the B's with $\gamma_B > 0$. Then, for every v and every two variables X and Y where Y is leaner than X,*

$$\int_{\mathcal{F}} X\,dv + \int_{\mathcal{F}} Y\,dv = \int_{\mathcal{F}} (X+Y)\,dv. \tag{5}$$

Note that the condition of the proposition is readily satisfied by $\mathcal{F}^{Ch}$. The reason is (see also the proof of Proposition 1) that every random variable essentially (ignoring indicators whose coefficients are zero) has a unique $\mathcal{F}^{Ch}$-allowable decomposition. Proposition 5 implies the comonotonic additivity of Choquet integral. Indeed, considering $\mathcal{F}^{Ch}$ and two comonotonic variables X and Y. Thus, both X and Y can be decomposed using indicators of events taken from the same chain, D. For very $\varepsilon > 0$, the variable $Z_\varepsilon = \sum_{A \in D} \frac{\varepsilon}{n} \mathbf{1}_A$ is smaller than or equal to ε and moreover, X is leaner than Z_ε and Y is leaner than $X + Z_\varepsilon$ (because the coefficients of all $A \in D$ are positive in the decompositions of Z_ε and of $X + Z_\varepsilon$). Proposition 5 implies that $\int_{\mathcal{F}^{Ch}} X\,dv + \int_{\mathcal{F}^{Ch}} Z_\varepsilon\,dv = \int_{\mathcal{F}^{Ch}} (X + Z_\varepsilon)\,dv$ and $\int_{\mathcal{F}^{Ch}} Y\,dv + \int_{\mathcal{F}^{Ch}}(X + Z_\varepsilon)\,dv = \int_{\mathcal{F}^{Ch}} (Y + X + Z_\varepsilon)\,dv$. Thus, $\int_{\mathcal{F}^{Ch}} Y\,dv + \int_{\mathcal{F}^{Ch}} X\,dv + \int_{\mathcal{F}^{Ch}} Z_\varepsilon\,dv = \int_{\mathcal{F}^{Ch}} (Y + X + Z_\varepsilon)\,dv$. As ε shrinks to 0 we obtain, $\int_{\mathcal{F}^{Ch}} Y\,dv + \int_{\mathcal{F}^{Ch}} X\,dv = \int_{\mathcal{F}^{Ch}} (Y + X)\,dv$, which is comonotonic additivity.

Proposition 5 implies also that whenever $\mathcal{F}$ consists of only one D, like $\mathcal{F}^{cav}$, its decomposition-integral respects the additivity property stated in Eq. (5). Therefore, if X and Y are leaner than each other (i.e., the same indicators possess positive coefficients in their optimal decompositions), then Eq. (5) holds true. In particular, the concave integral is linear over those variables that use the same indicators in their optimal decompositions.

The additivity of Choquet integral, as expressed in Eq. (5), does not depend on the underlying capacity. Two random variables are comonotone regardless of the capacity v, and for such variables, Eq. (5) would always be true. On the other hand, whether or not Eq. (5) applies to the concave integral, does depend on v. The reason for this difference between the integrals is that in Choquet integral the optimal decomposition does not depend on v (it always uses the same chain for every v), while it does depend on v when it comes to the concave integral.

5   Future Directions

• **Toward a characterizing of the decomposition-integral.** The decomposition-integral is homogeneous and therefore it is sufficient to consider only random variables in the unit simplex, $\Delta : \{X : N \to \mathbb{R}_+;\ \sum_{i \in N} X_i = 1\}$. For any given capacity v, the decomposition-integral, as a function over Δ, is piecewise-linear. That is, Δ can be partitioned into finitely many closed polygons, $G_1, ..., G_h$ (i.e., $\Delta = \cup_{j=1}^h G_j$) such that for every $j = 1, ..., h$ there is a vector $\alpha_j \in \mathbb{R}^N$ which satisfies[2] $\int_{\mathcal{F}} X dv = X \cdot \alpha_j$ for every $X \in G_j$.

A few natural question arises. Suppose that φ is a piecewise-linear and continuous function defined over Δ. Is there a capacity v and a set of collections $\mathcal{F}$ such that $\varphi(X) = \int_{\mathcal{F}} X dv$ for every $X \in \Delta$? A simple example shows that the answer is on the negative.

Example 3. Consider the case where $|N| = 2$ and for every $(X_1, X_2) \in \Delta$ (i.e., $X_1 + X_2 = 1$) define,

$$\varphi((X_1, X_2)) = \begin{cases} X_1 & \text{if } X_2 \leq 1/4 \\ 1.5 - X_1 & \text{if } 1/4 < X_2 \leq 1/2 \\ 2X_1 & \text{if } 1/2 < X_2 \leq 1. \end{cases}$$

The function φ is piecewise-linear and continuous, but not a decomposition-integral. The reason is that if φ is a decomposition-integral, due to that $\varphi((1,0)) = 1$, it must be that $v(\{1\}) = 1$. But the line $1.5 - X_1$ in the range $1/4 < X_2 \leq 1/2$ can be supported only by the collection $\{\{12\}, \{1\}\}$ while $v(\{12\}) = 1$ and $v(\{1\}) = 1/2$, which contradicts $v(\{1\}) = 1$.

What characterizes the piecewise-linear and continuous functions that are decomposition-integrals?

---

[2] $X \cdot \alpha_j$ denotes the inner product of X and α_j.

• **Measuring the degree of ambiguity.** One can define a dual integral to the concave integral: the convex integral (See Sect. 6 in Even and Lehrer [4] and Mesiar et al. [7]). The convex integral is defined as

$$\int^{vex} X dv = \min \left\{ \sum_{i=1}^{k} \alpha_i v(A_i); \; \sum_{i=1}^{k} \alpha_i \mathbf{1}_{A_i} \text{is } 2^N\text{-decomposition of } X \right\}.$$

$\int^{vex} X dv$ is a convex function of X. Obviously, $\int^{vex} X dv \leq \int^{cav} X dv$.
  Define,

$$A(v) = \max_{X \in \Delta} \int^{cav} X dv - \int^{vex} X dv.$$

$A(\cdot)$ is non-negative and a concave function over the set of capacities.

Proposition 6. v *is additive iff* $A(v) = 0$.

This proposition suggests that $A(v)$ might be a measurement to the extent of ambiguity embedded in v. In other words, $A(v)$ might measure how far v is from being additive. Can it be characterized?

• **Infinite spaces.** So far we were focusing on a finite N. The approach we used to define the concave integral can be applied to infinite spaces. Let $(\Omega, \mathcal{B})$ be a measurable space, where Ω is an arbitrary set and $\mathcal{B}$ is a σ-algebra of subsets of Ω. Let v be a capacity defined over the $\mathcal{B}$. That is, $v : \mathcal{B} \to \mathbb{R}_+$ is monotonic w.r.t. inclusion. Finally, Let $\mathcal{M}$ be the set of non-negative, bounded and $\mathcal{B}$-measurable functions defined over Ω. The concave integral for v is defined as,

$$\int^{cav} X dv = \sup \left\{ \sum_{i=1}^{k} \alpha_i v(A_i); \; \sum_{i=1}^{k} \alpha_i \mathbf{1}_{A_i} = X, A_i \in \mathcal{B}, \alpha_i \geq 0, i = 1, ..., k \right\}$$

for every $X \in \mathcal{M}$.
  To be more specific, suppose that $\Omega = [0, 1]$, $\mathcal{B}$ is the Borel σ-algebra of Ω, and λ is the Lebesgue. Although v attains only real values, without a further restriction on v, one may get that $\int^{cav} X dv = \infty$ for every $X \in \mathcal{M}$ such that $\int X d\lambda > 0$. This might happen, for instance, in the case where $v(B) = \sqrt{\lambda(B)}$ for every $B \in \mathcal{B}$.
  We say that a capacity v is *Lipschitz w.r.t. to* λ, if there is a constant $M > 0$ such that for every $B \in \mathcal{B}$, $v(B) \leq M\lambda(B)$. It is easy to show that if v is Lipschitz w.r.t. to λ, then for every $X \in \mathcal{M}$, $\int^{cav} X dv < \infty$. Furthermore, the concave integral is continuous w.r.t. the sup-norm, homogeneous and concave. These conditions are not sufficient to characterize the concave integral. Indeed, consider inf X. It satisfies all three conditions, but is not a concave integral w.r.t. any capacity.
  This leads us to the question of characterizing the concave integral over large spaces. Let φ be a continuous w.r.t. the sup-norm, homogeneous, and concave function define on $\mathcal{M}$. Under what conditions φ is a concave integral?

References

1. Choquet, G.: Theory of capacities. Ann Inst Fourier **5**, 131–295 (1955)
2. Ellsberg, D.: Risk, ambiguity, and the savage axioms. Q. J. Econ. **75**, 643–669 (1961)
3. Hadar, J., Russell, W.: Rules for ordering uncertain prospects. Am. Econ. Rev. **59**, 25–34 (1969)
4. Even, Y., Lehrer, E.: Decomposition-integral: unifying Choquet and the concave integrals. Econ. Theor. **56**, 33–58 (2014)
5. Lehrer, E.: A new integral for capacities. Econ. Theor. **39**, 157–176 (2009)
6. Lehrer, E.: Partially specified probabilities: decisions and games. Am. Econ. J. Microecon. **4**, 70–100 (2012)
7. Mesiar, R., Li, J., Pap, E.: Superdecomposition integrals. J. Fuzzy Sets Syst. **259**, 3–11 (2015)
8. Schmeidler, D.: Integral representation without additivity. Proc. Am. Math. Soc. **97**, 255–261 (1986)
9. Schmeidler, D.: Subjective probability and expected utility without additivity. Econometrica **57**, 571–587 (1989)
10. Shilkret, N.: Maxitive measure and integration. Indag. Math. **33**, 109–116 (1971)
11. Wang, S.S., Young, V.R., Panjer, H.H.: Axiomatic characterization of insurance prices. Insur. Math. Econ. **21**, 173–183 (1997)
12. Wang, Z., Klir, G.J.: Fuzzy Measure Theory. Springer, Boston (1992)

Super Level Measures: Averaging Before Integrating

Ondrej Hutník$^{(\boxtimes)}$

Institute of Mathematics, Faculty of Science, P. J. Šafárik University in Košice,
Jesenná 5, 040 01 Košice, Slovakia
ondrej.hutnik@upjs.sk

In many situations, such as the definition of the essential supremum, or definition of L_p-norms, or the weak type estimates of the Hardy-Littlewood operator, or the construction of certain nonadditive integrals, etc., the *upper level set* of a (measurable) function f of the form

$$F_\alpha = \{x \in X : |f(x)| > \alpha\},$$

frequently appears. This set is called the (strict) super level set as well as strict α-cut of f, etc. depending on the literature. Measuring this set using a monotone measure μ yields the non-increasing real-valued function $h_{\mu,f}(\alpha) := \mu(F_\alpha)$ known also as *decumulative distribution* of f with respect to μ, or *survival function*. We also use the term standard level measure in this text. For each $\alpha > 0$ the standard level measure may be written as follows

$$
\begin{aligned}
h_{\mu,f}(\alpha) &= \mu(\{x \in X : |f(x)| > \alpha\}) \\
&= \inf\left\{\mu(F) : F \in \mathbf{E_B}, \, (\forall x \in X \setminus F)\,|f(x)| \le \alpha\right\} \\
&= \inf\left\{\mu(F) : F \in \mathbf{E_B}, \, \sup_{x \in X \setminus F} |f(x)| \le \alpha\right\} \\
&= \inf\left\{m(F) : F \in \mathbf{E_B}, \, \operatorname{outsup}_{X \setminus F} \mathsf{s}_\infty(f)\langle\mathbf{E_B}\rangle \le \alpha\right\}.
\end{aligned}
\tag{1}
$$

Here, $\mathbf{E_B}$ is the σ-algebra of Borel sets of $X \ne \emptyset$, and s_∞ is an L_∞-based average of a Borel measurable function $f : X \to \mathbb{C}$ on $E \in \mathbf{E_B}$ given by

$$\mathsf{s}_\infty(f)(E) := \sup_{x \in E} |f(x)| = \sup |f|[E].$$

Finally, the *outer essential supremum* of f over a set $F \in \mathbf{E_B}$ has the form

$$\operatorname*{outsup}_{F} \mathsf{s}_\infty(f)\langle\mathbf{E_B}\rangle := \sup_{E \in \mathbf{E_B}} \mathsf{s}_\infty(f\mathbf{1}_F)(E).$$

Naturally, there are many other averaging procedures instead of L_∞-based average s_∞. This is the main idea for introducing a general mapping called a *size*. Indeed, size s is a mapping assigning a nonnegative number to each pair (f, E), where $f : X \to \mathbb{C}$ is a Borel measurable function and $E \in \mathbf{E_B}$, such that

© Springer Nature Switzerland AG 2019
R. Halaš et al. (Eds.): AGOP 2019, AISC 981, pp. 26–28, 2019.
https://doi.org/10.1007/978-3-030-19494-9_3

(i) if $|f| \le |g|$, then $\mathsf{s}(f)(E) \le \mathsf{s}(g)(E)$;

(ii) $\mathsf{s}(\lambda f)(E) = |\lambda|\,\mathsf{s}(f)(E)$ for each $\lambda \in \mathbb{C}$;

(iii) $\mathsf{s}(f+g)(E) \le C_{\mathsf{s}}\big(\mathsf{s}(f)(E) + \mathsf{s}(g)(E)\big)$ for some fixed $C_{\mathsf{s}} \ge 1$ depending only on s.

This approach involves averaging the functions on basic sets (usually on an appropriate collection $\mathbf{E} \subseteq \mathbf{E_B}$) such as the classical arithmetic mean, generalized arithmetic mean, weighting the integrals, the supremum of a function over a set, etc. A motivation for doing so can be seen in the efficiency of encoding the functions already in the classical Lebesgue theory of measure and integral. Indeed, classical coding describes functions as assignment of a value to every point of a basic set X. In the case of L_p-functions, the set of such assignments has a very large cardinality, which is only reduced after consideration of equivalence classes of L_p-functions. This detour over sets of large cardinality can be avoided by coding functions via their averages over dyadic cubes (alternatively, Euclidean balls). There are only countable many such averages, and by Lebesgue differentiation theorem these averages contain the complete information of the equivalence class of the L_p-function.

Introducing a new quantity to generalize the function $h_{\mu,f}(\alpha)$, formally replacing the supremum size s_∞ by a general size s on a collection $\mathbf{E} \subseteq \mathbf{E_B}$ in (1), we get the formula

$$\mu(\mathsf{s}(f)\langle \mathbf{E}\rangle > \alpha) := \inf\left\{\mu(F): \ F \in \mathbf{E_B}, \ \underset{X\setminus F}{\operatorname{outsup}}\,\mathsf{s}(f)\langle \mathbf{E}\rangle \le \alpha\right\}, \quad \alpha > 0.$$

This definition reminds the construction of well-known outer measure, but these notions coincide only in some cases (e.g. in the case of the Lebesgue outer measure and essential supremum identical to the standard supremum). From a practical point of view, the concept of super level measures and integrals with respect to them is a very suitable component connecting the theory of Carleson measures and the time-frequency analysis. Indeed, the first example of an outer measure space in which the definition of super level measure does not coincide with the level measure is the upper half-plane in the complex plane where the outer measure is generated by basic sets $\mathbf{E}_{\mathrm{tent}}$ (the so-called tents). The essentially bounded functions with respect to the outer measure in this upper half-plane are exactly the Carleson measures. In fact, this is the area where the concept of super level measures was born, see [2].

It is well-known that various aggregations of values of α and $h_{\mu,f}(\alpha)$ into one representative value yield various nonadditive integrals (including the prominent Choquet, Shilkret and Sugeno integrals). Once the super level measure is introduced, the definition of various integrals with respect to a (nonadditive) measure is immediate replacing formally the standard level measure with the super level measure. A novelty is that the new integrals depend on the size (= averaging procedure) to be chosen. For instance, the outer L_p-space norm for $1 \le p < +\infty$ is defined by

$$\|f\|_{L_p(X,\mu,\mathsf{s})} := \left(\int_0^\infty p\,\alpha^{p-1}\,\mu(\mathsf{s}(f)\langle \mathbf{E}\rangle > \alpha)\,\mathrm{d}\alpha \right)^{1/p}$$

mimicking the definition of L_p-spaces based on the standard Choquet integral. Also, weak L_p-space (Lorentz space) is defined by

$$\|f\|_{L_{p,\infty}(X,\mu,\mathsf{s})} := \sup_{\alpha>0} \big\{ \alpha^p\,\mu(\mathsf{s}(f)\langle \mathbf{E}\rangle > \alpha) \big\}.$$

Observe that both these definitions are closely related to non-additive integrals of Choquet and Shilkret (in the super level measure context), and the classical Choquet and Shilkret integrals are only their special cases.

In the talk we survey the recent progress in developing the super level measure concept described in [3]. Since this concept is technically and abstractly difficult in general, we describe an algorithm for super level measure computation [1] in the discrete setting (with a finite basic set X), and we exemplify certain integrals in such situation.

Acknowledgments. This work was supported by the Slovak Research and Development Agency under the contract No. APVV-16-0337.

References

1. Borzová, J., Halčinová, L., Šupina, J.: Size and standard super level measure. In: Medina, J., et al. (eds.) Information Processing and Management of Uncertainty in Knowledge-Based Systems: Theory and Foundations, IPMU 2018. Communications in Computer and Information Science, vol. 853, pp. 219–230. Springer, Cham (2018)
2. Do, Y., Thiele, C.: L^p theory for outer measures and two themes of Lennart Carleson united. Bull. Am. Math. Soc. **52**(2), 249–296 (2015)
3. Halčinová, L., Hutník, O., Kiseľák, J., Šupina, J.: Beyond the scope of super level measures. Fuzzy Sets Syst. **364**, 36–63 (2019). https://doi.org/10.1016/j.fss.2018.03.007

On Extreme Value Copulas with Given Concordance Measures

Piotr Jaworski[(✉)]

Institute of Mathematics, University of Warsaw, Warsaw, Poland
P.Jaworski@mimuw.edu.pl

Abstract. The dependence measures, like for example Kendall tau, Spearman rho, Blomquist beta or tail dependence coefficient, are the main numerical characterization of Bivariate Extreme Value Copulas. Such copulas are characterized by a function on the unit segment, called a Pickands dependence function, which is convex and comprised between two bounds. We identify the smallest possible compact sets containing the graphs of all Pickands dependence functions whose corresponding bivariate extreme-value copula has fixed values of given dependence measures. Moreover we provide the bounds for such bivariate extreme-value copulas.

Keywords: Extreme value copulas · Pickands functions · Kendall τ · Spearman ρ · Tail dependence coefficient

1 Introduction

Definition 1. *The function $C : [0,1]^2 \longrightarrow [0,1]$ is called a (bivariate) copula if the following three properties hold for all $u_1, u_2, v_1, v_2 \in [0,1]$, $u_1 \leq v_1, u_2 \leq v_2$*

$\quad$ (c1) $\ C(u_1, 0) = 0, \quad C(0, u_2) = 0;$

$\quad$ (c2) $\ C(u_1, 1) = u_1, \quad C(1, u_2) = u_2;$

$\quad$ (c3) $\ C(v_1, v_2) - C(u_1, v_2) - C(v_1, u_2) + C(u_1, u_2) \geq 0.$

We recall that every bivariate copula is a restriction to the unit square $[0,1]^2$ of the cumulative distribution function of a vector (U_1, U_2) of standard uniform random variables, i.e., for all $u_1, u_2 \in [0,1]$, $C(u_1, u_2) = \Pr(U_1 \leq u_1, U_2 \leq u_2)$ and for $i = 1, 2$, $\Pr(U_i \leq u_i) = u_i$. Such random variables U_1, U_2 are called representers of the copula C. They serve as a representative of the equivalence class of the pairs of random variables which are uniformly distributed on the unit interval $(0, 1)$ and admit C as their copula. The construction of *canonical* representers goes as follows:

Since copulas are continuous functions, to every copula C we may associate a probabilistic, Borel measure Q^C on the unit square given by

$$Q^C([u_1, v_1] \times [u_2, v_2]) = C(v_1, v_2) - C(u_1, v_2) - C(v_1, u_2) + C(u_1, u_2),$$

© Springer Nature Switzerland AG 2019
R. Halaš et al. (Eds.): AGOP 2019, AISC 981, pp. 29–46, 2019.
https://doi.org/10.1007/978-3-030-19494-9_4

where $u_1, u_2, v_1, v_2 \in [0,1]$, $u_1 \leq v_1, u_2 \leq v_2$. We define U_1 and U_2 on the probabilistic space $([0,1]^2, \mathcal{B}([0,1]^2), Q^C)$ as the coordinate functions $U_1(u_1, u_2) = u_1$ and $U_2(u_1, u_2) = u_2$. The copulas corresponding to three special cases: independent, comonotonic and anticomonotonic representers are denoted respectively by Π, M and W.

$$\Pi(u_1, u_2) = u_1 u_2, \quad M(u_1, u_2) = \min(u_1, u_2), \quad W(u_1, u_2) = (u_1 + u_2 - 1)^+.$$

In statistics, copulas are commonly used to model the dependence between the components of a random vector; see, e.g., Nelsen (2006), Durante and Sempi (2016) for an introduction to copulas and Genest and Favre (2007), Jaworski et al. (2010), Joe et al. (2015) for reviews of statistical modeling techniques using copulas.

To understand the nature of copula models, let H denote the joint distribution function of the vector $X = (X_1, X_2)$ and let F_1, F_2 be its univariate margins, i.e., for all $x_1, x_2 \in \mathbb{R}$, $H(x_1, x_2) = \Pr(X_1 \leq x_1, X_2 \leq x_2)$ and for $i = 1, 2$, $F_i(x_i) = \Pr(X_i \leq x_i)$. Assume that the functions F_1, F_2 are continuous, as is often the case when the variables X_1, X_2 are measurements. Sklar (1959) showed that in this case, there exists a unique copula $C : [0,1]^2 \to [0,1]$ such that, for all $x_1, x_2 \in \mathbb{R}$,

$$H(x_1, x_2) = C(F_1(x_1), F_2(x_2)). \tag{1}$$

Thus if H is known, its margins F_1, F_2 can be deduced from it and the copula C induces the dependence between them. In fact, C is simply the restriction to $[0,1]^2$ of the joint distribution function of the vector $(U_1, U_2) = (F_1(X_1), F_2(X_2))$. In practice, however, H is often unknown and a model for it can be constructed by selecting F_1, F_2 and C of specific forms in Eq. (1).

In statistical practice, it is often assumed that an unknown copula C belongs to a given parametric class $\{C_\theta : \theta \in \Theta\}$. There is then an interest in estimating the parameter θ from data. A standard approach, useful when the parameter space Θ is an interval in $\mathbb{R}$, is to compute the rank correlation, say ρ_n, in a random sample of size n from (X_1, X_2). The theoretical analog, called Spearman's rho, is given by

$$\text{corr}\{F_1(X_1), F_2(X_2)\} = \rho(C) = -3 + 12 \int_0^1 \int_0^1 C(u_1, u_2) du_1 du_2.$$

An estimate of θ is then obtained by solving the equation $\rho(C) = \rho_n$. Another approach consists of computing Kendall's coefficient of concordance, τ_n, whose population value is

$$\tau(C) = -1 + 4 \int_0^1 \int_0^1 C(u_1, u_2) dC(u_1, u_2).$$

In general, the inversion of Spearman ρ and Kendall τ both lead to consistent estimators of the dependence parameter θ if $C \in \{C_\theta : \theta \in \Theta\}$ and $\Theta \subset \mathbb{R}$ as $\rho_n \to \rho$ and $\tau_n \to \tau$ almost surely as $n \to \infty$. For details, see, e.g., Genest and

Favre (2007). Alternatively, instead of Spearman ρ and Kendall τ other measures of dependence may be used, like the tail coefficient or Spearman's foot-rule.

Copula modeling is of particular interest in assessing the joint probability of occurrence of rare events with potentially catastrophic consequences; see, e.g., McNeil et al. (2015). It is then known from multivariate extreme-value theory that the copula of the max-stable distribution is itself max-stable, i.e., such that for all $u_1, \ldots, u_d \in [0,1]$ and $m \in \mathbb{N}$, $C(u_1, \ldots, u_d) = C^m(u_1^{1/m}, \ldots, u_d^{1/m})$; see, e.g., de Haan and Ferreira (2006). In the bivariate case, Pickands (1981) showed that every max-stable (or extreme-value, or BEV) copula can be written in terms of a convex function $A : [0,1] \rightarrow [0, 1/2]$ such that, for all $t \in [0,1]$, $A(t) \geq \max(t, 1-t)$. Specifically, one has, for all $u_1, u_2 \in (0,1)$,

$$C(u_1, u_2) = \exp\left[\ln(u_1 u_2) A\left\{\ln(u_2)/\ln(u_1 u_2)\right\}\right] = (uv)^{A(\ln(v)/\ln(uv))}. \qquad (2)$$

The mapping A is called a Pickands dependence function and in modeling the dependence between $d = 2$ extreme risks, the choice of copula C in Eq. (1) amounts to selecting an appropriate function A, see for example Gudendorf and Segers (2010), Genest and Nešlehová (2012). Suppose, that one has observed that for some dependence measure κ, $\kappa(C) = \kappa_*$ in a sample. It is then of interest to know how broad is the choice of Pickands dependence functions with this specific value of κ. For Spearman ρ and Kendall τ, this question was answered in Kamnitui et al. (2019). In this paper we deal with general continuous dependence measures, like for example, besides τ and ρ, the tail coefficient or Spearman's foot rule. Furthermore we study the interdependence between such measures.

2   Notation

Let $\mathcal{A}$ denotes the set of Pickands dependence functions. In more details:

Definition 2. $\mathcal{A}$ is a family of all convex functions $A : [0,1] \longrightarrow [0.5, 1]$ such that $1 \geq A(t) \geq \max(t, 1-t)$.

Note that the set of Pickands dependence functions has "nice" algebraic and analytic characterizations:

1. The five-tuple $(\mathcal{A}, \max, GCM, \mathbf{1}, \max(t, 1-t))$, where GCM stands for the greatest convex minorant, is a bounded lattice.
2. $\mathcal{A}$ is ordered by point-wise inequalities and this ordering is compatible with lattice operations.
3. $\mathcal{A}$ is a convex, bounded, complete subset of the Banach space $C([0,1])$ with supremum norm.

For a function A from $\mathcal{A}$ we denote by C_A the corresponding copula

$$C_A(u, v) = (uv)^{A(\ln(v)/\ln(uv))}, \qquad (u,v) \in (0,1]^2 \setminus \{(1,1)\}.$$

The bijective mapping $\Psi : \mathcal{A} \longrightarrow BEV$, $\quad \Psi(A) = C_A$, is order reverting and continuous with respect to supremum norm. Indeed[1]

$$\forall t \in [0,1] \ A_1(t) \le A_2(t) \iff \forall (u,v) \in [0,1]^2 \ C_{A_1}(u,v) \ge C_{A_2}(u,v).$$

$$\|C_{A_1} - C_{A_2}\|_\infty \le 2\|A_1 - A_2\|_\infty.$$

Furthermore $\Psi(\mathbf{1})(u,v) = uv = \Pi(u,v)$, $\quad \Psi(\max(t, 1-t))(u,v) = \min(u,v) = M(u,v)$. Note that Ψ is inducing the lattice structure on BEV.

$$
\begin{aligned}
\Psi(\max(A_1, A_2))(u,v) &= \exp\big(\ln(uv) \max\big(A_1(\ln(u)/\ln(uv)), A_2(\ln(u)/\ln(uv))\big)\big) \qquad (3) \\
&= \min\big(\exp\big(\ln(uv)A_1(\ln(u)/\ln(uv))\big), \exp\big(\ln(uv)A_2(\ln(u)/\ln(uv))\big)\big) \\
&= \min\big(\Psi(A_1), \Psi(A_2)\big)(u,v).
\end{aligned}
$$

In a similar way we get $\Psi(GCM(A_1, A_2)) = \inf\{C \in BEV : C \ge \max(\Psi(A_1), \Psi(A_2))\}$.

3   Dependence Measures

Let κ be a continuous mapping $\kappa : BEV \longrightarrow [0,1]$, which is compatible with concordance ordering:

$$(\forall u,v \in [0,1] \ C_1(u,v) \le C_2(u,v)) \implies \kappa(C_1) \le \kappa(C_2), \quad \kappa(\Pi) = 0, \quad \kappa(M) = 1.$$

Such a dependence measure can be expressed in terms of Pickands functions. We abbreviate $\kappa(A) := \kappa(\Psi(A))$. Furthermore the composed function is reversing the ordering of Pickands functions.

$$(\forall t \in [0,1] \ A_1(t) \le A_2(t)) \implies \kappa(A_1) \ge \kappa(A_2).$$

We shall deal with two sources of dependence measures κ. First, it might be a restriction of the concordance measure defined for all copulas, like Spearman ρ, Kendall τ, Blomqvist β or Spearman foot-rule ϕ (for definitions see for example Nelsen (2006)).

When C_A is of the form (2) for some A, it was shown by Ghoudi et al. (1998) that

$$\rho(C_A) = \rho(A) = -3 + 12 \int_0^1 \frac{1}{(1 + A(t))^2}\, dt, \quad \tau(C_A) = \tau(A) = \int_0^1 \frac{t(1-t)}{A(t)}\, dA'(t),$$

where A' denotes the right-hand derivative of A on $[0,1)$, which always exists because A is convex, and $A'(1)$ is defined as the supremum of $A'(t)$ on $(0,1)$. The formulas for next two measures are straightforward:

$$\text{Blomqvist } \beta : \beta(A) = \beta(C_A) = 4\, C_A\left(\frac{1}{2}, \frac{1}{2}\right) - 1 = 4^{1 - A(0.5)} - 1.$$

---

[1] See Kamnitui et al. (2019) Proposition 3 for more accurate estimate of the norm.

Spearman foot-rule: $\phi(A) = \phi(C_A) = 6 \int_0^1 C_A(t,t)dt - 2 = 4\dfrac{1 - A(0.5)}{1 + 2A(0.5)}$.

Second, it might be a restriction of a tail dependence measure. We recall the notion of the leading tail term of the copula called also the tail dependence function, for details see Jaworski (2004, 2006, 2009). The upper tail leading term of a copula C is given by a formula

$$L_C(x,y) = \lim_{t\to 0+} \frac{t(x+y) + C(1-tx, 1-ty) - 1}{t} = \lim_{t\to 0+} \frac{C(1-tx, 1-ty) - W(1-tx, 1-ty)}{t},$$

where $x, y \geq 0$, provided that the limit exists. Note that for all BEV copulas the upper tail leading term is well defined (Jaworski 2004) and equals

$$L_{C_A}(x,y) = (x+y)\left(1 - A\left(\frac{x}{x+y}\right)\right).$$

Furthermore, note that BEV copulas can be characterized in terms of the upper tail leading term.

$$C_A(u_1, u_2) = \exp\left(ln(u_1 u_2) - L_{C_A}(-\ln(u_1), -\ln(u_2))\right).$$

Specifically, we will deal with two measures: the upper tail index λ and the maximum upper tail index ω.

$$\lambda(C_A) = \lim_{t\to 1^-} \frac{1 - 2t + C_A(t,t)}{1-t} = L_A(1,1) = 2(1 - A(0.5)).$$

$$\omega(C_A) = 2 \limsup_{(x,y)\to(1,1)} \frac{C_A(x,y) - W(x,y)}{2 - x - y} = 2\max(L_A(t, 1-t) : t \in [0,1])$$

$$= 2 - 2\min(A(t) : t \in [0,1]).$$

While λ is a well-known object (compare Nelsen (2006), Joe et al. (2015)), ω seems to be absent in the literature.

4   Main Results

Our goal is to determine for given dependence measure $\kappa_* \in (0,1)$:

1. the lower bound L for Pickands functions A with $\kappa(A) \leq \kappa_*$;
2. the upper bound U for Pickands functions A with $\kappa(A) \geq \kappa_*$;
3. the set $\Omega_\kappa(\kappa^*)$ being the union of graphs of Pickands functions A with $\kappa(A) = \kappa_*$

$$\Omega_\kappa(\kappa^*) = \{(s,a) : A(s) = a, \kappa(A) = \kappa_*\}.$$

The following two families of piece-wise linear Pickands dependence functions will be used extensively in subsequent derivations.

1. We denote by $\mathcal{A}_m$ the set of V-shaped Pickands functions.

$$\mathcal{A}_m = \{A_m(t,a): \ t \in (0,1), \ \max(t, 1-t) \le a \le 1\}, \qquad \text{where}$$

$$A_m(t,a)(s) = \max\left(1 - s\frac{1-a}{t}, 1 - (1-s)\frac{1-a}{1-t}\right) = \begin{cases} 1 - s\frac{1-a}{t} & \text{for } s \in [0,t], \\ 1 - (1-s)\frac{1-a}{1-t} & \text{for } s \in (t,1]. \end{cases} \tag{4}$$

Note that the BEV copula corresponding to $\mathcal{A}_m$ belongs to the Marshall-Olkin family

$$\Psi(A_m(t,a)) = MO(\alpha, \beta), \quad \alpha = \frac{1-a}{t}, \quad \beta = \frac{1-a}{1-t}.$$

Indeed

$$\Psi(A_m(t,a))(u,v) = \exp\left(\ln(uv)\max\left(1 - \alpha\frac{\ln u}{\ln(uv)}, 1 - \beta\frac{\ln v}{\ln(uv)}\right)\right)$$
$$= \exp\left(\min\left(\ln(uv) - \alpha\ln(u), \ln(uv) - \beta\ln(v)\right)\right)$$
$$= \min\left(u^{1-\alpha}v, uv^{1-\beta}\right) = MO(\alpha, \beta)(u,v).$$

2. We denote by $\mathcal{A}_M$ the set of ∨-shaped Pickands functions.

$$\mathcal{A}_M = \left\{A_M(t_1, t_2): \ 0 \le t_1 \le \frac{1}{2} \le t_2 \le 1\right\}, \qquad \text{where}$$

$$A_M(t_1, t_2)(s) = \max\left(s, 1-s, (1-t_1)\frac{t_2 - s}{t_2 - t_1} + t_2\frac{s - t_1}{t_2 - t_1}\right) \tag{5}$$
$$= \begin{cases} 1 - s & \text{for } s \in [0, t_1], \\ (1-t_1)\frac{t_2 - s}{t_2 - t_1} + t_2\frac{s - t_1}{t_2 - t_1} & \text{for } s \in (t_1, t_2), \\ s & \text{for } s \in [t_2, 1]. \end{cases}$$

The BEV copula corresponding to $\mathcal{A}_M$ belongs to the two parameter family of copulas C having a monomial diagonal $\delta(t) = C(t,t) = t^p$, $p \in [1,2]$.

$$\Psi(A_M(t_1, t_2))(u,v) = \min\left(u, v, u^\alpha v^\beta\right), \quad \alpha = \frac{(1-t_1)(2t_2 - 1)}{t_2 - t_1}, \quad \beta = \frac{(1-2t_1)t_2}{t_2 - t_1}.$$

Indeed

$$\Psi(A_M(t_1, t_2))(u,v) = \exp\left(\ln(uv)\max\left(\frac{\ln u}{\ln(uv)}, \frac{\ln v}{\ln(uv)}, \alpha\frac{\ln u}{\ln(uv)} + \beta\frac{\ln v}{\ln(uv)}\right)\right)$$
$$= \exp\left(\min\left(\ln(u), \ln(v), \alpha\ln(u) + \beta\ln(v)\right)\right) = \min\left(u, v, u^\alpha v^\beta\right).$$

The interest in these two families comes from the fact that, since the Pickands functions are convex and take value 1 at the ends of the unit interval we can bound them from the top and from the bottom by piece-wise linear functions

A_m and A_M. Indeed, let A be a Pickand function. We fix a point $t_0 \in (0,1)$, put $a = A(t_0)$ and select

$$\dot{a} \in \left[\frac{\partial A(t_0^-)}{\partial t}, \frac{\partial A(t_0^+)}{\partial t} \right].$$

We construct a piece-wise linear Pickands functions A_m and A_M which bound A from top and from bottom (Fig. 1). We choose

$$A_m(s) = A_m(t_0, a)(s) = \max \left(1 - s\frac{1-a}{t_0}, 1 - (1-s)\frac{1-a}{1-t_0} \right),$$

$$A_M(s) = A_M(t_1, t_2)(s) = \max \left(s, 1 - s, \dot{a}(s - t_0) + a \right),$$

where for $\dot{a} \neq \pm 1$, t_1 and t_2 are given by

$$\dot{a}(t_1 - t_0) + a = 1 - t_1 \text{ and } \dot{a}(t_2 - t_0) + a = t_2.$$

for $\dot{a} = -1$ we put $t_1 = 0, t_2 = 1/2$ and for $\dot{a} = 1$ we put $t_1 = 1/2, t_2 = 1$.
    Finally we get

$$\forall s \in [0,1] \quad A_m(t_0, a)(s) \geq A(s) \geq A_M(t_1, t_2)(s) \quad \text{and} \quad \kappa(A_m(t_0, a)) \leq \kappa(A) \leq \kappa(A_M(t_1, t_2)).$$

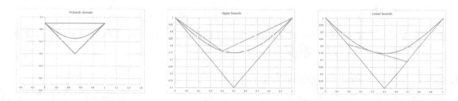

Fig. 1. Graphs illustrating the Pickands domain, upper bound A_m and lower bound A_M.

The conditions $\kappa(A) \leq \kappa_*$ or $\kappa(A) \geq \kappa_*$ give us an extra lower and upper bounds on A, L and U. Furthermore these bounds can be expressed in terms of A_m and A_M.

Theorem 1. *Let κ be a continuous dependence function. Then for every Pickands function A and $\kappa_* \in (0,1)$ it holds:*

1. *If $\kappa(A) \leq \kappa_*$ then for every $s \in [0,1]$ $A(s) \geq L(s)$, where the function $L : [0,1] \to [0.5,1]$ is given by*

$$L(s) = \min(A(s) : A \in \mathcal{A}_m, \kappa(A) = \kappa_*).$$

2. *If $\kappa(A) \geq \kappa_*$ then for every $s \in [0,1]$ $A(s) \leq U(s)$, where the function $U : [0,1] \to [0.5,1]$ is given by*

$$U(s) = \max(A(s) : A \in \mathcal{A}_M, \kappa(A) = \kappa_*).$$

3. *The above bounds are strict and*

$$\Omega_\kappa(\kappa_*) = \{(s,a) \in [0,1]^2 : L(s) \le a \le U(s)\}.$$

Note that the upper bound U is a convex function, hence a Pickands one ($U \in \mathcal{A}$). Thus the function

$$Q_U(u,v) = \exp\left(ln(uv)\, U\left(\frac{\ln u}{\ln(uv)}\right)\right), \quad (u,v) \in (0,1]^2 \setminus \{(1,1)\}$$

extends to a BEV copula (a point-wise minimum of some Marshall-Olkin copulas).

But the lower bound L need not to be convex. It depends on κ. Hence the function

$$Q_L(u,v) = \exp\left(ln(uv)\, L\left(\frac{\ln u}{\ln(uv)}\right)\right), \quad (u,v) \in (0,1]^2 \setminus \{(1,1)\}$$

might extend only to a quasi-copula (for definition and properties see Nelsen (2006), Durante and Sempi (2016)). The proof of Theorem 1 and more detailed discussion are provided in Sect. 7.1.

The above allows us to give the following characterization in terms of copulas or quasi-copulas:

Corollary 1. *Let κ be a continuous dependence function. Then for every BEV copula C and $\kappa_* \in (0,1)$ it holds:*

1. *If $\kappa(C) \le \kappa_*$ then for every $u,v \in [0,1]$*

$$C(u,v) \le Q_L(u,v) = \max\{\min(u,v,u^\alpha v^\beta) : \alpha,\beta \in [0,1],\, \alpha+\beta \in [1,2],\, \kappa(\min(u,v,u^\alpha v^\beta)) = \kappa_*\}.$$

2. *If $\kappa(C) \ge \kappa_*$ then for every $u,v \in [0,1]$*

$$C(u,v) \ge Q_U(u,v) = \min\{D(u,v) : D \in MO,\ \kappa(D) = \kappa_*\}.$$

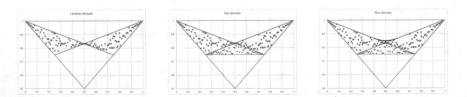

Fig. 2. Graphs illustrating the Ω domain for λ, β and ϕ (left), τ and ω (center) and ρ (right).

5   Special Cases

In this section we investigate the examples of the continuous dependence measures listed in Sect. 3. It shows that up to some parametrization of κ_* the sets $\Omega_\kappa(\cdot)$, for $\kappa = \beta, \phi, \lambda, \tau, \omega, \rho$, are increasing (Fig. 2). For $\alpha \in [0, 1]$

$$\Omega_\beta \left(2^\alpha - 1\right) = \Omega_\phi \left(\frac{2\alpha}{3 - \alpha}\right) = \Omega_\lambda(\alpha) \subset \Omega_\tau(\alpha) = \Omega_\omega\left(\frac{2\alpha}{1 + \alpha}\right) \subset \Omega_\rho\left(\frac{3\alpha}{2 + \alpha}\right).$$

We describe this in more details in the following propositions.

Proposition 1. *For $\kappa \in \{\beta, \phi, \lambda\}$ we have*

$$\Omega_\kappa(\kappa_*) = \left\{(s, a) : \min\left(A_m(1 - t_*, t_*)(s), A_m(t_*, t_*)(s)\right) \le a \le A_m\left(\frac{1}{2}, \frac{3t_* - 1}{2t_*}\right)\right\}, \quad where$$

$$t_* = \begin{cases} \frac{\ln 2}{\ln 2 + \ln(1 + \kappa_*)} & for \ \kappa = \beta, \\ \frac{2 + \kappa_*}{2 + 4\kappa_*} & for \ \kappa = \phi, \\ \frac{1}{1 + \kappa_*} & for \ \kappa = \lambda. \end{cases}$$

In terms of copulas the above can be restated as:

Proposition 2. *Let C be a BEV copula such that $\beta(C) = 2^\alpha - 1$ or $\phi(C) = \frac{2\alpha}{3-\alpha}$ or $\lambda(C) = \alpha$, $\alpha \in [0, 1]$, then for all $u, v \in [0, 1]$*

$$MO(\alpha, \alpha)(u, v) \le C(u, v) \le \max(MO(1, \alpha)(u, v), MO(\alpha, 1)(u, v)).$$

Note that the upper bound is only a quasi-copula.

Proposition 3. *For $\kappa \in \{\tau, \omega\}$*

$$\Omega_\kappa(\kappa_*) = \left\{(s, a) : A_M(1 - t_*, t_*)(s) \le a \le A_m\left(\frac{1}{2}, \frac{3t_* - 1}{2t_*}\right)\right\}, \quad where$$

$$t_* = \begin{cases} \frac{1}{1 + \kappa_*} & for \ \kappa - \tau, \\ 1 - \frac{\kappa_*}{2} & for \ \kappa = \omega. \end{cases}$$

In terms of copulas the above can be restated as:

Proposition 4. *Let C be a BEV copula such that $\tau(C) = \alpha$ or $\omega(C) = \frac{2\alpha}{1+\alpha}$, $\alpha \in [0, 1]$, then for all $u, v \in [0, 1]$*

$$MO(\alpha, \alpha)(u, v) \le C(u, v) \le \min(u, v, u^{1-\alpha}v^{1-\alpha}).$$

The last case $\kappa = \rho$ gives the set Ω bigger than previous ones.

Proposition 5.

$$\Omega_\rho(\rho_*) = \left\{(s, a) : A_M(1 - t_*, t_*)(s) \le a \le U_\rho(\rho_*)(s)\right\}, \quad where \ t_* = \frac{3 - \rho_*}{3 + \rho_*} \quad and$$

$$U_\rho(\rho_*)(t) = \begin{cases} 1 - \dfrac{2\rho_*}{3 - \rho_*} t & \text{if } 0 \leq t < \frac{3-\rho_*}{6+\rho_*}, \\[2mm] 1 - \dfrac{6 + 2\rho_* - 2\sqrt{\Delta}}{15 + \rho_*} & \text{if } \frac{3-\rho_*}{6+\rho_*} \leq t \leq \frac{3+2\rho_*}{6+\rho_*}, \\[2mm] 1 - \dfrac{2\rho_*}{3 - \rho_*}(1 - t) & \text{if } \frac{3+2\rho_*}{6+\rho_*} < t \leq 1. \end{cases}$$

$$\Delta = 9(1 - \rho_*) + (15 + \rho_*)(2t - 1)^2.$$

The function $U_\rho(\rho_*) \in \mathcal{A}$ is symmetric with respect to $1/2$, convex and continuously differentiable on $(0, 1)$.

In terms of copulas the above can be restated as:

Proposition 6. *Let C be a BEV copula such that $\rho(C) = \frac{3\alpha}{2+\alpha}$, $\alpha \in [0, 1]$, then for all $u, v \in [0, 1]$*

$$C_U(u, v) \leq C(u, v) \leq \min(u, v, u^{1-\alpha}v^{1-\alpha}), \quad \text{where}$$

$$C_U(u, v) = \begin{cases} vu^{1-\alpha} & \text{if } u^{2+3\alpha} \geq v^2, \\[2mm] \exp\left(\frac{9-\alpha}{15+9\alpha}\ln(uv) \right. \\ \quad \left. - \frac{2\sqrt{2+\alpha}}{15+9\alpha}\sqrt{12(\ln^2 u + \ln^2 v) - 6(1 + 3\alpha)\ln u \ln v}\right) & \text{if } u^{2+3\alpha} < v^2 \wedge v^{2+3\alpha} < u^2, \\[2mm] uv^{1-\alpha} & \text{if } v^{2+3\alpha} \geq u^2. \end{cases}$$

Propositions 5 and 3 for $\kappa = \tau$ are proved in Kamnitui et al. (2019). The remaining proofs are included in Sect. 7.2.

6 Constraints

Let C be a bivariate extreme value copula. We will briefly describe how different the dependence measures can be.

1. If $\omega(C) = \omega_* \in [0, 1]$, then:

$$\tau(C), \lambda(C) \in \left[\frac{\omega_*}{2 - \omega_*}, \omega_*\right], \quad \rho(C) \in \left[\frac{3\omega_*}{4 - \omega_*}, \frac{3\omega_*(8 - 5\omega_*)}{(4 - \omega_*)^2}\right],$$

$$\beta(C) \in \left[\exp\left(\frac{\omega_*}{2 - \omega_*}\ln 2\right) - 1, 2^{\omega_*} - 1\right], \quad \phi(C) \in \left[\frac{\omega_*}{3 - 2\omega_*}, \frac{2\omega_*}{3 - \omega_*}\right].$$

2. If $\lambda(C) = \lambda_* \in [0, 1]$, then:

$$\tau(C) \in \left[\frac{\lambda_*}{2 - \lambda_*}, \lambda_*\right], \quad \rho(C) \in \left[\frac{3\lambda_*}{4 - \lambda_*}, \frac{3\lambda_*(8 - 5\lambda_*)}{(4 - \lambda_*)^2}\right],$$

$$\omega(C) \in \left[\lambda_*, \frac{2\lambda_*}{1 + \lambda_*}\right], \quad \beta(C) = 2^{\lambda_*} - 1, \quad \phi(C) = \frac{2\lambda_*}{3 - \lambda_*}.$$

3. If $\tau(C) = \tau_* \in [0, 1]$ then:

$$\beta(C) \in \left[2^{\tau_*} - 1, \exp\left(\frac{2\tau_*}{1 + \tau_*} \ln 2\right) - 1\right], \quad \omega(C), \lambda(C) \in \left[\tau_*, \frac{2\tau_*}{1 + \tau_*}\right], \quad \phi(C) \in \left[\frac{2\tau_*}{3 - \tau_*}, \frac{4\tau_*}{3 + \tau_*}\right].$$

4. If $\rho(C) = \rho_* \in [0, 1]$ then:

$$\beta(C) \in \left[\exp\left(\frac{4\rho_*}{3 + \rho_* + 3\sqrt{1 - \rho_*}} \ln 2\right) - 1, \exp\left(\frac{4\rho_*}{3 + \rho_*} \ln 2\right) - 1\right],$$

$$\omega(C), \lambda(C) \in \left[\frac{4\rho_*}{3 + \rho_* + 3\sqrt{1 - \rho_*}}, \frac{4\rho_*}{3 + \rho_*}\right], \quad \phi(C) \in \left[\frac{8\rho_*}{9 - \rho_* + 9\sqrt{1 - \rho_*}}, \frac{8\rho_*}{9 - \rho_*}\right].$$

Note that all the above bounds are strict. They are attained for $A = A_m(1/2, a)$, $A = A_m(a, a)$ or $A = A_M(1 - a, a)$, where $a \in [1/2, 1]$. The detailed calculations are provided in Sect. 7.3. The bounds for λ given τ or ρ are discussed also in Kamnitui et al. (2019).

7 Proofs and Auxiliary Results

7.1 Proof of Theorem 1

We start with correctness of the definition of L and U bounds. First we observe that the mappings A_m and A_M which parametrize the piece-wise linear Pickands functions

$$A_m : D_m = \{(t, a) \in [0, 1]^2 : \min(t, 1 - t) \le a\} \longrightarrow (\mathcal{A}, ||\cdot||_\infty), \quad (t, a) \longmapsto A_m(t, a),$$

$$A_M : D_M = [0, 1/2] \times [1/2, 1] \longrightarrow (\mathcal{A}, ||\cdot||_\infty), \quad (t_1, t_2) \longmapsto A_M(t_1, t_2),$$

are continuous. Hence also the compositions of the continuous dependence measure κ and the parametrizations A_m and A_M are continuous.

$$\kappa_m : D_m \longrightarrow [0, 1], \quad \kappa_m(t, a) = \kappa(A_m(t, a)).$$

$$\kappa_M : D_M \longrightarrow [0, 1], \quad \kappa_M(t_1, t_2) = \kappa(A_M(t_1, t_2)).$$

Therefore the level sets $\kappa_m^{-1}(\{\kappa_*\})$ and $\kappa_M^{-1}(\{\kappa_*\})$ are nonempty and compact (i.e. closed and bounded). Since the mappings Φ_m and Φ_M for a fixed parameter $s \in [0, 1]$

$$\Phi_m^s : D_m \longrightarrow \left[\frac{1}{2}, 1\right], \quad \Phi_m^s(t, a) = A_m(t, a)(s),$$

$$\Phi_M^s : D_M \longrightarrow \left[\frac{1}{2}, 1\right], \quad \Phi_M^s(t_1, t_2) = A_m(t_1, t_2)(s),$$

are continuous, the images of the level sets are compact. Hence both minimum and maximum are attainable. Moreover the bounds L and U are given by

$$L(s) = \min\{x \in \Phi_m^s(\kappa_m^{-1}(\{\kappa_*\}))\}, \quad U(s) = \max\{x \in \Phi_M^s(\kappa_M^{-1}(\{\kappa_*\}))\}.$$

Next we provide another characterizations of L and U.

Lemma 1. *For every* $(t, a) \in D_m$

$$\kappa(A_m(t, a)) \leq \kappa_* \iff a \geq L(t).$$

Proof. We fix t from $(0, 1)$ and select τ and a such that

$$A_m(\tau, a)(t) = L(t), \quad \kappa(A_m(\tau, a)) = \kappa_*.$$

Since the piece-wise linear $A_m(t, L(t))$ is greater than the convex function $A_m(\tau, a)$ and for $a \geq L(t)$, $A_m(t, a)$ is greater than $A_m(t, L(t))$, we get

$$\kappa(A_m(t, a)) \leq \kappa(A_m(t, L(t))) \leq \kappa(A_m(\tau, a)) = \kappa_*.$$

Which concludes the proof of the implication "$\Leftarrow$".

To prove the next implication we observe that since $L(s)$ is a minimum, we have for all $(t, a) \in D_m$

$$\kappa(A_m(t, a)) = \kappa_* \implies a \geq L(t).$$

The same is valid when $\kappa(A_m(t, a)) < \kappa_*$. We consider a path consisting of A_m functions joining $A_m(t, a)$ with $A_m(1/2, 1/2)$. First we take

$$A(\xi) = A_m(t, \xi), \quad \xi \in (\max(t, 1 - t), a],$$

then

$$A(\xi) = \begin{cases} A_m(\xi, \xi), & \xi \in [1/2, t] & \text{for } t \geq \frac{1}{2}, \\ A_m(1 - \xi, \xi), & \xi \in [1/2, 1 - t] & \text{for } t < \frac{1}{2}. \end{cases}$$

At one end we have $\kappa(A_m(t, a)) < \kappa_*$, on the other $\kappa(A_m(1/2, 1/2)) = \kappa(M) = 1$. Hence for some ξ_*, $\kappa(A(\xi_*))$ is equal to κ_*. Since $L(t)$ is a minimum we get

$$L(t) \leq A(\xi_*)(t) = \max(\xi_*, t, 1 - t) \leq a.$$

$$\square$$

The similar results are valid also for U and A_M.

Lemma 2.

1. *For every* $(t_1, t_2) \in D_M$

$$\kappa(A_M(t_1, t_2)) \geq \kappa_* \implies \forall s \in [0, 1] \ A_M(t_1, t_2)(s) \leq U(s).$$

2. *For every* $t \in [0, 1]$ *and* a *such that* $\max(t, 1 - t) \leq a \leq U(t)$, *there exists* $(t_1, t_2) \in D_M$ *such that*

$$A_M(t_1, t_2)(t) = a \quad \text{and} \quad \kappa(A_M(t_1, t_2)) \geq \kappa_*.$$

Proof.

Point 1. We consider a "monotonic" path consisting of A_M functions joining $A_M(t_1, t_2)$ with $A_M(0, 1)$. First we take

$$A(\xi) = A_M(\xi, 1), \quad \xi \in [0, t_1],$$

then
$$A(\xi) = A_M(t_1, 1 + t_1 - \xi), \quad \xi \in (t_1, 1 - t_2 + t_1].$$

At one end we have $\kappa(A_M(t_1, t_2) \geq \kappa_*$, on the other $\kappa(A_M(0, 1)) = \kappa(\Pi) = 0$. Hence for some ξ_*, $\kappa(A(\xi_*))$ is equal to κ_*. Since $U(t)$ is a maximum we get for any $s \in (0, 1)$
$$U(s) \geq A(\xi_*)(s) \geq A_M(t_1, t_2)(s).$$

Point 2. Since $U(t)$ is the maximum there exists $(s_1, s_2) \in D_M$ such that
$$A_M(s_1, s_2)(t) = U(t) \quad \text{and} \quad \kappa(A_M(s_1, s_2)) = \kappa_*.$$

We consider a "monotonic" path
$$A(\xi) = A_M(\xi/2 + (1 - \xi)s_1, \xi/2 + (1 - \xi)s_2); \quad \xi \in [0, 1].$$

Note that
$$A(0)(t) = U(t) \geq a \quad \text{and} \quad A(1)(t) = \max(t, 1 - t) \leq a.$$

Hence there exists ξ_* such that
$$A(\xi_*)(t) = a.$$

We put
$$t_1 = \xi_*/2 + (1 - \xi_*)s_1, \quad t_2 = \zeta_*/2 + (1 - \xi_*)s_2.$$

Since $s_1 \leq t_1 \leq t_2 \leq s_2$, $A_M(s_1, s_2)$ dominates $A_M(t_1, t_2)$ and
$$\kappa(A_M(t_1, t_2)) \geq A_M(s_1, s_2)) \geq \kappa_*.$$

$\square$

Proof of the Theorem

Point 1 of Theorem 1 follows from Lemma 1.

We fix $s \in (0, 1)$. Since every Pickands function A is dominated by a piecewise linear Pickands function $A_m(s, A(s))$, we have
$$\kappa(A_m(s, A(s))) \leq \kappa(A) \leq \kappa_*.$$

Hence $A(s) \geq L(s)$.

The proof of point 2 of Theorem 1 is similar, it follows from the first point of Lemma 2.

We fix $s \in (0, 1)$. Since every Pickands function A dominates some piece-wise linear Pickands function $A_M(t_1, t_2))$, such that
$$A_M(t_1, t_2)(s) = A(s),$$

we have
$$\kappa(A_M(t_1, t_2)) \geq \kappa(A) \geq \kappa_*.$$

Hence $A(s) \leq U(s)$.

The proof of point 3 of Theorem 1 follows from the previous two points and both Lemmas 1 and 2.

Let A be a Pickands functions with $\kappa(A) = \kappa_*$. We fix $s \in (0,1)$. From the previous points we get

$$L(s) \leq A(s) \leq U(s).$$

Next we fix a, $L(s) \leq a \leq U(s)$. For a piece-wise linear function $A_m(s,a)$ we have $\kappa(A_m(s,a)) \leq \kappa_*$. Next we select $(t_1, t_2) \in D_M$ such that

$$A_M(t_1, t_2)(t) = a \quad \text{and} \quad \kappa(A_M(t_1, t_2)) \geq \kappa_*.$$

We join these two functions by a path consisting of Pickands functions

$$A(\xi) = \xi A_M(t_1, t_2) + (1 - \xi)A_m(s,a), \quad \xi \in [0,1].$$

Since $\kappa(A(0)) \leq \kappa_*$ and $\kappa(A(1)) \geq \kappa_*$, there exists ξ_* such that $\kappa(A(\xi_*)) = \kappa_*$. Since for each ξ $A(\xi)(s) = a$, the point (s,a) belongs to the graph of the Pickands function with κ equal κ_*. Hence the point (s,a) belongs to the set $\Omega_\kappa(\kappa_*)$, being the union of graphs. □

7.2 Proofs of Propositions 1,...,6

It shows that for $A_m(t,a)$, i.e. for Marshall-Olkin copulas, Spearman ρ, Kendall τ and the maximum upper tail index do not depend on t (for ρ and τ see Nelsen (2006) Examples 5.5 and 5.7 or Kamnitui et al. (2019)):

$$\rho(A_m(t,a)) = 3\frac{1-a}{1+a}, \quad \tau(A_m(t,a)) = \frac{1-a}{a}, \quad \omega(A_m(t,a)) = 2(1 - \min(A_m(t,a))) = 2(1-a).$$

Since

$$A_m(t,a)\left(\frac{1}{2}\right) = 1 - \frac{1-a}{2\max(t, 1-t)},$$

we get

$$\lambda(A_m(t,a)) = \frac{1-a}{\max(t, 1-t)}, \quad \beta(A_m(t,a)) = \exp\left(\ln(2)\frac{1-a}{\max(t, 1-t)}\right) - 1,$$

$$\phi(A_m(t,a)) = 4\frac{1-a}{3\max(t, 1-t) - 1 + a}.$$

For $A_M(t_1, t_2)$ we get (for ρ and τ see Kamnitui et al. (2019)):

$$\rho(A_M(t_1, t_2)) = 3\frac{2 - 3t_1 + 2t_2 + 5t_1 t_2}{(2 - t_1)(1 + t_2)}, \quad \tau(A_M(t_1, t_2)) = \frac{1 - 3t_1 - t_2 + 4t_1 t_2}{t_2 - t_1},$$

$$\lambda(A_M(t_1, t_2)) = \frac{1}{2}\frac{-1 + t_1 + 3t_2 - 4t_1 t_2}{t_2 - t_1}, \quad \omega(A_M(t_1, t_2)) = 2(1 - \min(t_2, 1 - t_1)).$$

For $t_1 + t_2 = 1$ the formulas simplify. Namely for $t \in [1/2, 1]$ we have

$$\lambda(A_M(1 - t, t)) = \omega(A_M(1 - t, t)) = \tau(A_M(1 - t, t)) = 2(1 - t), \quad \rho(A_M(1 - t, t)) = \frac{3(1 - t)(5t - 1)}{(1 + t)^2}.$$

Proof of Proposition 1.

We observe that Pickands functions A with fixed value of respectively the tail index λ, Blomqvist β or Spearman foot-rule ϕ, have a fixed value at $1/2$, say a_*. Hence all such A are dominated by $A_m(1/2, a_*)$.

On the other side the parameters of the piece-wise linear Pickands functions $A_m(t, a)$ with fixed value a_*, $a_* \in [1/2, 1]$, at $1/2$ are subject to the following constraints:

$$0 \le t \le \frac{1}{2}, \quad 1 - t \le a \le 1, \quad 1 - \frac{1}{2}\frac{1-a}{1-t} = a_* \quad \text{or} \quad \frac{1}{2} \le t \le 1, \quad t \le a \le 1, \quad 1 - \frac{1}{2}\frac{1-a}{t} = a_*.$$

These constraints are equivalent to

$$0 \le t \le 1, \quad 1 - t \le a \le 1, \quad a = 1 - 2(1 - a_*)\min(t, 1 - t).$$

Hence the lower bound L equals

$$L(s) = \max(s, 1-s, 1-2(1-a_*)\max(s, 1-s)) = \min(A_m(1-t_*, t_*), A_m(t_*, t_*)),$$

where $t_* = \frac{1}{3-2a_*}$. Since

$$a_* = \frac{3t_* - 1}{2t_*},$$

we have

$$A_m(1/2, u_*) = A_m\left(1/2, \frac{3t_* - 1}{2l_*}\right).$$

To conclude the proof one has to observe that for $\kappa(A) = \kappa_*$, $\kappa \in \{\lambda, \beta, \phi\}$,

$$a_* = \begin{cases} 1 - \frac{1}{2}\kappa_* & \text{for } \kappa = \lambda, \\ 1 - \frac{\ln(1+\kappa_*)}{2\ln 2} & \text{for } \kappa = \beta, \\ \frac{4-\kappa_*}{4+\kappa_*} & \text{for } \kappa = \phi. \end{cases}$$

Hence

$$t_* = \begin{cases} \frac{\ln 2}{\ln 2 + \ln(1+\kappa_*)} & \text{for } \kappa = \beta, \\ \frac{2+\kappa_*}{2+4\kappa_*} & \text{for } \kappa = \phi, \\ \frac{1}{1+\kappa_*} & \text{for } \kappa = \lambda. \end{cases}$$

$\square$

Proof of Proposition 3, $\kappa = \omega$.

We observe that all Pickands functions A such that the upper tail maximum index is fixed, say $\omega(A) = \omega_*$, $\omega_* \in [0, 1]$ are tangent to the segment $[1 - t_*, t_*] \times \{t_*\}$, where $t_* = 1 - \frac{1}{2}\omega_*$. Hence they dominate piece-wise linear Pickands function $A_M(1 - t_*, t_*)$. On the other side the parameters of the piece-wise linear Pickands functions $A_M(t_1, t_2)$ with $\omega(A_M(t_1, t_2)) = \omega_*$, are subject to the following constraints:

$$0 \le t_1 \le \frac{1}{2} \le t_2 \le 1, \quad \min(1 - t_1, t_2) = t_*.$$

The Pickands functions $A_M(1 - t_*, t_2)$ with $t_2 \geq t_*$, are dominated by $A_M(1 - t_*, 1)$ and $A_M(t_1, t_*)$ with $1 - t_1 \geq t_*$, are dominated by $A_M(1, t_*)$. Hence the upper bound U equals

$$U(s) = \max(A_M(1 - t_*, t_2)(s), A_M(1, t_*)(s)) = A_m\left(\frac{1}{2}, \frac{3t_* - 1}{2t_*}\right).$$

$\square$

Propositions 5 and 3 $\kappa = \tau$ follow from Kamnitui et al. (2019), where the upper and lower bounds for the sets Ω_ρ and Ω_τ are calculated.

Propositions 2, 4 and 6 are obtained by substitution $t_* = \frac{1}{1+\alpha}$ and application of the Pickands mapping Ψ (compare Corollary 1). We get for BEV copula C with fixed dependence measure κ, $\kappa \in \{\lambda, \beta, \phi, \omega, \tau, \rho\}$

$$\Psi(U)(u, v) = \exp\left(\ln(uv)U\left(\frac{\ln u}{\ln(u, v)}\right)\right) \leq C(u, v) \leq \exp\left(\ln(uv)L\left(\frac{\ln u}{\ln(u, v)}\right)\right).$$

7.3   Proofs of "constraints"

Step 1. The upper tail index λ.

Let κ be any continuous dependence measure. Since λ depends only on the value of the Pickands function A at $1/2$

$$\lambda(A) = 2\left(1 - A\left(\frac{1}{2}\right)\right).$$

we get for A, with fixed κ, say $\kappa(A) = \kappa_*$,

$$2\left(1 - U\left(\frac{1}{2}\right)\right) \leq \lambda(A) \leq 2\left(1 - L\left(\frac{1}{2}\right)\right),$$

where L and U are lower and upper bounds of $\Omega_\kappa(\kappa_*)$. Since $L(1/2)$ and $U(1/2)$ are are attained by some Pickands functions with $\kappa = \kappa_*$, the bounds are strict. For $\kappa = \omega, \tau$ or ρ we have:

$$2 - 2L(1/2) = 2 - 2A_M(1 - t_*, t_*)(1/2) = 2 - 2t_* = \begin{cases} \frac{2\kappa_*}{1+\kappa_*} & \text{for } \kappa = \tau, \\ \kappa_* & \text{for } \kappa = \omega, \\ \frac{4\kappa_*}{3+\kappa_*} & \text{for } \kappa = \rho. \end{cases}$$

The second bound is more complicated. For $\kappa = \omega$ and τ we have:

$$2 - 2U(1/2) = 2 - 2A_m\left(1/2, \frac{3t_* - 1}{2t_*}\right)(1/2) = \frac{1}{t_*} - 1 = \begin{cases} \kappa_* & \text{for } \kappa = \tau, \\ \frac{\kappa_*}{2-\kappa_*} & \text{for } \kappa = \omega. \end{cases}$$

For $\kappa = \rho$ we have:

$$2 - 2U(1/2) = 2\frac{6 + 2\rho_* - 6\sqrt{1 - \rho_*}}{15 + \rho_*} = \frac{4\rho_*}{3 + \rho_* + 3\sqrt{1 - \rho_*}}.$$

Finally we obtain the following constraints (compare Kamnitui et al. (2019) Sect. 6)

$$\frac{\omega}{2-\omega} \le \lambda \le \omega, \quad \tau \le \lambda \le \frac{2\tau}{1+\tau}, \quad 2\frac{6+2\rho-6\sqrt{1-\rho}}{15+\rho} = \frac{4\rho_*}{3+\rho_*+3\sqrt{1-\rho_*}} \le \lambda \le \frac{4\rho}{3+\rho}.$$

Applying the inverse functions we get the bounds for τ, ρ and ω given λ. Since β, ϕ and λ are functionally dependent, the constraints with β and ϕ follows from the above.

Step 2. The maximum upper tail index ω.

We recall from the proof of Proposition 3 that the Pickands function A with $\omega(A) = \omega_*$ is bounded by piece-wise linear Pickands functions with the same ω

$$A_M(\omega_*/2, 1-\omega_*/2)(s) \le A(s) \le A_m(t, 1-\omega_*/2)(s)),$$

where A has a minimum at the point t. Hence

$$\tau \le \omega \le \frac{2\tau}{1+\tau}, \quad 2\frac{6+2\rho-6\sqrt{1-\rho}}{15+\rho} = \frac{4\rho_*}{3+\rho_*+3\sqrt{1-\rho_*}} \le \omega \le \frac{4\rho}{3+\rho}.$$

Applying the inverse functions we get the bounds for τ and ρ given ω.

References

Durante, F., Sempi, C.: Principles of Copula Theory. CRC Press, Boca Raton (2016)

de Haan, L., Ferreira, A.: Extreme Value Theory: An Introduction. Springer, New York (2006)

Genest, C., Favre, A.-C.: Everything you always wanted to know about copula modeling but were afraid to ask. J. Hydrol. Eng. **12**, 347–368 (2007)

Genest, C., Nešlehová, J.: Copula modeling for extremes. In: El-Shaarawi, A.H., Piegorsch, W.W. (eds.) Encyclopedia of Environmetrics, vol. 2, 2nd edn, pp. 530–541. Wiley, Chichester (2012)

Ghoudi, K., Khoudraji, A., Rivest, L.-P.: Propriétés statistiques des copules de valeurs extrêmes bidimensionnelles. Canad. J. Statist. **26**, 187–197 (1998)

Gudendorf, G., Segers, J.: Extreme value copulas. In: Jaworski, P., Durante, F., Härdle, W.K., Rychlik, T. (eds.), Copula Theory and Its Applications: Proceedings of the Workshop Held in Warsaw, 25–26 September 2009, pp. 127–145. Springer, New York (2010)

Jaworski, P.: On uniform tail expansions of bivariate copulas. Applicationes Mathematicae **31**(4), 397–415 (2004)

Jaworski, P.: On uniform tail expansions of multivariate copulas and wide convergence of measures. Applicationes Mathematicae **33**(2), 159–184 (2006)

Jaworski, P.: Tail behaviour of copulas. In: Jaworski, P., Durante, F., Härdle, W.K., Rychlik, T. (eds.), Copula Theory and Its Applications: Proceedings of the Workshop Held in Warsaw, 25–26 September 2009, pp. 161–186. Springer, New York (2010)

Jaworski, P., Durante, F., Härdle, W.K., Rychlik, T. (eds.): Copula Theory and Its Applications: Proceedings of the Workshop Held in Warsaw, 25–26 September 2009. Springer, New York (2010)

Joe, H.: Dependence Modeling with Copulas. CRC Press, Boca Raton (2015)

Kamnitui, N., Genest, C., Jaworski, P., Trutschnig, W.: On the size of the class of bivariate extreme-value copulas with fixed value of Spearman's rho or Kendall's tau. J. Math. Anal. Appl. **472**, 920–936 (2019)

McNeil, A.J., Frey, R., Embrechts, P.: Quantitative Risk Management: Concepts, Techniques and Tools. Princeton University Press, Princeton, (2015)

Nelsen, R.B.: An Introduction to Copulas, 2nd edn. Springer, New York (2006)

Pickands, J.: Multivariate extreme value distributions. In: Proceedings of the 43rd Session of the International Statistical Institute, Volume 2, Buenos Aires, vol. 49, pp. 859–878, 894–902 (1981)

Sklar, A.: Fonctions de répartition à n dimensions et leurs marges. Publ. Inst. Statist. Univ. Paris **8**, 229–231 (1959)

Some Remarks About Polynomial Aggregation Functions

Sebastia Massanet[1,2](✉) [iD], Juan Vicente Riera[1,2] [iD], and Joan Torrens[1,2] [iD]

[1] SCOPIA Research Group, Department of Mathematics and Computer Science, University of the Balearic Islands, Crta. Valldemossa, km. 7.5, 07122 Palma, Spain
{s.massanet,jvicente.riera,jts224}@uib.es
[2] Balearic Islands Health Research Institute (IdISBa), 07010 Palma, Spain

Abstract. There exist a great quantity of aggregation functions at disposal to be used in different applications. The choice of one of them over the others in each case depends on many factors. In particular, in order to have an easier implementation, the selected aggregation is required to have an expression as simple as possible. In this line, aggregation functions given by polynomial expressions were investigated in [22]. In this paper we continue this investigation focussing on binary aggregation functions given by polynomial expressions only in a particular sub-domain of the unit square. Specifically, splitting the unit square by using the classical negation, the aggregation function is given by a polynomial of degree one or two in one of the sub-domains and by 0 (or 1) in the other sub-domain. This is done not only in general, but also requiring some additional properties like idempotency, commutativity, associativity, neutral (or absorbing) element and so on, leading to some families of binary polynomial aggregation functions with a non-trivial 0 (or 1) region.

Keywords: Aggregation functions · Polynomial functions · Classical negation

1 Introduction

Although the use of some particular aggregation functions, such as for instance the arithmetic mean, dates from long ago, the systematic study of mathematical functions that could be used in merging several input data into a representative output value is quite more recent. The boom of the study of aggregation functions can be probably dated in the 1980s and from then they have been extensively studied (see [4,5,7,12]). The great quantity of application fields where these operators play a key role has been the main reason for which the theoretical study of aggregation functions have experienced this important growth in last decades.

The usual definition of n-ary aggregation function (increasing functions $f : [0,1]^n \to [0,1]$ with $f(0,\ldots,0) = 0$ and $f(1,\ldots,1) = 1$) is rather general. This fact leads usually to require additional properties to the aggregation

© Springer Nature Switzerland AG 2019
R. Halaš et al. (Eds.): AGOP 2019, AISC 981, pp. 47–59, 2019.
https://doi.org/10.1007/978-3-030-19494-9_5

functions depending on each concrete study in the literature. In this sense, the following classification of n-ary aggregation functions is generally accepted (see [5]): the class of *conjunctive* aggregation functions (those that take values below the minimum), *disjunctive* aggregation functions (those that take values over the maximum), *averaging* aggregation functions (those that take values between the minimum and the maximum) and *mixed* aggregation functions (the remaining ones, that is, those with different types of behaviour on different parts of the domain). Many examples in each one of these classes are well-known. For instance,

- Conjunctive (respectively, disjunctive) aggregation functions include t-norms, copulas and quasi-copulas (respectively, t-conorms, co-copulas and duals of quasi-copulas). These kinds of aggregations have been used mainly as logical operators with consequent applications in fuzzy logic and approximate reasoning, but also in image processing, probability and statistics, and economy, among others (see for instance [1,13–15,23]). In these topics the relation between aggregation functions and fuzzy implication functions becomes of great interest as it can be seen in many references (see [2,3,8,11,18,24,25]).
- Averaging aggregation functions include all weighted means, OWA's, Choquet and Sugeno integrals and many generalizations. They are mainly used in all processes related to aggregation of information, decision making, consensus, optimization, image analysis, and so on (see [4,5,7,12,26]).
- Finally, mixed aggregation functions include uninorms and nullnorms among others, and they have been proved to be useful in many of the applications already mentioned for the other types of aggregations like fuzzy logic and approximate reasoning, image processing, decision making and so on (see for instance [6,10,16,27] and the recent survey on uninorms [17]).

Among the theoretical aspects on aggregation functions, one of special interest is the study of their expressions in order to define aggregation functions whose expression is as simple as possible in order to make easier their implementation and computation in applications. In this direction, a first step could be to study aggregation functions whose expression is given by polynomial or rational functions of different degrees. This was already done in some particular cases. See for instance [1], where all the rational Archimedean continuous t-norms are characterized leading to the well-known Hamacher class (in particular, the only polynomial t-norm is the product t-norm $T_P(x,y) = xy$). In [9], all the rational uninorms were characterized as those whose expression is given by

$$U_e(x,y) = \begin{cases} \frac{(1-e)xy}{(1-e)xy+e(1-x)(1-y)} & \text{if } (x,y) \in [0,1]^2 \setminus \{(0,1),(1,0)\}, \\ 0 \text{ (or 1)} & \text{otherwise.} \end{cases}$$

In this case, there do not exist any polynomial uninorm since they are never continuous. A similar study for polynomial and rational fuzzy implication functions was done in the successive papers [19–21], where many different examples are shown.

Recently, an initial study of polynomial aggregation functions in general of degrees one and two was presented in [22]. Whereas weighted arithmetic means are the only polynomial aggregation functions of degree one, many different families appear in the case of degree 2. Such study was firstly done for binary aggregation functions and after that, some of these results were extended to n-ary aggregations (see [22]). Following this line of research, in this paper we want to deal with the case of binary aggregation functions given by polynomial expressions only in some partial domains of the unit square. The idea is that many aggregation functions have a non trivial 0 region (recall for instance the Lukasiewicz or the nilpotent minimum t-norm) or dually a non trivial 1 region (recall the dual t-conorms of the previous mentioned t-norms).

Thus, this paper is organized as follows. After some preliminaries in Sect. 2, we will focus in Sect. 3 firstly on binary aggregation functions with a non trivial 0 region limited by the classical negation N_c and given by a polynomial of degrees 1 or 2 over N_c. In this study we will analyse some of their common additional properties like commutativity, associativity, idempotency and neutral element. Then in Sect. 4, using the duality, all these results can be adequately applied to the study of binary aggregation functions with a non trivial 1 region limited by the classical negation N_c and given by a polynomial of degrees 1 or 2 under N_c. We end the paper with Sect. 5 devoted to conclusions and future work.

2   Preliminaries

Let us recall some concepts and results that will be used throughout this paper. First, we give the definition of an aggregation function.

Definition 1 ([5,7]).   *An n-ary aggregation function is a function of $n > 1$ arguments that maps the (n-dimensional) unit cube onto the unit interval $f : [0,1]^n \to [0,1]$, with the properties*

(i) $f(\underbrace{0,0,\ldots,0}_{n-times}) = 0$ *and* $f(\underbrace{1,1,\ldots,1}_{n-times}) = 1$.

(ii) $\mathbf{x} \leq \mathbf{y}$ *implies* $f(\mathbf{x}) \leq f(\mathbf{y})$ *for all* $\mathbf{x}, \mathbf{y} \in [0,1]^n$.

In particular, a 2-ary aggregation function will be called a binary aggregation function.

As we have already mentioned in the introduction, many families of aggregation functions have been introduced in the literature. One of such families which will play an important role in this paper is the family of weighted arithmetic means.

Definition 2 ([5,7]).   *The* weighted arithmetic mean *is the n-ary function given by*

$$M_w(\mathbf{x}) = w_1 x_1 + w_2 x_2 + \ldots + w_n x_n = \sum_{i=1}^{n} w_i x_i$$

where $\boldsymbol{w} = (w_1, \ldots, w_n)$ is the so-called weighting vector satisfying $w_i \in [0,1]$ for all $1 \leq i \leq n$ and $\sum_{i=1}^{n} w_i = 1$.

Some additional properties of binary aggregation functions which will be used in this work are:

– The *idempotency*,
$$f(x,x) = x, \quad \text{for all } x \in [0,1]. \tag{ID}$$

– The *symmetry*,
$$f(x,y) = f(y,x), \tag{SYM}$$
for all $(x,y) \in [0,1]^2$.

– The *associativity*,
$$f(f(x,y),z) = f(x,f(y,z)), \quad \text{for all } x,y,z \in [0,1]. \tag{ASS}$$

– It is said that $a \in (0,1)$ is a *zero divisor* when
$$f(a,y) = f(y,a) = 0, \tag{ZD(a)}$$
for some $y > 0$.

– It is said that $a \in (0,1)$ is a *one divisor* when
$$f(a,y) = f(y,a) = 1, \tag{OD(a)}$$
for some $y < 1$.

– The *left neutral element property* with a fixed $e \in [0,1]$,
$$f(e,y) = y, \quad \text{for all } y \in [0,1]. \tag{L $-$ NE(e)}$$

– The *right neutral element property* with a fixed $e \in [0,1]$,
$$f(x,e) = x, \quad \text{for all } x \in [0,1]. \tag{R $-$ NE(e)}$$

– The *neutral element property* with a fixed $e \in [0,1]$,
$$f(e,x) = f(x,e) = x, \quad \text{for all } x \in [0,1]. \tag{NE(e)}$$

– The *left absorbing element property* with a fixed $a \in [0,1]$,
$$f(a,y) = a, \quad \text{for all } y \in [0,1]. \tag{L $-$ AE(a)}$$

– The *right absorbing element property* with a fixed $a \in [0,1]$,
$$f(x,a) = a, \quad \text{for all } x \in [0,1]. \tag{R $-$ AE(a)}$$

– The *absorbing element property* with a fixed $a \in [0,1]$,
$$f(a,x) = f(x,a) = a, \quad \text{for all } x \in [0,1]. \tag{AE(a)}$$

Finally, we will call *conjunctors* to those binary aggregation functions with absorbing element 0 and *disjunctors* to those aggregation functions with absorbing element 1. Note that conjunctive (disjunctive) aggregation functions are trivially conjunctors (disjunctors) but not vice versa.

3 Polynomial Binary Aggregation Functions with a Non Trivial 0 Region

In this section, we want to deal with binary aggregation functions given by polynomial expressions only in some partial domains of the unit square. As we have already commented, the idea is that in many cases the aggregation function has a non trivial 0 region as for instance the Łukasiewicz or the nilpotent minimum t-norms. Thus, we want to focus first on aggregation functions with a non trivial 0 region limited by the classical negation N_c and given by a polynomial of degree one or two over N_c. Note that this ensures in particular the existence of zero divisors. Formally, we will study functions of the form

$$f(x,y) = \begin{cases} 0 & \text{if } y \leq 1 - x, \\ P(x,y) & \text{if } y > 1 - x, \end{cases} \tag{1}$$

where $P(x,y)$ is a polynomial of degree 1 or 2.

Remark 1. Of course, any other region could be considered as the zero-region of the aggregation function. For instance, it could be delimited by any other polynomial negation of the form $N(x) = 1 - x^n$. However, in this paper we will focus on the classical negation N_c which in particular will allow us to retrieve the Łukasiewicz t-norm as one of the searched aggregation functions.

First of all note that there exist a lot of aggregation functions of the form (1) as the following examples show.

Example 1. All the following cases are aggregation functions of the form (1):

(i) It is clear that if we take as $P(x,y)$ any binary polynomial aggregation function of degree one or two from those characterized in [22] we trivially obtain aggregation functions of the desired form. However, we will see along the paper that there are many other examples different from these ones, as for instance the next example.

(ii) The Łukasiewicz t-norm $T_{\mathbf{L}}$ is also an aggregation function of the form (1), just taking as $P(x,y)$ the polynomial $P(x,y) = x + y - 1$.

(iii) Note that, although we will deal only with polynomials of degree one or two, it can be deduced from the first item that there are examples with $P(x,y)$ of any degree. Take for instance

$$f(x,y) = \begin{cases} 0 & \text{if } y \leq 1 - x, \\ \frac{x^n + y^n}{2} & \text{if } y > 1 - x. \end{cases}$$

Let us divide the following results in two different sections, one devoted to the case when $P(x,y)$ is a polynomial of degree smaller than or equal to one, and the other devoted to the case when $P(x,y)$ is a polynomial of degree two.

3.1   The Case of Degree 0 or 1

We investigate in this section which functions of the form (1) are in fact aggregation functions where $P(x, y)$ is a polynomial of degree 0 or 1. That is, we will deal with functions of the form

$$f(x, y) = \begin{cases} 0 & \text{if } y \leq 1 - x, \\ ax + by + c & \text{if } y > 1 - x, \end{cases} \tag{2}$$

where a, b, c are real numbers. In this case it is easy to characterize all possibilities as it is done in the next result.

Theorem 1. *Let f be a binary function of the form (2). The following statements are equivalent:*

(i) *f is a binary aggregation function.*
(ii) *There exist $a, b \in [0, 1]$ such that $f = f_{a,b}$ where $f_{a,b} : [0, 1]^2 \to [0, 1]$ is given by*

$$f_{a,b}(x, y) = \begin{cases} 0 & \text{if } y \leq 1 - x, \\ ax + by + 1 - a - b & \text{if } y > 1 - x. \end{cases} \tag{3}$$

The theorem above includes the following particular cases.

Example 2. (i) The case $a = b = 0$ gives the only aggregation function of the form (2) with $P(x, y)$ a polynomial of degree zero. That is, the function:

$$f_{0,0}(x, y) = \begin{cases} 0 & \text{if } y \leq 1 - x, \\ 1 & \text{if } y > 1 - x. \end{cases}$$

(ii) The first and second projections with a 0 region limited by N_c are also obtained in the cases $a = 1, b = 0$, and $a = 0, b = 1$, respectively.
(iii) The Łukasiewicz t-norm is obtained taking $a = b = 1$.
(iv) Taking $b = 1 - a$ we obtain the family of weighted arithmetic means with a 0 region limited by N_c, that is,

$$f_{a,1-a}(x, y) = \begin{cases} 0 & \text{if } y \leq 1 - x, \\ ax + (1 - a)y & \text{if } y > 1 - x. \end{cases} \tag{4}$$

Note that the last item in the previous example corresponds with those given in Example 1-(i) when $P(x, y)$ is of degree 1 (see Theorem 11 in [22]). Consequently, even in the case of degree one, we obtain many other examples than the trivial ones given in Example 1-(i).

From Theorem 1, it is easy to characterize all binary aggregation functions of the form (3) fulfilling some additional properties. We collect some of them in the following proposition.

Proposition 1. *Let $f = f_{a,b}$ be a binary aggregation function of the form (3). The following statements are true:*

(i) $f_{a,b}$ is continuous if and only if $a = b = 1$, that is, if and only if $f_{a,b} = f_{1,1}$ is the Łukasiewicz t-norm.

(ii) $f_{a,b}$ satisfies (**SYM**) if and only if $a = b$. That is, when $f_{a,b} = f_{a,a}$ is given by

$$f_{a,a}(x,y) = \begin{cases} 0 & \text{if } y \leq 1 - x, \\ a(x+y) + 1 - 2a & \text{if } y > 1 - x. \end{cases}$$

(iii) $f_{a,b}$ satisfies (**ID**) in its positive region, if and only if $b = 1 - a$. That is, when $f_{a,b} = f_{a,1-a}$ is given by Eq. (4).

(iv) $f_{a,b}$ satisfies (**L-NE(e)**) if and only if the neutral element is $e = 1$ and $b = 1$. That is, when $f_{a,b} = f_{a,1}$ is given by

$$f_{a,1}(x,y) = \begin{cases} 0 & \text{if } y \leq 1 - x, \\ ax + y - a & \text{if } y > 1 - x. \end{cases}$$

(v) $f_{a,b}$ satisfies (**R-NE(e)**) if and only if and only if the neutral element is $e = 1$ and $a = 1$. That is, when $f_{a,b} = f_{1,b}$ is given by

$$f_{1,b}(x,y) = \begin{cases} 0 & \text{if } y \leq 1 - x, \\ x + by - b & \text{if } y > 1 - x. \end{cases}$$

(vi) $f_{a,b}$ satisfies (**NE(e)**) if and only if the neutral element is $e = 1$ and $f_{a,b} = f_{1,1}$ is the Łukasiewicz t-norm.

(vii) $f_{a,b}$ satisfies (**ASS**) if and only if $f_{a,b} = f_{1,1}$ is the Łukasiewicz t-norm.

3.2   The Case of Degree 2

In this section, those functions of the form (1) where $P(x, y)$ is a polynomial of degree two are studied. In particular, we are interested in characterizing which functions of the form

$$f(x,y) = \begin{cases} 0 & \text{if } y \leq 1 - x \\ ax^2 + by^2 + cxy + dx + ey + f & \text{if } y > 1 - x, \end{cases} \tag{5}$$

where a, b, c, d, e, f are real numbers such that $a^2 + b^2 + c^2 > 0$, are aggregation functions. This case is more complex than the previous one and it leads to a huge number of possibilities for the values of a, b, c, d, e, f. Due to this great quantity of cases and the lack of space, we will present in the next example just some families of binary aggregation functions of the form (5) where $P(x, y)$ is a polynomial of degree two.

Example 3. (i) The family

$$f(x,y) = \begin{cases} 0 & \text{if } y \leq 1 - x, \\ by^2 + bxy + dx + ey + 1 - 2b - d - e & \text{if } y > 1 - x, \end{cases}$$

where $-1 < b < 0$, $-b \leq d \leq 1 - d$ and $-3b \leq e \leq 1 - 2b$ is a family of binary aggregation functions with a 0-region limited by N_c and given by a polynomial of degree 2 over N_c.

(ii) The family

$$f(x,y) = \begin{cases} 0 & \text{if } y \le 1-x, \\ cxy + 2\sqrt{-c}x - cy + 1 - 2\sqrt{-c} & \text{if } y > 1-x, \end{cases}$$

where $-4 \le c \le -1$ is also a family of these aggregation functions which depends only on one parameter.

By requiring the fulfilment of some additional properties, we can reduce the number of possibilities for the values of a, b, c, d, e, f making feasible to present the complete families of aggregation functions of the form (5) satisfying these properties. Let us start studying the continuity and the existence of a neutral element independently.

Proposition 2. *Let f be a binary function of the form (5). The following statements are equivalent:*

(i) f is a continuous aggregation function.
(ii) f is given by $f(x,y) =$

$$\begin{cases} 0 & \text{if } y \le 1-x, \\ ax^2 + by^2 + (a+b)xy + (1-2a-b)x + (1-a-2b)y + a+b-1 & \text{if } y > 1-x, \end{cases}$$

where $-1 \le a, b \le 1$ and $a^2 + b^2 > 0$.

Proposition 3. *Let f be a binary function of the form (5). The following statements are equivalent:*

(i) f is an aggregation function satisfying (NE(e)).
(ii) $e = 1$ and f is given by

$$f(x,y) = \begin{cases} 0 & \text{if } y \le 1-x, \\ cxy + (1-c)x + (1-c)y + c - 1 & \text{if } y > 1-x, \end{cases}$$

where $0 < c \le 1$.

Remark 2. Propositions 2 and 3 again clearly show that the study of these aggregation functions does not reduce to truncate by 0 the family of polynomial functions, i.e., there are more possibilities than the aggregation functions presented in Example 1-(i). Indeed, there are no continuous aggregation functions within the family presented in Example 1-(i). Moreover, since the unique polynomial aggregation function of degree 2 having 1 as neutral element is $f(x,y) = xy$, the product truncated by 0 must be obtained in Proposition 3 (as it is, just take $c = 1$). However, for any value $0 < c < 1$, other aggregation functions are obtained.

Other additional properties can be also fully characterized but there still remain too many possibilities for the parameter values. Thus, let us consider pairs of additional properties in order to reduce the number of possible values. We will begin with the symmetry property jointly with other additional properties.

Proposition 4. *Let f be a binary function of the form (5). The following statements are equivalent:*

(i) f is a continuous aggregation function satisfying **(SYM)**.
(ii) f is given by

$$f(x,y) = \begin{cases} 0 & \text{if } y \leq 1 - x, \\ a(x^2 + y^2) + 2axy + (1 - 3a)(x + y) + 2a - 1 & \text{if } y > 1 - x, \end{cases}$$

where $a \in [-1, 0[\cup]0, 1]$.

From Proposition 3, the following result is straightforward.

Proposition 5. *Let f be a binary function of the form (5). If f satisfies* **(NE(e))** *then f satisfies also* **(SYM)**.

Let us study now the idempotency in the positive region jointly with other additional properties.

Proposition 6. *Let f be an aggregation function of the form (5). If f is continuous or satisfies* **(NE(e))**, *then f is not idempotent in its positive region.*

Contrarily, there exist solutions when we join **(ID)** in the positive region with **(SYM)**.

Proposition 7. *Let f be a binary function of the form (5). The following statements are equivalent:*

(i) f is an aggregation function satisfying **(SYM)** *and* **(ID)** *in its positive region.*
(ii) f is given by

$$f(x,y) = \begin{cases} 0 & \text{if } y \leq 1 - x, \\ a(x^2 + y^2) - 2axy + \frac{1}{2}(x + y) & \text{if } y > 1 - x, \end{cases}$$

where $a \in [-\frac{1}{2}, 0[\cup]0, \frac{1}{2}]$.

From Corollary 17 in [22], it can be checked that the aggregation functions of the form (5) satisfying **(SYM)** and **(ID)** in its positive region are exactly those polynomial aggregation functions of degree 2 truncated by 0.

Besides the results presented in this section, more combinations of additional properties can be analysed but they are not included due to the lack of space.

4   Polynomial Binary Aggregation Functions with a Non Trivial 1-region

Analogously to the study carried out in Sect. 3 for polynomial binary aggregation functions with a non trivial 0-region, a similar study can be done for polynomial binary aggregation functions with a non trivial 1-region. This family of aggregation functions includes the well-known Łukasiewicz or the nilpotent maximum t-conorms. This construction ensures the existence of one divisors and taking again the classical negation N_c as limit of the 1-region, the expression of these operators can be explicitly given as

$$f(x,y) = \begin{cases} P(x,y) & \text{if } y < 1-x, \\ 1 & \text{if } y \geq 1-x, \end{cases} \tag{6}$$

where $P(x,y)$ is a polynomial of degree 1 or 2.

This family of aggregation functions is huge and there exist a lot of aggregation functions of the form (6) as the following examples show.

Example 4. All the following cases are aggregation functions of the form (6):

(i) A construction method of aggregation functions of the form (6) is straightforward from any binary polynomial aggregation function of degree one or two from those characterized in [22]. It is based on assigning a polynomial aggregation function as $P(x,y)$ and impose the value 1 all over N_c. This construction method does not provide the whole family of aggregation functions of the form (6) since for instance, it does not provide any continuous aggregation function.

(ii) The Łukasiewicz t-conorm $S_{\mathbf{L}}$ is an aggregation function of the form (6), just taking as $P(x,y)$ the polynomial $P(x,y) = x + y$. Note that this operator is continuous.

(iii) The following family of aggregation functions of the form (6) proves the existence of operators of this kind for any degree n

$$f(x,y) = \begin{cases} 1 - \frac{(1-x)^n + (1-y)^n}{2} & \text{if } y < 1-x, \\ 1 & \text{if } y \geq 1-x. \end{cases}$$

As the reader can perceive from the previous examples is that all of them are the dual operators (with respect to N_c) of the corresponding ones in Sect. 4. Indeed, both forms are connected through the duality as the following result states.

Proposition 8. $f(x,y)$ *is a binary function of the form (1) if and only if $1 - f(1-x, 1-y)$ is a binary function of the form (6).*

This result allows us to rewrite easily all the results presented in Sect. 4 to results involving aggregation functions of the form (6). For the sake of clarity, we show for instance the characterization of all aggregation functions of the form (6) with degree less or equal to 1.

Theorem 2. *Let f be a binary function of the form (6) with $P(x,y)$ a polynomial of degree less or equal to 1. The following statements are equivalent:*

(i) f *is a binary aggregation function.*

(ii) *There exist* $a, b \in [0, 1]$ *such that* $f = f_{a,b}$ *where* $f_{a,b} : [0, 1]^2 \to [0, 1]$ *is given by*

$$f_{a,b}(x, y) = \begin{cases} ax + by & \text{if } y < 1 - x, \\ 1 & \text{if } y \geq 1 - x. \end{cases} \tag{7}$$

5 Conclusions and Future Work

This paper continues the line of research initiated in [22] about the construction of aggregation functions whose expressions are as simple as possible. Specifically, aggregation functions which have a 0 (or 1)-region delimited by N_c and polynomial function as expression in the other sub-domain have been deeply analysed. Indeed, we have characterized all binary aggregation functions with 0 (or 1)-region delimited by N_c and defined by a polynomial of degree less than or equal to 1 in the other sub-domain. Moreover, other additional desirable properties such as the idempotency in the non-constant region, symmetry, continuity or the existence of a neutral element among others, have been studied for the members of this family of aggregation functions. Furthermore, a similar study for polynomials of degree 2 has been also performed. In this case, due to the huge number of possibilities, some particular families have been presented as well as some additional properties, often in pairs, have been studied. Finally, using the duality, every result on this class of aggregation functions with the 0 region can be translated to the class of aggregation functions with a 1-region.

As future work, we want to complete the study of the additional properties when an underlying polynomial of degree 2 is considered. Also, some general results for polynomials of degree n could also be proved similarly to as it was done in [22]. Finally, another fuzzy negation to delimitate the constant region could be also considered.

Acknowledgments. This paper has been partially supported by the Spanish Grant TIN2016-75404-P AEI/FEDER, UE.

References

1. Alsina, C., Frank, M.J., Schweizer, B.: Associative Functions: Triangular Norms and Copulas. Kluwer Academic Publishers, Dordrecht (2000)
2. Baczyński, M., Jayaram, B.: Fuzzy Implications. Studies in Fuzziness and Soft Computing, vol. 231. Springer, Heidelberg (2008)
3. Baczyński, M., Jayaram, B., Massanet, S., Torrens, J.: Fuzzy implications: past, present, and future. In: Kacprzyk, J., Pedrycz, W. (eds.) Springer Handbook of Computational Intelligence, pp. 183–202. Springer, Heidelberg (2015)
4. Beliakov, G., Bustince Sola, H., Calvo Sánchez, T.: A Practical Guide to Averaging Functions. Studies in Fuzziness and Soft Computing, vol. 329. Springer, Heidelberg (2016)

5. Beliakov, G., Pradera, A., Calvo, T.: Aggregation Functions: A Guide for Practitioners. Studies in Fuzziness and Soft Computing, vol. 221. Springer, Heidelberg (2007)
6. Calvo, T., De Baets, B., Fodor, J.: The functional equation of Frank and Alsina for uninorms and nullnorms. Fuzzy Sets Syst. **120**, 385–394 (2001)
7. Calvo, T., Mayor, G., Mesiar, R.: Aggregation Operators: New Trends and Applications. Studies in Fuzziness and Soft Computing. Physica-Verlag, Heidelberg (2002)
8. Durante, F., Klement, E.P., Mesiar, R., Sempi, C.: Conjunctors and their residual implicators: characterizations and construction methods. Mediterr. J. Math. **4**, 343–356 (2007)
9. Fodor, J.C.: On rational uninorms. In: Proceedings of the First Slovakian–Hungarian Joint Symposium on Applied Machine Intelligence, Herlany, Slovakia, pp. 139–147 (2003)
10. Fodor, J., Yager, R.R., Rybalov, A.: Structure of uninorms. Int. J. Uncertain. Fuzziness **5**, 411–427 (1997)
11. Gottwald, S.: A Treatise on Many-Valued Logic. Research Studies Press, Baldock (2001)
12. Grabisch, M., Marichal, J.-L., Mesiar, R., Pap, E.: Aggregation Functions (Encyclopedia of Mathematics and Its Applications), 1st edn. Cambridge University Press, New York (2009)
13. Kerre, E.E., Huang, C., Ruan, D.: Fuzzy Set Theory and Approximate Reasoning. Wu Han University Press, Wu Chang (2004)
14. Kerre, E.E., Nachtegael, M.: Fuzzy Techniques in Image Processing. Studies in Fuzziness and Soft Computing, vol. 52. Springer, New York (2000)
15. Klement, E.P., Mesiar, R., Pap, E.: Triangular Norms. Kluwer Academic Publishers, Dordrecht (2000)
16. Mas, M., Mayor, G., Torrens, J.: t-Operators. Int. J. Uncertain. Fuzziness Knowl.-Based Syst. **7**(1), 31–50 (1999)
17. Mas, M., Massanet, S., Ruiz-Aguilera, D., Torrens, J.: A survey on the existing classes of uninorms. J. Intell. Fuzzy Syst. **29**, 1021–1037 (2015)
18. Mas, M., Monserrat, M., Torrens, J., Trillas, E.: A survey on fuzzy implication functions. IEEE Trans. Fuzzy Syst. **15**(6), 1107–1121 (2007)
19. Massanet, S., Riera, J.V., Ruiz-Aguilera, D.: On fuzzy polynomial implications. In: Laurent, A., et al. (eds.) Information Processing and Management of Uncertainty in Knowledge-Based Systems. Communications in Computer and Information Science, vol. 442, pp. 138–147. Springer, Heidelberg (2014)
20. Massanet, S., Riera, J.V., Ruiz-Aguilera, D.: On (OP)-polynomial implications. In: Alonso, J.M., Bustince, H., Reformat, M. (eds.) Proceedings of the 2015 Conference of the International Fuzzy Systems Association and the European Society for Fuzzy Logic and Technology (IFSA-EUSFLAT 2015), pp. 1208–1215. Atlantis Press (2015)
21. Massanet, S., Riera, J.V., Ruiz-Aguilera, D.: On rational fuzzy implication functions. In: Proceedings of IEEE World Congress on Computational Intelligence (IEEE WCCI), pp. 272–279 (2016)
22. Massanet, S., Riera, J.V., Torrens, J.: Aggregation functions given by polynomial functions. In: Proceedings of IEEE International Conference on Fuzzy Systems (FUZZ-IEEE 2017), Naples, Italy, pp. 1–6 (2017). https://doi.org/10.1109/FUZZ-IEEE.2017.8015631
23. Nelsen, R.B.: An Introduction to Copulas. Springer, New York (2006)

24. Pradera, A., Beliakov, G., Bustince, H., De Baets, B.: A review of the relationships between implication, negation and aggregation functions from the point of view of material implication. Inf. Sci. **329**, 357–380 (2016). Special issue on Discovery Science

25. Pradera, A., Massanet, S., Ruiz-Aguilera, D., Torrens, J.: The non-contradiction principle related to natural negations of fuzzy implication functions. Fuzzy Sets Syst. **359**, 3–21 (2019)

26. Torra, V., Narukawa, Y.: Modeling Decisions: Information Fusion and Aggregation Operators. Cognitive Technologies. Springer, Heidelberg (2007)

27. Yager, R.R., Rybalov, A.: Uninorm aggregation operators. Fuzzy Sets Syst. **80**, 111–120 (1996)

Other Speakers

Aggregation Through Composition: Unification of Three Principal Fuzzy Theories

Irina Perfilieva[1]([✉]), Anand P. Singh[1], and S. P. Tiwari[2]

[1] Institute for Research and Applications of Fuzzy Modeling, NSC IT4Innovations,
30. dubna 22, 701 03 Ostrava 1, Czech Republic
irina.perfilieva@osu.cz, anandecc@gmail.com
[2] Indian Institute of Technology (ISM), 826004 Dhanbad, India
sptiwarimaths@gmail.com

Abstract. This paper shows that the theories of fuzzy rough sets, F-transforms and fuzzy automata can be unified in the framework of fuzzy relational structures. Specifically, the key concepts in such theories are represented as lattice-based aggregations in the form of compositions with suitable fuzzy relations. Furthermore, it is shown that the principal parts of morphisms between all considered fuzzy relational structures coincide.

Keywords: Fuzzy relation · Fuzzy rough set · F-transform ·
Fuzzy automata

1 Introduction

After the introduction of fuzzy sets by Zadeh [19] in 1965, very rapid and extensive development of methods, techniques appeared using this concept. Some of the well-known concepts in fuzzy theory are fuzzy rough sets, fuzzy transforms and fuzzy automata. The concept of fuzzy rough set was introduced by Dubois and Prade [2] and further extensively studied in the fuzzy literature. On the other hand, fuzzy transform (F-transform, in short), firstly proposed by Perfilieva [7] has now been significantly developed and shown to be useful in many areas. The main idea of the F-transform is to factorize (or fuzzify) the precise values of independent variables by a closeness relation, and precise values of dependent variables are averaged to an approximate value. The theory of F-transform has been extended from real valued functions to lattice-valued functions (cf., [7,8]), from fuzzy sets to parametrized fuzzy sets and from the single variable to the two (or more variables). This theory has shown to be useful in many applications: denoising, scheduling, trading, time series, numerical solutions of partial differential equations, data analysis and neural networks. The extension of the F-transform to the functions of two variables shows powerful applications in signal and image processing, particularly, image compression, edge detection,

R. Halaš et al. (Eds.): AGOP 2019, AISC 981, pp. 63–74, 2019.
https://doi.org/10.1007/978-3-030-19494-9_6

image reconstruction, image fusion, image compression. In addition to the above, a research in the direction of theoretical study of lattice based F-transform has been carried out as well (cf., [9,16,18]).

Finally, fuzzy automata was firstly proposed by Wee and Santos [15] after the introduction of Zadeh's fuzzy set theory [19]. After few decades Malik, Mordeson and Sen introduced the concept of fuzzy finite state machine (a concept similar to fuzzy automaton). Further generalizations has been made by extending the truth values of fuzzy automata from the closed interval [0, 1] to complete residuated lattices, lattice ordered monoids and some other kinds of lattices (c.f., [12,13]).

However, some researches have thought that most of the concepts studied in these theories are simple translation of the results from the classical theories. Contrary to it, in [1], it has been justified that it is possible to use powerful tools of the theory of fuzzy sets in the study of fuzzy automata. One such tool is the concept of fuzzy relations, which have been shown to be useful in the intensive development of several areas of fuzzy mathematics. In past few decades, several authors studied above mentioned theories independently, but none of the studies has been made to unify them in the framework of fuzzy relations. In this paper, we specify these relationship in more precise and consistent way. A closer observations to the above discussed theories leads to the conclusion that the basis and development of all such theories are based on the composition of fuzzy relations with fuzzy sets. Specifically, we show that the basic concepts in fuzzy rough sets, F-transforms and fuzzy automata can be described as the direct and subdirect images of a fuzzy sets under some suitable fuzzy relations.

The remainder of the paper is structured as follows. In Sect. 2, basic concepts of residuated lattices and the notions of fuzzy sets and fuzzy relations are recalled. The fuzzy lower and fuzzy upper approximation operators are defined in sect. 3 and shown that they can be considered as direct and subdirect images of fuzzy sets. Section 4, recalls the notion of F-transforms and shows its relationship with direct and subdirect images of fuzzy sets under suitable fuzzy relations and discuss how the morphism of F-transform can be obtained from the morphism of fuzzy relational structures. In Sect. 5, we introduce and study fuzzy relation based fuzzy automata and show its equivalence with fuzzy transition function based fuzzy automata. Furthermore, we represent the key notion in fuzzy automata as the direct and subdirect image of fuzzy sets. Section 6 concludes the paper.

2   Preliminaries

In this section, we recall some basic concepts related to residuated lattice, fuzzy sets and fuzzy relations which we need in the subsequent sections. We begin with the following concept of a residuated lattice. For details on residuated lattices, we refer to [3].

Definition 1. *A* **residuated lattice** *L is an algebra $(L, \wedge, \vee, \otimes, \rightarrow, 0, 1)$ such that*

(i) $(L, \wedge, \vee, 0, 1)$ *is a bounded lattice with the least element* 0 *and the greatest element* 1;

(ii) $(L, \otimes, 1)$ *is a commutative monoid; and*

(iii) $\forall a, b, c \in L$;

$$a \otimes b \leq c \ iff \ a \leq b \rightarrow c,$$

i.e. $(\rightarrow, \otimes)$ *is an adjoint pair on* L.

A residuated lattice $(L, \wedge, \vee, \otimes, \rightarrow, 0, 1)$ is **complete**, if it is complete as a lattice.

Throughout this paper, we work with the complete residuated lattice $L = (L, \wedge, \vee, \otimes, \rightarrow, 0, 1)$. The fuzzy sets, considereed in this paper, take membership values in L. For a nonempty set X, L^X denotes the collection of all fuzzy subsets (L-valued functions) of X.

Definition 2. *Let* X *be a nonempty set. Then for all* $\lambda, \mu \in L^X$ *and* $x \in X$,

(i) $(\lambda \otimes \mu)(x) = \lambda(x) \otimes \mu(x)$;

(ii) $(\lambda \rightarrow \mu)(x) = \lambda(x) \rightarrow \mu(x)$;

(iii) $\lambda \leq \mu \Leftrightarrow \lambda(x) \leq \mu(x)$.

Now, let $f : X \rightarrow Y$ be a map. Then according to Zadeh's extension principle f can be extended to the operators $f^{\rightarrow} : L^X \rightarrow L^Y$ and $f^{\leftarrow} : L^Y \rightarrow L^X$ such that $\lambda \in L^X, \mu \in L^Y, y \in Y$

$$f^{\rightarrow}(\lambda)(y) = \bigvee_{x, f(x) = y} \lambda(x), \quad f^{\leftarrow}(\mu) = \mu \cap f.$$

Fuzzy relation as a measure of indistinguishability has shown to be useful in various fields like fuzzy control, approximate reasoning, fuzzy cluster analysis, etc. Here we recall the concept of a fuzzy relation as it was introduced in [14].

Definition 3. *Let* X *and* Y *be a nonempty sets. A* **fuzzy relation** R *from* X *to* Y *is a map* $R : X \times Y \rightarrow L$.

We denote by $L^{X \times Y}$ the set of all fuzzy relations from X to Y. If $X = Y$, R is called a fuzzy relation on X. A fuzzy relation R is called **reflexive** if $\forall x \in X$, $R(x, x) = 1$.

A set X equipped with a fuzzy relation R is denoted by (X, R) and is called a **fuzzy relational structure**.

Definition 4. *For two fuzzy relational structures* (X, R) *and* (Y, S), *a* **morphism** *is a map* $f : X \rightarrow Y$, *such that for all* $x, y \in X$, $R(x, y) \leq S(f(x), f(y))$.

The following two compositions of fuzzy relations will be used in the sequel.

Definition 5. *Let* $R \in L^{X \times Y}$ *and* $S \in L^{Y \times Z}$. *Then* **sup-$\otimes$ composition** *of* R *and* S *is given by*

$$R \circ_{\otimes} S(x, z) = \bigvee_{y \in Y} (R(x, y) \otimes S(y, z)), \quad x \in X, z \in Z. \tag{1}$$

The direct image of fuzzy sets under a fuzzy relation can be derived from the above given compositions of fuzzy relations.

Definition 6. *Let* $\lambda \in L^X$ *and* $R \in L^{X \times Y}$. *Then the* **direct image** *of* λ *under* R *is given by*

$$R \circ_{\otimes} \lambda(x) = \bigvee_{y \in Y} (R(x,y) \otimes \lambda(y)), \quad x \in X. \tag{2}$$

Definition 7. *Let* $R \in L^{X \times Y}$ *and* $S \in L^{Y \times Z}$. *Then* **inf-$\rightarrow$ composition** *of* R *and* S *is given by*

$$R \triangleleft S(x,z) = \bigwedge_{y \in Y} (R(x,y) \rightarrow S(y,z)), \quad x \in X, z \in Z. \tag{3}$$

The subdirect image of a fuzzy set under a fuzzy relation can be derived from the above considered compositions of fuzzy relations.

Definition 8. *Let* $\lambda \in L^X$ *and* $R \in L^{X \times Y}$. *Then the* **subdirect image** *of* λ *under* R *is given by*

$$R \triangleleft \lambda(x) = \bigwedge_{y \in Y} (R(x,y) \rightarrow \lambda(y)), \quad x \in X. \tag{4}$$

3   Fuzzy Rough Sets

Fuzzy rough sets, introduced by Dubois and Prade [2], have been extensively studied in the literature: theoretically and in connection with applications. The theory is based on the concept of fuzzy upper and lower approximation operators. In this section, we show that both approximation operators can be obtained as direct and subdirect images of a fuzzy set under a given fuzzy relation. The morphisms between fuzzy approximation spaces also come from fuzzy relational structures. Below, we recall the definitions as they appeared in [14].

Definition 9. [14] *A pair* (X, R) *is called a* **fuzzy approximation space** *if* X *is a nonempty set and* R *is a fuzzy relation on* X.

Definition 10. [14] *Let* (X, R) *be a fuzzy approximation space. The pair* $(\underline{R}(\lambda), \overline{R}(\lambda))$ *of lower and upper approximations of a fuzzy set* $\lambda \in L^X$ *is a fuzzy rough set in* (X, R) *where*

$$\underline{R}(\lambda)(x) = \bigwedge_{y \in X} (R(x,y) \rightarrow \lambda(y)),$$

$$\overline{R}(\lambda)(x) = \bigvee_{y \in X} (R(x,y) \otimes \lambda(y)).$$

The two operators $\underline{R}, \overline{R} : L^X \longrightarrow L^X$ are called a **lower fuzzy approxima-tion operator** and ane **upper fuzzy approximation operator**, respectively. Let us emphasize that the order of operators is essential.

Proposition 1. *Let* (X, R) *be a fuzzy approximation space. Then for approxi-mation operators* $(\underline{R}, \overline{R})$ *and for all* $\lambda \in L^X$,

(i) *The upper fuzzy approximation operator* $\overline{R}$ *is the direct image of a fuzzy set* λ *under the fuzzy relation* R, *i.e.*

$$\overline{R}(\lambda)(x) = R \circ_\otimes \lambda(x).$$

(ii) *The lower fuzzy approximation operator* $\underline{R}$ *is the subdirect image of a fuzzy set* λ *under the fuzzy relation* R, *i.e.*

$$\underline{R}(\lambda)(x) = R \triangleleft \lambda(x).$$

Proof. The proof is based on definitions of direct and subdirect image of a fuzzy set under a fuzzy relation.

For given two fuzzy approximation spaces, the following is the notion of morphism between them.

Definition 11. [11] *The* **morphism** $f : (X, R) \rightarrow (Y, S)$ *between two fuzzy approximation spaces* (X, R) *and* (Y, S) *is given by*

$$f^\leftarrow(\underline{S}(\lambda)) \le \underline{R}(f^\leftarrow(\lambda)) \quad \forall \lambda \in L^Y.$$

Theorem 1. *Let* $f : (X, R) \rightarrow (Y, S)$ *be a morphism between two fuzzy rela-tional structures* (X, R) *and* (Y, S). *Then* f *is a morphism between fuzzy approx-imation spaces* (X, R) *and* (Y, S).

Proof. Let $f : (X, R) \rightarrow (Y, S)$ be a morphism between two fuzzy relational structures (X, R) and (Y, S). Then $R(x, y) \le S(f(x), f(y)), \forall x, y \in X$. Now, let $\lambda \in L^Y$ and $x \in X$. Then,

$$\underline{R}(f^\leftarrow(\lambda))(x) = \bigwedge_{y \in X} R(x, y) \rightarrow f^\leftarrow(\lambda)(y)$$

$$= \bigwedge_{y \in X} R(x, y) \rightarrow \lambda(f(y))$$

$$\ge \bigwedge_{y \in X} S(f(x), f(y)) \rightarrow \lambda(f(y))$$

$$\ge \bigwedge_{z \in Y} S(f(x), z) \rightarrow \lambda(z)$$

$$= f^\leftarrow(\underline{S}(\lambda))(x).$$

4   F-transform

The theory of F-transforms, proposed by Perfilieva [7], has been significantly developed aftewards. It has been shown that F-transforms are useful in various applications, as e.g., denoising, scheduling, trading and neural network approaches. In this section, we show how the F-transform can be connected with the theory of fuzzy relations. The basic concepts of F-transform, recalled here, can be found in [7].

In the past decades, the notion of fuzzy partition has been proposed in several papers (cf., [7,10]). In the most cases, this notion was connected with a finite collection of fuzzy sets that are defined on the set of reals $\mathbb{R}$ or its Cartesian product. In [9], we proposed to define a fuzzy partition of an arbitrary non-empty universe by a particular collection (not necessarily finite) of fuzzy subsets on it. Below, we repeat this definition.

Definition 12. *A collection Π_X of normal fuzzy sets $\{A_\xi : \xi \in \Lambda\}$ in X is a **fuzzy partition** of X, if the corresponding collection of ordinary sets $\{core(A_\xi) : \xi \in \Lambda\}$ is a partition of X. A pair (X, Π_X) where Π_X is a fuzzy partition of X is called a **space with a fuzzy partition**.*

Let $\Pi_X = \{A_\xi : \xi \in \Lambda\}$ be a fuzzy partition of X. With this partition, we associate the following surjective index-function $i_\Pi : X \to \Lambda$:

$$i_\Pi(y) = \xi \iff y \in core(A_\xi). \tag{5}$$

Then Π can be uniquely represented by the reflexive L-fuzzy relation R_Π on X, such that

$$R_\Pi(x, y) = A_{i_\Pi(y)}(x). \tag{6}$$

It is not difficult to show that this characterization is correct.

Definition 13. *Let λ be an L-valued function on X and $\Pi_X = \{A_\xi : \xi \in \Lambda\}$ be a fuzzy partition of X. Then*

(i) the **direct** $F^\uparrow$**-transform** *of λ with respect to fuzzy partition $\Pi_X = \{A_\xi : \xi \in \Lambda\}$ is a collection of lattice elements $\{F_\xi^\uparrow : \xi \in \Lambda\}$, where*

$$F_\xi^\uparrow[\lambda] = \bigvee_{x \in X} (A_\xi(x) \otimes \lambda(x)), \qquad \xi \in \Lambda.$$

(ii) The **direct** $F^\downarrow$**-transform** *of λ w.r.t. fuzzy partition $\Pi_X = \{A_\xi : \xi \in \Lambda\}$ is a collection of lattice elements $\{F_\xi^\downarrow : \xi \in \Lambda\}$, where*

$$F_\xi^\downarrow[\lambda] = \bigwedge_{y \in X} (A_\xi(y) \to \lambda(y)), \quad \xi \in \Lambda.$$

We denote by $F^\uparrow[\lambda] = \{F_\xi^\uparrow : \xi \in \Lambda\}$, the direct $F^\uparrow$-transform of λ and $F_\xi^\uparrow[\lambda]$ its ξ-th component. Similarly, $F^\downarrow[\lambda] = \{F_\xi^\downarrow : \xi \in \Lambda\}$ and $F_\xi^\downarrow[\lambda]$ the, direct $F^\downarrow$-transform of λ and its ξ-th component, respectively.

Proposition 2 ([9]). *Let $\Pi_X = \{A_\xi : \xi \in \Lambda\}$ be a fuzzy partition of X, represented by a fuzzy relation R_Π, and R_Π^T be the transpose of R_Π. Then,*

(i) in the fuzzy approximation space (X, R_Π), the upper approximation $\overline{R}_\Pi(\lambda)$ of $\lambda \in L^X$ is determined by the $F^\uparrow$-transform of f w.r.t. Π_X, i.e., for any $y \in X$, there exists $\xi_y \in \Lambda$, such that

$$\overline{R}_\Pi(\lambda)(y) = F_{\xi_y}^\uparrow[\lambda], \text{and} \tag{7}$$

(ii) in the fuzzy approximation space (X, R_Π^T), the lower approximation of $\lambda \in L^X$ is determined by the $F^\downarrow$-transform of λ w.r.t. Π_X, i.e. for any $x \in X$, there exists $\xi_x \in \Lambda$, such that

$$\underline{R}_\Pi^T(\lambda)(x) = F_{\xi_x}^\downarrow[\lambda]. \tag{8}$$

Corollary 1. *Let the assumptions of Proposition 2 be fulfilled. Then,*

(i) for any $y \in X$, there exists $\xi_y \in \Lambda$, such that the corresponding upper F-transform component $F_{\xi_y}^\uparrow[\lambda]$ can be computed, using the direct image of λ under fuzzy relation R_Π, i.e.

$$F_{\xi_y}^\uparrow[\lambda] = (R_\Pi \circ_\otimes \lambda)(y).$$

(ii) for any $x \in X$, there exists $\xi_x \in \Lambda$, such that the corresponding lower F-transform $F_{\xi_x}^\downarrow[\lambda]$ can be computed, using the subdirect image of a fuzzy set λ under the reflexive fuzzy relation R_Π, i.e.

$$F_{\xi_x}^\downarrow[\lambda] = (R_\Pi^T \triangleleft \lambda)(x).$$

For given two spaces with fuzzy partitions, the following is the notion of morphism between them.

Definition 14 [5]. *For two spaces with fuzzy partitions $\Pi_X = (X, \{A_\xi : \xi \in \Lambda\})$ and $\Pi_Y = (Y, \{B_\psi : \psi \in \Omega\})$ an **FP-map** is a pair of maps (f, g), where $f : X \to Y$ and $g : \Lambda \to \Omega$ are maps such that for each $x \in X$, $\forall \xi \in \Lambda$, $A_\xi(x) \le B_{g(\xi)}(f(x))$.*

The criterion [17] of whether fuzzy partition is represented by reflexive fuzzy relation is repeated below.

A reflexive fuzzy relation R on X represents fuzzy partition Π_X of X, if and only if

$$(\forall y \in X) \, (\forall z \in core(R(\cdot, y))) \quad R(\cdot, y) = R(\cdot, z). \tag{9}$$

Proposition 3. *Let $f : (X, R) \to (Y, S)$ be a morphism between fuzzy relational structures, where reflexive fuzzy relations R, S fulfill property (9). Let moreover, R, S represent fuzzy partitions Π_X, Π_Y of X, Y respectively. Then there exists a map $g : \Lambda \to \Omega$ such that (f, g) is a morphism (FP-map) between spaces with fuzzy partitions (X, Π_X) and (Y, Π_Y).*

Proof. Let $f : (X, R) \rightarrow (Y, S)$ be a morphism between fuzzy relational structures (X, R) and (Y, S). Then $R(x, y) \leq S(f(x), f(y)), \forall x, y \in X$. Because fuzzy relation R is reflexive and fulfills property (9), the cores of fuzzy sets $\{R(\cdot, y) \mid y \in X\}$ constitute a partition of X with the corresponding to it equivalence relation $\equiv_R$ on X. Moreover, R determines a fuzzy partition of X $\Pi_X = \{A_\xi : \xi \in \Lambda\}$, where

(i)   for every $\xi \in \Lambda$, there exists $y \in X$, such that $A_\xi = R(\cdot, y)$,
(ii) if $y_1 \equiv_R y_2$, and $A_\xi = R(\cdot, y_1)$ then $A_\xi = R(\cdot, y_2)$,
(iii) for every $y \in X$, there exists $\xi_y \in \Lambda$, such that $R(\cdot, y) = A_{\xi_y}$.

Therefore, we can define an index function $i_{\Pi_X} : X \rightarrow \Lambda$ such that

$$i_{\Pi_X}(y) = \xi, \text{ if } y \in core(A_\xi).$$

Let us apply the similar reasoning as above to fuzzy relation S on Y and denote Π_Y the corresponding fuzzy partition of Y with the index function $k_{\Pi_Y} : Y \rightarrow \Omega$. Now define map $g : \Lambda \rightarrow \Omega$ such that, $g(\xi) = k_{\Pi_Y}(f(y))$, where $\xi = i_{\Pi_X}(y)$. Let $y \in X$ and $i_{\Pi_X}(y) = \xi$. Then, $y \in core(A_\xi)$ so that $A_\xi(x) = R(x, y) \leq S(f(x), f(y)) = B_{k_{\Pi_Y}(f(y))}(f(x)) = B_{g(\xi)}(f(x))$, (since $f(y) \in core(B_{g(\xi)})$). Therefore, (f, g) is a morphism (FP-map) between spaces with fuzzy partitions (X, Π_X) and (Y, Π_Y).

5   Fuzzy Automata

The theory of fuzzy automata, introduced by Wee and Santos [15], has drawn attention of a number of researchers. In the literature, fuzzy automata have been studied either (i) as recognizers of fuzzy languages, or (ii) as algebraic structures. In this section, we propose a new definition of fuzzy automaton (as an algebraic structure), where a transition function is replaced by a system of fuzzy relations. We prove the equivalence of both approaches, where the original one is based on the notion of transtiton function.

In the following, we give a glimpse to the basic concepts fuzzy automaton and associated fuzzy languages.

5.1   Fuzzy Automaton Based on Fuzzy Transition Function

Below, we recall some basic concepts of fuzzy automaton, based on a fuzzy transition function.

Definition 15. *A* **fuzzy automaton** *is a 5-tuple $M = (Q, X, \delta, \sigma_0, \sigma_1)$, where Q and X are nonempty finite sets, called the set of states and the set of inputs, respectively, $\delta : Q \times X \times Q \rightarrow L$ is a fuzzy subset of $Q \times X \times Q$, called a fuzzy transition function, and $\sigma_0, \sigma_1 : Q \rightarrow L$ are fuzzy subsets of Q, called fuzzy initial and final state, respectively.*

For a fixed set X, X^* denotes the free monoid generated by X. We denote by e the identity element of X^* (empty word).

It can be easily defined, how fuzzy transition function $\delta : Q \times X \times Q \to L$ can be extended to $\delta^* : Q \times X^* \times Q \to L$, such that $\forall p, q \in Q$, $\forall u \in X^*$, and $\forall x \in X$,

$$\delta^*(q, e, p) = \begin{cases} 1 & \text{if } q = p \\ 0 & \text{if } q \neq p, \text{ and} \end{cases}$$

$$\delta^*(p, ux, q) = \bigvee \{\delta^*(p, u, r) \otimes \delta(r, x, q) : r \in Q\}.$$

Also, in [12], it has been observed that

$$\delta^*(p, uv, q) = \bigvee \{\delta^*(p, u, r) \otimes \delta^*(r, u, q) : r \in Q\},$$

$\forall p, q \in Q, \forall u, v \in X^*$.

Definition 16. *A **right fuzzy language**, associated with a state $q \in Q$ of fuzzy automaton $M = (Q, X, \delta, \sigma_0, \sigma_1)$, is a fuzzy subset f_q of X^*, given by $f_q : X^* \to L$, such that for all $u \in X^*$, $f_q(u) = \vee\{\sigma_0(q) \otimes \delta^*(q, u, p) : q \in Q\}$.*

Remark 1. It can be seen that a right fuzzy language associated with a state $q \in Q$ of a fuzzy automaton $M = (Q, X, \delta, \sigma_0, \sigma_1)$ can be obtained as a result of sup-$\otimes$-composition: $f_q(u) = \sigma_0 \circ_\otimes \delta_u^*$, $u \in X^*$.

Further, an algebraic study of fuzzy automata was initiated, aiming at understanding a behaviour of a system in a fuzzy environment. The important contribution to this development was made in [4]. It is worthing to note that in these studies, the notion of fuzzy initial/final state was eliminated, and replaced by the notion of fuzzy source, fuzzy successor and fuzzy core. Below, we give a modified definition of a fuzzy automaton.

Definition 17. *A **fuzzy automaton** is a triple $M = (Q, X, \delta)$, where Q, X and δ are as in the Definition 15.*

We now introduce the following concept of homomorphism between fuzzy automata.

Definition 18. *A **homomorphism** from a fuzzy automaton (Q, X, δ) to a fuzzy automaton (Q', X, δ') is a map $f : Q \to Q'$, such that*

$$\forall (q, u, p) \in Q \times X \times Q, \quad \delta'(f(q), u, f(p)) \geq \delta(q, u, p).$$

Definition 19. [12] *Let (Q, X, δ) be a fuzzy automaton and $\lambda \in L^Q$ a fuzzy current state. Then **fuzzy source**, **fuzzy successor** and **fuzzy core** of λ are respectively defined as the following fuzzy states:*

$$\sigma(\lambda)(q) = \bigvee \{\lambda(p) \otimes \delta(q, u, p) : p \in Q, u \in X\},$$

$$s(\lambda)(q) = \bigvee \{\lambda(p) \otimes \delta(p, u, q) : p \in Q, u \in X\}, \text{ and}$$

$$\mu(\lambda)(q) = \bigwedge \{\lambda(p) \to \delta(q, u, p) : p \in Q, u \in X\}.$$

5.2   Fuzzy Relation-Based Fuzzy Automaton

In this section, we introduce a new definition of fuzzy automaton, where a transition function is replaced by a system of fuzzy relations. We show that both definitions of a fuzzy automaton are equivalent. Moreover, we show that the derived notions of fuzzy source, fuzzy successor and fuzzy core can be expressed using the language of fuzzy relations and their compositions.

Definition 20. *A fuzzy relation-based fuzzy automaton is a triple* $\mathcal{R} = (Q, X, \{R_u, u \in X\})$, *where* Q *and* X *are nonempty finite sets, called the set of states and the set of inputs, respectively, and for every input* $u \in X$, $R_u : Q \times Q \to L$ *is a fuzzy (transition) relation on* Q.

Let X^* denote a free monoid generated by X, and e be the identity element of X^*. We define the extended fuzzy transition relation $\{R_u^*\}_{u \in X^*} : Q \times Q \to L$ as follows:

(i) if $u = e$,

$$R_e^*(q, p) = \begin{cases} 1 & \text{if } q = p \\ 0 & \text{if } q \neq p, \end{cases}$$

(ii) if $u = (v, x)$, where $v \in X^*$, $x \in X$,

$$R_{vx}^*(q, p) = \bigvee \{R_v^*(q, r) \otimes R_x(r, q) : r \in Q\}.$$

It is easy to see that, $\forall q, p \in Q, \forall u, v \in X^*$, we have

$$R_{uv}^*(q, p) = \bigvee \{R_u^*(q, r) \otimes R_v^*(r, p) : r \in Q\}.$$

In the next proposition, we prove that both the definitions of fuzzy automata are **equivalent**.

Proposition 4. *Let* (Q, X, δ) *be a transition-based fuzzy automaton. Then, there exists a fuzzy relation-based fuzzy automaton* $(Q, X, \{R_u, u \in X\})$ *such that for all* $u \in X$, *and for all* $q, p \in Q$,

$$\delta(q, u, p) = R_u(q, p), \tag{10}$$

and vice versa.

Proof. Let (Q, X, δ) be a fuzzy automaton and $\delta : Q \times X \times Q \to L$ be a fuzzy transition function. For all $u \in X$, and for all $q, p \in Q$, we define fuzzy relation R_u on Q, using (10). It is easy to see that $(Q, X, \{R_u, u \in X\})$ is a fuzzy relation-based fuzzy automaton. The opposite claim can be proved similarly.

We now introduce the following concept of homomorphism between fuzzy relation-based fuzzy automata.

Definition 21. *A* **homomorphism** *between fuzzy relation-based fuzzy automata* $(Q, X, \{R_u, u \in X\}$ *and* $(Q, X, \{S_u, u \in X\})$ *is a map* $f : Q \to Q'$ *such that for all* $u \in X$, *and for all relations* R_u *and* S_u,

$$S_u(f(q), f(p)) \geq R_u(q, p), \text{ for all } q, p \in Q.$$

Remark 2. With every fuzzy relation-based fuzzy automaton $(Q, X, \{R_u, u \in X\})$ we associate the system of fuzzy relational structures $\{(Q, R_u), u \in X\}$ where X is an index set.

It is not difficult to show that the following proposition holds.

Proposition 5. *Let $f : Q \to Q'$ be a homomorphism between fuzzy relation-based fuzzy automata $(Q, X, \{R_u, u \in X\})$ and $(Q', X, \{S_u, u \in X\})$. Let moreover, $\{(Q, R_u), u \in X\}$ and $\{(Q', S_u), u \in X\}$ be the corresponding associated fuzzy relational structures. Then, for every $u \in X$, f is a morphism between fuzzy relational structures (Q, R_u) and (Q', S_u).*

On the other hand, let $\{(Q, R_u), u \in X\}$ and $\{(Q', S_u), u \in X\}$ be two fuzzy relational structures, connected by the same index set X. Let moreover, for every $u \in X$, $f : Q \to Q'$ be a morphism between (Q, R_u) and (Q', S_u). Then, f is a homomorphism between fuzzy relation-based fuzzy automata $(Q, X, \{R_u, u \in X\}$ and $(Q', X, \{S_u, u \in X\})$, determined by the corresponding structures $\{(Q, R_u), u \in X\}$ and $\{(Q', S_u), u \in X\}$.

Definition 22. *Let $(Q, X, \{R_u, u \in X\})$ be a fuzzy relation-based fuzzy automaton and $\lambda \in L^Q$ a fuzzy state. The **fuzzy source**, **fuzzy successor** and **fuzzy core** of λ are respectively defined as follows:*

$$\sigma(\lambda)(q) = \bigvee\{\lambda(p) \otimes R_u(q, p) : p \in Q, u \in X\},$$

$$s(\lambda)(q) = \bigvee\{\lambda(p) \otimes R_u(p, q) : p \in Q, u \in X\}, \text{ and}$$

$$\mu(\lambda)(q) = \bigwedge\{\lambda(p) \to R_u^T(p, q) : p \in Q, u \in X\}.$$

Now, we have the following.

Proposition 6. *For a given fuzzy relation-based fuzzy automaton $(Q, X, \{R_u, u \in X\})$, and $\lambda \in L^Q$, the fuzzy source, fuzzy successor and fuzzy core, respectively, can be given by*

$$\sigma(\lambda)(q) = \bigvee\{R_u \circ_\otimes \lambda(q) : u \in X\},$$

$$s(\lambda)(q) = \bigvee\{R_u^T \circ_\otimes \lambda(q) : u \in X\},$$

$$\mu(\lambda)(q) = \bigwedge\{\lambda \triangleleft R_u^T(q) : u \in X\}.$$

6   Conclusion

In this paper, we analyzed the theories of fuzzy rough sets, F-transforms and fuzzy automata and showed that all of them are particular cases of fuzzy relational structures. In all these theories, the aggregation (in the form of lattice-based composition with an underlying fuzzy relation) is a main transforming tool. We proved that the principal parts of morphisms between all considered fuzzy relational structures coincide.

Acknowledgment. This work is supported by University of Ostrava grant lRP201824 "Complex topological structures". The additional support was also provided by the Czech Science Foundation (GAČR) through the project of No. 18-06915S.

References

1. Ćirić, M., Ignjatović, J.: Fuzziness in automata theory: why? How? Stud. Fuzziness Soft Comput. **298**, 109–114 (2013)
2. Dubois, D., Prade, H.: Rough fuzzy sets and fuzzy rough sets. Int. J. Gen. Syst. **17**, 191–209 (1990)
3. Hájek, P.: Metamathematics of Fuzzy Logic. Kluwer Academic Publishers, Boston (1998)
4. Jin, J., Li, Q., Li, Y.: Algebraic properties of L-fuzzy finite automata. Inf. Sci. **234**, 182–202 (2013)
5. Močkoř, J., Holčapek, M.: Fuzzy objects in spaces with fuzzy partitions. Soft Comput. **21**, 7269–7284 (2017)
6. Perfilieva, I., Valasek, R.: Fuzzy transforms in removing noise. Adv. Soft Comput. **2**, 221–230 (2005)
7. Perfilieva, I.: Fuzzy transforms: theory and applications. Fuzzy Sets Syst. **157**, 993–1023 (2006)
8. Perfilieva, I.: Fuzzy transforms: a challenge to conventional transforms. Adv. Image Electron Phys. **147**, 137–196 (2007)
9. Perfilieva, I., Singh, A.P., Tiwari, S.P.: On the relationship among F-transform, fuzzy rough set and fuzzy topology. Soft Comput. **21**, 3513–3523 (2017)
10. Perfilieva, I., Holčapek, M., Kreinovich, V.: A new reconstruction from the F-transform components. Fuzzy Sets Syst. **288**, 3–25 (2016)
11. Qiao, J., Hua, B.Q.: A short note on L-fuzzy approximation spaces and L-fuzzy pretopological spaces. Fuzzy Sets Syst. **312**, 126–134 (2017)
12. Qiu, D.: Automata theory based on quantum logic: some characterizations. Inf. Comput. **190**, 179–195 (2004)
13. Qiu, D.: Automata theory based on complete residuated lattice-valued logic(I). Sci. China **44**, 419–429 (2001)
14. Radzikowska, A.M., Kerre, E.E.: Fuzzy rough sets based on residuated lattices. Lecture Notes in Computer Science, vol. 3135, pp. 278–296 (2005)
15. Santos, E.S.: Maximin automata. Inf. Control. **12**, 367–377 (1968)
16. Singh, A.P., Tiwari, S.P.: Lattice F-transform for functions in two variables. J. Fuzzy Set Valued Anal. **3**, 185–195 (2016)
17. Singh, A. P., Tiwari, S. P., Perfilieva, I.: F-transforms, L-fuzzy partitions and L-fuzzy pretopological spaces: an operator oriented view. Fuzzy Sets Syst. (Submitted)
18. Sussner, P.: Lattice fuzzy transforms from the perspective of mathematical morphology. Fuzzy Sets Syst. **288**, 115–128 (2016)
19. Zadeh, L.A.: Fuzzy sets. Inf. Control **8**, 338–353 (1965)

Pseudo-Additions and Shift Invariant Aggregation Functions

Andrea Stupňanová[1]([✉])[iD], Doretta Vivona[2][iD], and Maria Divari[2]

[1] Faculty of Civil Engineering, Slovak University of Technology in Bratislava,
Bratislava, Slovak Republic
andrea.stupnanova@stuba.sk
[2] Faculty of Civil and Industrial Engineering, Sapienza - University of Rome,
Rome, Italy
doretta.vivona@sbai.uniroma1.it, maria.divari@alice.it

Abstract. Shift invariant aggregation functions are related to the shifts based on the standard addition $+$ and their complete characterization is well known. We discuss the aggregation functions invariant with respect to a pseudo-addition $\oplus$. Our study has two directions. In the first one, we discuss $\oplus$-shift invariant aggregation functions with respect to a fixed pseudo-addition $\oplus$. In the second one, for a fixed aggregation function A, we discuss pseudo-additions $\oplus$ such that A is $\oplus$-shift invariant.

Keywords: Aggregation function · Pseudo addition · Shift invariantness

1 Introduction

Shift invariantness of n-dimensional real functions appears naturally in several branches of mathematics and applied fields. So, for example, for a function $F : \mathbb{R} \to \mathbb{R}$, it can be seen as the distributivity of the common addition $+$ of reals over F, i.e.,

$$F(x_1 + z, \ldots, x_n + z) = F(x_1, \ldots, x_n) + z \tag{1}$$

In measurement theory [11], shift invariant functions $F : I^n \to \mathbb{R}$ (where I is some real interval) are called difference scale invariant function, see also [9, Subsect. 7.3]. In general information theory [15,17], the shift invariantness (1) of a function $F : [0, \infty]^2 \to [0, \infty]$ considered for characterizing information measures is called a compatibility equation. For more details we recommend [4–6].

In several situations the standard addition $+$ is not appropriate to model real problems. Therefore, several models based on some pseudo-addition $\oplus$ were

This work was supported by the Slovak Research and Development Agency under the contract no. APVV-17-0066 and grant VEGA 1/0682/16 for the first author and MIUR (ITALY) for the second one.

R. Halaš et al. (Eds.): AGOP 2019, AISC 981, pp. 75–82, 2019.
https://doi.org/10.1007/978-3-030-19494-9_7

introduced, studied and successfully applied, see, e.g. [13,17,19]. These facts have inspired our study of ⊕-shift invariant aggregation functions.

Our contribution is organized as follows. In the next section, some preliminaries are given and characterization of shift invariant aggregation functions is recalled. In Sect. 3, after recalling the notion of a pseudo-addition ⊕ we introduce and discuss, for some fixed pseudo-additions ⊕, the classes of ⊕-shift invariant binary aggregation functions. In Sect. 4, for some particular binary aggregation function A, we study pseudo-additions ⊕ making A a ⊕-shift invariant aggregation function. Finally, some concluding remarks are added.

2   Shift Invariant Aggregation Functions

In this contribution, we deal with aggregation functions defined on $[0,1]$, for more details see [2,3,7,9].

Definition 1. *A mapping $A : [0,1]^n \to [0,1]$ is called an (n-ary) aggregation function if and only if it is monotone non-decreasing and satisfy the boundary conditions $A(0,\ldots,0) = 0$, $A(1,\ldots,1) = 1$.*

Definition 2. *An aggregation function $A : [0,1]^n \to [0,1]$ is called shift invariant if and only if it satisfies the equality*

$$A(x_1 + c, \ldots, x_n + c) = A(x_1, \ldots, x_n) + c \qquad (2)$$

for any $\mathbf{x} = (x_1, \ldots, x_n) \in [0,1]^n$ and $c \in [0, 1 - \mathrm{Max}(\mathbf{x})]$.

Observe that due to the boundary condition $A(0,\ldots,0) = 0$, for any shift invariant aggregation function A it holds $A(c,\ldots,c) = c$, $c \in [0,1]$, i.e., shift invariantness of an aggregation function A implies its idempotency. The next characterizations can be found in [12,14], see also [9, Subsect. 7.3].

Proposition 1. *A binary aggregation function $A : [0,1]^2 \to [0,1]$ is a shift invariant aggregation function if and only if there are monotone non-decreasing 1-Lipschitz functions $f, g : [0,1] \to [0,1]$, $f(0) = g(0) = 0$, so that*

$$A(x,y) = \begin{cases} y + f(x - y) & \text{if } x \geq y, \\ x + g(y - x) & \text{if } x \leq y. \end{cases} \qquad (3)$$

Obviously, $A(x,0) = f(x)$ and $A(0,y) = g(y)$.

Proposition 2. *Let $n \geq 2$. An aggregation function $A : [0,1]^n \to [0,1]$ is shift invariant if and only if*

$$A(\mathbf{x}) = \mathrm{Min}(\mathbf{x}) + A(\mathbf{x} - \mathbf{1} \cdot \mathrm{Min}(\mathbf{x})),$$

and

$$0 \leq A(\mathbf{x}) - A(\mathbf{y}) \leq \max\{x_1 - y_1, \ldots, x_n - y_n\}$$

for all $\mathbf{x}, \mathbf{y} \in [0,1]^n$ such that $x_i = y_i = 0$ for some $i \in \{1, \ldots, n\}$ and $\mathbf{x} \geq \mathbf{y}$.

Note that for $n = 2$, Proposition 2 is just Proposition 1.

Proposition 2 also gives a hint how to construct shift invariant aggregation functions. Indeed, for any aggregation function $B : [0,1]^n \to [0,1]$ with minimal Chebyshev norm (i.e., $\|B\|_\infty = 1$), one can define

$$B_s(\mathbf{x}) = \text{Min}(\mathbf{x}) + B(\mathbf{x} - \mathbf{1} \cdot \text{Min}(\mathbf{x})), \tag{4}$$

what is the shift invariant aggregation function.

Example 1. (i) Each weighted arithmetic mean $W_\mathbf{w}$, $W_\mathbf{w}(\mathbf{x}) = \sum\limits_{i=1}^{n} w_i x_i$, is shift invariant. Similarly, each Choquet integral $\mathbf{Ch}_m$ with respect to a fuzzy measure m,

$$\mathbf{Ch}_m(\mathbf{x}) = \sum_{i=1}^{n} (x_{\sigma(i)} - x_{\sigma(i-1)})\, m(\{\sigma(i), \dots, \sigma(n)\}),$$

where $\sigma : \{1, \dots, n\} \to \{1, \dots, n\}$ is a permutation such that $x_{\sigma(1)} \le \cdots \le x_{\sigma(n)}$, and $x_{\sigma(0)} = 0$ by convention, is shift invariant.

(ii) Consider $B : [0,1]^2 \to [0,1]$ given by

$$B(x_1, x_2) = \sqrt[p]{\frac{x_1{}^p + x_2{}^p}{2}},$$

where $p \ge 1$. Then $\|B\|_\infty = 1$, and, based on (4), $B_s : [0,1]^2 \to [0,1]$ given by

$$B_s(x_1, x_2) = \frac{2^{\frac{1}{p}} - 1}{2^{\frac{1}{p}}} \text{Min}(x_1, x_2) + \frac{1}{2^{\frac{1}{p}}} \text{Max}(x_1, x_2)$$

is a shift invariant aggregation function (observe that B_s is an OWA operator, i.e., Choquet integral with respect to a symmetric fuzzy measure).

(iii) As a solution of compatibility equation related to Shannon entropy, Aczél et al. [1] have obtained the following family of shift invariant aggregation functions (the same formula can be introduced on $[0, \infty]$, too):

$$A_\alpha(x, y) = -\alpha \log \left(\frac{e^{-\frac{x}{\alpha}} + e^{-\frac{y}{\alpha}}}{2} \right)$$

for $\alpha \in]0, \infty[$. Note that due to Proposition 1, A_α is related to function $f_\alpha = g_\alpha$ given by

$$f_\alpha(x) = -\alpha \log \left(\frac{1 + e^{-\frac{x}{\alpha}}}{2} \right).$$

3   Pseudo-Additions and $\oplus$-Shift Invariant Aggregation Functions

Definition 3. *An operation $\oplus : [0, \infty]^2 \to [0, \infty]$ is called a pseudo-addition whenever it is a continuous associative aggregation function on $[0, \infty]$ with a neutral element $e = 0$.*

Note that $\oplus$ is an I-semigroup and it was completely described by Mostert and Shields as an ordinal sum of Archimedean pseudo-additions [16]. Our main interest is in aggregation of inputs from $[0, 1]$, and then we have 3 prototypical pseudo-additions which we consider in the rest of this contribution.

Example 2. The next operations are pseudo-additions:

(i) The only idempotent pseudo-addition $\oplus = \vee$ (maximum operator);
(ii) Archimedean pseudo-addition $\oplus_g$ generated by an additive generator g : $[0, \infty] \to [0, \infty]$, g is continuous strictly increasing, $g(0) = 0$, and

$$x \oplus_g y = g^{-1}(\min\{g(\infty), g(x) + g(y)\}).$$

○ if g is the identity, i.e., $g(x) = g_1(x) = x$, then $\oplus_{g_1} = +$ is the standard addition;
○ if $g(x) = g_p(x) = x^p$, $p \in\,]0, \infty[$, then $x \oplus_{g_p} y = (x^p + y^p)^{\frac{1}{p}}$;
○ if $g(x) = \frac{2x}{x+1}$, then $x \oplus_g y = \begin{cases} \frac{2xy+x+y}{1-xy} & \text{if } xy < 1, \\ \infty & \text{otherwise .} \end{cases}$

(iii) If $1 \oplus 1 = 1$, then $\oplus|[0, 1]^2 = S$ is a continuous t-conorm [10].
○ if S is a strict t-conorm then there is an increasing bijection $t : [0, 1] \to [0, \infty]$ such that, for $x, y \in [0, 1]$,

$$x \oplus_t y = S(x, y) = t^{-1}(t(x) + t(y)).$$

If $t(x) = -\log(1 - x)$, $S = S_P$ is the probabilistic sum, $S_P(x, y) = x + y - xy$;
○ if S is a nilpotent t-conorm, then there is an increasing bijection $u : [0, 1] \to [0, 1]$ such that, for $x, y \in [0, 1]$,

$$x \oplus_u y = S(x, y) = u^{-1}(\min\{1, u(x) + u(y)\}).$$

If $u(x) = x$, then $S = S_L$ is the Lukasiewicz t-conorm (truncated sum), $S_L(x, y) = \min\{1, x + y\}$.

Based on a fixed pseudo-addition $\oplus$ we modify Definition 2 as follows.

Definition 4. *Let* $\oplus : [0, \infty]^2 \to [0, \infty]$ *be a fixed pseudo-addition and* $A : [0, 1]^n \to [0, 1]$ *an aggregation function. Then* A *is called* $\oplus$-*shift invariant if and only if it satisfies the equality*

$$A(x_1 \oplus c, \ldots, x_n \oplus c) = A(x_1, \ldots, x_n) \oplus c \qquad (5)$$

for any $\mathbf{x} = (x_1, \ldots, x_n) \in [0, 1]^n$ *and* $c \in [0, 1]$ *such that* $x_1 \oplus c, \ldots, x_n \oplus c \in [0, 1]$.

Note that the equality (5) can be seen as a distributivity of $\oplus$ over A. For some particular pseudo-additions introduced in Example 2 we have the next results.

(i) If $\oplus = \vee$, then the $\vee$-shift invariantness is called also max-homogeneity [9]. Typical examples of $\vee$-shift invariant aggregation functions are Sugeno integrals [8,18]. In binary case, we have the next modification of Proposition 1.

Proposition 3. *An aggregation function $A : [0,1]^2 \to [0,1]$ is $\vee$-shift invariant if and only if there are monotone non-decreasing functions $f, g : [0,1] \to [0,1]$, $f(0) = g(0) = 0, f \leq \mathrm{id}, g \leq \mathrm{id}$, (id is identity function) and*

$$A(x,y) = \begin{cases} y \vee f(x) & \text{if } x \geq y \\ x \vee g(y) & \text{if } x \leq y. \end{cases} \tag{6}$$

Note that if $f(x) = \min\{x, a\}, g(y) = \min\{y, b\}$, then A is the Sugeno integral with respect to fuzzy measure $m : 2^{\{1,2\}} \to [0,1], m \sim (a, b)$ where $m(\{1\}) = a$ and $m(\{2\}) = b$.

(*ii*) In the case of an Archimedean pseudo-addition $\oplus_g$ the $\oplus_g$-shift invariantness is a property isomorphic to the standard shift invariantness.

Proposition 4. *Let $\oplus_g : [0,\infty]^2 \to [0,\infty]$ be an Archimedean pseudo-addition generated by an additive generator $g : [0,1] \to [0,\infty]$, such that $g(1) = 1$. Denote $\varphi = g|[0,1]$. Then $\varphi : [0,1] \to [0,1]$ is an automorphism, and an aggregation function $A : [0,1]^n \to [0,1]$ is $\oplus_g$-shift invariant if and only if the φ^{-1}-transform of A, $A_{\varphi^{-1}} : [0,1]^n \to [0,1]$ given by*

$$A_{\varphi^{-1}}(x_1, \ldots, x_n) = \varphi\left(A(\varphi^{-1}(x_1), \ldots, \varphi^{-1}(x_n))\right),$$

is a shift invariant aggregation function.

Due to Propositions 4 and 1 (for binary case) and Proposition 2 (for $n \geq 2$), we have a complete characterization of $\oplus_g$-shift invariant aggregation functions.

As a prototypical example, we recall the weighted quasi-arithmetic mean $A(x_1, \ldots, x_n) = \varphi^{-1}\left(\sum_{i=1}^n w_i\, \varphi(x_i)\right)$. Note that then

$$A_{\varphi^{-1}}(x_1, \ldots, x_n) = \varphi\left(\varphi^{-1}\left(\sum_{i=1}^n w_i\, \varphi(\varphi^{-1}(x_i))\right)\right) = \sum_{i=1}^n w_i\, x_i,$$

i.e., $A_{\varphi^{-1}}$ is a weighted arithmetic mean, which is shift invariant.

(*iii*) A similar characterization is valid in the case when $\oplus|[0,1]^2 = S$ is a strict t-conorm. Then, instead of $\oplus$-shift invariantness we will use notation S-shift invariantness. (Similar notation we use for nilpotent t-conorms, too).

Proposition 5. *Let $S : [0,1]^2 \to [0,1]$ be a strict t-conorm generated by an additive generator $t : [0,1] \to [0,\infty]$. Then an aggregation function $A : [0,1]^n \to [0,1]$ is S-shift invariant if and only if the function $A_{t^{-1}} : [0,\infty]^n \to [0,\infty]$ given by*

$$A_{t^{-1}}(x_1, \ldots, x_n) = t\left(A(t^{-1}(x_1), \ldots, t^{-1}(x_n))\right),$$

is a shift invariant aggregation function on $[0,\infty]$.

Note that shift invariant aggregation functions on $[0,\infty]$ have a similar characterization as those on $[0,1]$ described in Proposition 2. For more details see [9].

Example 3. Consider the arithmetic mean $M : [0, \infty]^n \to [0, \infty]$, $M(x_1, \ldots, x_n) = \frac{1}{n} \sum_{i=1}^{n} x_i$, and $t(x) = -\log(1 - x)$ (related to the probabilistic sum S_P). Then $A_{t^{-1}} = M$ if and only if $A = M_t$, i.e., for $\mathbf{x} \in [0, 1]^n$, we have

$$A(\mathbf{x}) = t^{-1}\left(M(t(x_1), \ldots, t(x_n))\right) = 1 - \left(\prod_{i=1}^{n}(1 - x_i)\right)^{\frac{1}{n}},$$

i.e., $A = G^d$ is the dual of standard geometric mean and it is S_P-shift invariant.

As a problem for further study we propose S-shift invariant aggregation functions, where $S : [0, 1]^2 \to [0, 1]$ is a nilpotent t-conorm.

Example 4. Consider the Lukasiewicz t-conorm S_L, (i.e., $x \oplus y = S_L(x, y) = \min\{1, x + y\}$) and the arithmetic mean $M : [0, 1]^2 \to [0, 1]$. Then for $x = 0.4$, $y = 0.8$ and $c = 0.4$, we have $M(x \oplus c, y \oplus c) = M(0.8, 1) = 0.9$. On the other hand $M(x, y) \oplus c = S_L(0.6, 0.4) = 1$, i.e., M is not S_L-shift invariant.

Obviously, we have also positive examples. In particular, for any pseudo-addition $\oplus$ the next aggregation functions are $\oplus$-shift invariant:

(1) projections $P_1, \ldots, P_n$, where $P_i(\mathbf{x}) = x_i$, $\mathbf{x} \in [0, 1]^n$;
(2) order statistics $OS_1, \ldots, OS_n$, where $OS_i(\mathbf{x}) = x_{\sigma(i)}$ for any permutation $\sigma : \{1, \ldots, n\} \to \{1, \ldots, n\}$ such that $x_{\sigma(1)} \leq \cdots \leq x_{\sigma(n)}$.

4   Pseudo-Additions that Correspond to Some Fixed Aggregation Functions

In this section, for a fixed aggregation function $A : [0, 1]^n \to [0, 1]$ we are interested in characterization of pseudo-additions $\oplus : [0, \infty]^2 \to [0, \infty]$ such that A is $\oplus$-shift invariant. As already mentioned above, if $A \in \{P_1, \ldots, P_n, OS_1, \ldots, OS_n\}$, then A is $\oplus$-shift invariant for any pseudo-addition $\oplus$. We conjecture that this claim foes not hold for any aggregation functions A which is not a projection neither an order statistics.

Consider, for example, the arithmetic mean $M : [0, 1]^n \to [0, 1]$,

$$M(\mathbf{x}) = \frac{1}{n} \sum_{i=1}^{n} x_i.$$

Obviously, M is shift invariant (with respect the classical addition $+$), i.e.,

$$+ \in \mathcal{O}_M = \{\oplus | M \text{ is } \oplus-\text{shift invariant}\}.$$

On the other hand, M is not $\vee$-shift invariant, $\vee \notin \mathcal{O}_M$.

If $\oplus = \oplus_g$ is an Archimedean pseudo-addition, then M is $\oplus_g$-shift invariant if and only if $g = \mathrm{id}$ (i.e., $\oplus_g = +$) or g is given by $g(x) = \log(1 + \lambda x)$, where $\lambda > 0$. Note that then $\oplus = \oplus_\lambda$ is the λ-addition introduced by Sugeno [18], and

$$x \oplus_\lambda y = x + y + \lambda xy.$$

Hence $\oplus_\lambda \in \mathcal{O}_M$ for any $\lambda \in [0, \infty]$. On the other hand if $1 \oplus 1 = 1$, and $\oplus|[0,1]^2 = S$ is a nilpotent t-conorm, then $\oplus \notin \mathcal{O}_M$. Finally, if $\oplus|[0,1]^2$ is a strict t-conorm, it can be shown that M is $\oplus$-invariant if and only if $S = S_P$.

The same results are valid for an arbitrary non-trivial weighted arithmetic mean W (i.e., W is not a projection),

$$\{\oplus_\lambda | \lambda \in [0, \infty]\} \cup \{\oplus | \oplus |[0,1]^2 = S_P\} \subseteq \mathcal{O}_W$$

and

$$(\{\vee\} \cup \{\oplus | 1 \oplus 1 = 1, \text{ and } \oplus |[0,1]^2 \neq S_P\}) \cap \mathcal{O}_W = \emptyset.$$

Observe that similar results are valid for non-trivial weighted quasi-arithmetic means, too.

5 Concluding Remarks

We have introduced and studied $\oplus$-shift invariant aggregation functions , where $\oplus$ is a pseudo-addition. For particular fixed pseudo-additions $\oplus$, we have characterized aggregation functions which are $\oplus$-shift invariant. On the other hand, for some particular aggregation functions A, we have studied the set $\mathcal{O}_A$ of pseudo-additions $\oplus$ such that A is $\oplus$-shift invariant. Our results open several new problems. For example, a complete description of the set $\mathcal{O}_M$ is missing, though we have shown several positive examples when $\oplus \in \mathcal{O}_M$, as well as several negative examples when $\oplus \notin \mathcal{O}_M$.

Our results could be of interest in several domains, in particular in measurement theory and in information measures theory.

References

1. Aczél, J., Forte, B., Ng, C.T.: L'équation fonctionnelle triangulaire et la théorie de l'information sans probabilité. C. R. Acad. Sci. Paris Sér. A-B, **275**, A727–A729 (1972)
2. Beliakov, G., Bustince Sola, H., Calvo Sánchez, T.: A Practical Guide to Averaging Functions. Springer, Heidelberg (2016)
3. Beliakov, G., Pradera, A., Calvo, T.: Aggregation Functions: A Guide for Practitioners. Springer, Heidelberg (2007)
4. Benvenuti, P.: Sulle misure di informazione compositive con traccia compositiva unversale. Rendiconti di Matematica **3–4**(2), 481–505 (1969). Serie IV
5. Benvenuti, P.: Sur l'independence dans l'information. Colloquies Internationaux du C.N.R.S. Theorie De l'Information, vol. 276, pp. 49–55 (1974)

6. Benvenuti, P., Divari, M., Pandolfi, M.: Su un sistema di equazioni funzionali provenienti dalla teoria soggettiva dell'informazione. Rendiconti di Matematica **39**(5), 529–540 (1972). Serie VI
7. Calvo, T., Kolesárová, A., Komorníková, M., Mesiar, R.: Aggregation operators: properties, classes and construction methods. In: Aggregation Operators. Studies in Fuzziness and Soft Computing, vol. 97, pp 3–104. Physica, Heidelberg (2002)
8. Couceiro, M., Marichal, J.-L.: Characterizations of discrete Sugeno integrals as polynomial functions over distributive lattices. Fuzzy Sets Syst. **161**(5), 694–707 (2010)
9. Grabisch, M., Marichal, J.-L., Mesiar, R., Pap, E.: Aggregation Functions. Cambridge University Press, Cambridge (2009)
10. Klement, E.P., Mesiar, R., Pap, E.: Triangular Norms. Kluwer Academic Publishers, Dordrecht (2000)
11. Krantz, D.H., Luce, R.D., Suppes, P., Tversky, A.: Foundations of Measurement. Vol. I: Additive and Polynomial Representations. Academic Press, New York (1971)
12. Lázaro, J., Rückschlossová, T., Calvo, T.: Shift invariant binary aggregation operators. Fuzzy Sets Syst. **142**(1), 51–62 (2004)
13. Maslov, V.-P., Samborskii, S.-N.: Idempotent analysis (in place of an introduction). Idempotent Analysis. Advances in Soviet Mathematics, vol. 13, pp. vii–xi. American Mathematical Society, Providence (1992)
14. Mesiar, R., Rückschlossová, T.: Characterization of invariant aggregation operators. Fuzzy Sets Syst. **142**(1), 63–73 (2004)
15. Mesiar, R., Rybárik, J.: Pseudo-arithmetical operations. Tatra Mount. Math. Publ. **2**, 185–192 (1993)
16. Mostert, P.-S., Shield, L.: On the structure of semigroups on a compact manifold with boundary. Ann. Math. **65**, 117–143 (1957)
17. Pap, E.: Null-additive Set Functions. Mathematics and its Applications, vol. 337. Kluwer Academic Publishers Group, Dordrecht (1995)
18. Sugeno, M.: Theory of fuzzy integrals and its applications. Ph.D. thesis, Tokyo Institute of Technology (1974)
19. Vivona, D., Divari, M.: Pseudo-analysis: measures of general conditional information. Adv. Sci. Technol. Eng. Syst. J. **2**(2), 36–40 (2016)

Importation Algebras

Vikash Kumar Gupta$^{(\boxtimes)}$ and Balasubramaniam Jayaram

Department of Mathematics, Indian Institute of Technology Hyderabad,
Sangareddy, Telangana, India
{MA17RESCH01002,jbala}@iith.ac.in

Abstract. In recent years, many works have appeared that propose order from basic fuzzy logic connectives. However, all of them assume the connectives to possess some kind of monotonicity, which succinctly implies that the underlying set is already endowed with an order. In this work, given a set $\mathbb{P} \neq \emptyset$, we define an algebra based on an implicative-type function I without assuming any order-theoretic properties, either on $\mathbb{P}$ or I. Terming it the importation algebra, since the law of importation becomes one of the main axioms in this algebra, we show that such algebras can impose an order on the underlying set $\mathbb{P}$. We show that in the case $\mathbb{P} = [0, 1]$ we can obtain new order-theoretic structures on it even when I is not a fuzzy implication and that one can recover the usual order on $[0, 1]$ even from fuzzy implications that do not have the classical ordering property. Finally, we show a similar approach can lead us to obtaining order from conjunctive type connectives too.

Keywords: Ordered sets · Law of importation · Importation algebra · Fuzzy implications

1 Introduction

In recent years, many works have appeared that propose order from basic fuzzy logic connectives. For instance, given a bounded lattice $(\mathbb{P}, \leq, 0, 1)$, Karaçal *et al.* [5] have proposed an order on $\mathbb{P}$ based on a given t-norm on $\mathbb{P}$. Taking cue from this Ertugrul *et al.* [4], Asici [2] and Kesiciouglu and Mesiar [6] have proposed partial orders based on uninorms, nullnorms and implications, respectively.

From the above works, the following observations can be made:

- Note that, however, there already exists an order on the underlying set $\mathbb{P}$.
- Clearly, the operations come from either the conjunctive or implicative type. However, not all the properties of the employed fuzzy logic connectives play a role in obtaining an order. For instance, the commutativity of a T or a U does not play a role in obtaining the order w.r.t. the proposed definitions.

On the one hand, the above observations give us the first motivation for this work. We consider an underlying set $\mathbb{P}$ that has no pre-existing order defined

© Springer Nature Switzerland AG 2019
R. Halaš et al. (Eds.): AGOP 2019, AISC 981, pp. 83–94, 2019.
https://doi.org/10.1007/978-3-030-19494-9_8

on it. Further, we consider only those properties on the operation that lead us to obtaining an order on $\mathbb{P}$, without presupposing it to come from any particular class of connectives. On the other hand, if we consider $\mathbb{P} = [0, 1]$ and restrict ourselves to fuzzy implications, only R-implications obtained from left-continuous t-norms lead us to richer order-theoretic structures on $[0, 1]$, viz., a residuated lattice, and allow us to recover the underlying order. Can one obtain similar structures and also recover the order from fuzzy implications outside of the above class? This forms the other motivation for this work.

In this paper, we define an algebra based on (I, C) where I and C are implication-type and conjunction-type connectives, respectively. Also, we like to impose minimal conditions on the pair (I, C) and study the algebraic and order-theoretic structures they impose on the underlying set $\mathbb{P}$.

The rest of this paper is organised as follows. In Sect. 2, we propose our Importation algebra and discuss the minimality and mutual independence of the axioms. In Sect. 3 we show how an order can be obtained from an importation algebra and discuss some results that relate the different order-theoretic properties with some known functional equations. In Sect. 4, we discuss importation algebras when the underlying set is the unit interval $[0, 1]$. In Sect. 5 we discuss the conditions under which the original order, if any on the underlying set, can be recovered through the newly proposed order. Our investigations show that we can recover the order on $[0, 1]$ from very many families of fuzzy implications. Finally, in Sect. 6 we present further possible interesting explorations that are underway on this topic.

2   Importation Algebra - Definition and Examples

In this section, we begin by proposing a new algebra, called the *Importation Algebra*. After giving a few examples in the finite setting, we also show that the axioms of this algebra are mutually independent.

Henceforth, we assume that $\mathbb{P}$ is only a non-empty set with no other structure on it.

2.1   Importation Algebra

Definition 1. *Let $\mathbb{P} \neq \emptyset$. An **Importation algebra** $(\mathbb{P}, I, C, \mathbf{1})$ is a $(2, 2, 0)$ algebra satisfying the following conditions:*

– $\mathbf{1}$ *is the left neutral element of I, i.e.,*

$$I(\mathbf{1}, x) = x \ , \ \text{for every } x \in \mathbb{P}. \tag{NP}$$

– *I satisfies the following quasi-neutrality: For any $x, y, z \in \mathbb{P}$,*

$$I(x, I(y, z)) = z \Longrightarrow I(y, z) = z \ . \tag{QN}$$

– *The pair (I, C) satisfies the law of importation, i.e.,*

$$I(C(x, y), z) \ = I(x, I(y, z)) \ . \tag{LI}$$

2.2 Examples of Importation Algebra

If $F : \mathbb{P} \times \mathbb{P} \longrightarrow \mathbb{P}$ is a binary operation on $\mathbb{P}$ then by $\mathbb{U}_F$ we denote the set of all left neutral elements of F, i,e., $\mathbb{U}_F = \{p \in \mathbb{P} \mid F(p,y) = y \text{ for all } y \in \mathbb{P}\}$. For an $x \in \mathbb{P}$, let $U_x^F = \{p \in \mathbb{P} \mid F(p,x) = x\}$. Then $\mathbb{U}_F = \bigcap_{x \in \mathbb{P}} U_x^F$.

In the following examples it can be verified that the given I_i satisfies **(NP)** and **(QN)** and that the pair (I_i, C_i) satisfies **(LI)** on the corresponding $\mathbb{P}_i$ and hence $(\mathbb{P}_i, I_i, C_i, \mathbf{1})$ becomes an importation algebra, for $i = 1, 2, 3$ (Table 1).

Table 1. Functions I_i, C_i for $i = 1, 2, 3$ in Example 2

I_1	p	q	1
p	p	q	1
q	p	q	1
1	p	q	1

C_1	p	q	1
p	p	p	p
q	q	q	q
1	1	1	1

I_2	p	q	r	1
p	1	1	1	1
q	r	1	1	1
r	q	1	1	1
1	p	q	r	1

C_2	p	q	r	1
p	p	p	p	p
q	p	p	p	q
r	p	p	p	r
1	p	q	r	1

I_3	p	q	r	1
p	r	r	1	1
q	p	q	r	1
r	1	1	1	1
1	p	q	r	1

C_3	p	q	r	1
p	r	p	r	p
q	p	q	r	q
r	r	r	r	r
1	r	q	r	1

Example 2. (i) Let $\mathbb{P}_1 = \{p, q, \mathbf{1}\}$. Note that $\mathbb{U}_{I_1} = \mathbb{P}_1$, i.e., every element of $\mathbb{P}_1$ is a left neutral element of I_1 and that C_1 is not commutative.

(ii) Let $\mathbb{P}_2 = \{p, q, r, \mathbf{1}\}$. Note that $\mathbb{U}_{I_2} = \{\mathbf{1}\}$ and that $\mathbb{U}_{C_2} = \emptyset$, i.e., C_2 has no left neutral element.

(iii) Let $\mathbb{P}_3 = \{p, q, r, \mathbf{1}\}$. Note that $\mathbb{U}_{I_3} = \{q, \mathbf{1}\}$. Given an I on $\mathbb{P}$ the C with which it satisfies **(LI)** may not be unique. For instance, it can be easily verified that, in the above importation algebra, if $C_3(\mathbf{1}, q) = \mathbf{1}$ or $C_3(\mathbf{1}, \mathbf{1}) = q$ (see the values in the box), then too the pair (I_3, C_3) satisfies **(LI)**.

2.3 Mutual Independence of the Axioms (NP), (QN) and (LI)

Let $\mathbb{P} = \{p, q, \mathbf{1}\}$. In each of the algebras given in $(\mathbb{P}, I'_i, C'_i, \mathbf{1})$, where the functions I'_i, C'_i are as defined in Table 2, all the axioms of importation algebra except one are satisfied, showing that **(NP)**, **(QN)** and **(LI)** are mutually independent.

Table 2. Functions I'_i, C'_i for $i = 1, 2, 3$ in Example 3

I'_1	p	q	1
p	1	1	1
q	1	1	1
1	1	1	1

C'_1	p	q	1
p	p	q	p
q	q	p	q
1	p	q	p

I'_2	p	q	1
p	q	p	1
q	p	q	1
1	p	q	1

C'_2	p	q	1
p	q	p	p
q	p	q	q
1	p	q	q

I'_3	p	q	1
p	1	1	1
q	p	q	1
1	p	q	1

C'_3	p	q	1
p	p	q	p
q	p	q	q
1	p	q	1

Example 3. (i) Since $\mathbb{U}_{I'_1} = \emptyset$, I'_1 does not satisfy **(NP)** for any $p \in \mathbb{P}$. However, it can be shown that I'_1 satisfies **(LI)** with C'_1 and also **(QN)**.

(ii) *Clearly I_2' satisfies (**NP**). Further, it can be verified that the pair (I_2', C_2') satisfies (**LI**). However, $I_2'(p, I(p,q)) = q$ but $I(p,q) \neq q$, i.e., the axiom (**QN**) is not satisfied.*

(iii) *The pair (I_3', C_3') does not satisfy (**LI**). For instance, $I_3'\left(C_3'(p,q), p\right) = I_3'(q,p) = p \neq 1 = I_3'\left(p, I_3'(q,p)\right) = I_3'(p,p)$. However, it can be shown that I_3' satisfies both (**QN**) and (**NP**).*

3 Order Based on Importation Algebras

The equivalence between a lattice as an algebraic system and as an order-theoretic structure is well-known, i.e., from the partial order $\leq$ one can obtain the algebraic operations of meet $\wedge$ and join $\vee$ uniquely and vice-versa.

In this section, we show that any importation algebra imposes a partial order on the underlying set. Further, we present some results which show how the functional equations or properties satisfied by the operations involved lead to some interesting order-theoretic properties on the obtained poset and also among the operations w.r.t. the obtained new order.

3.1 Order on Importation Algebras

Proposition 4. *Let $(\mathbb{P}, I, C, \mathbf{1})$ be an importation algebra. The relation $\preceq$ on $\mathbb{P}$ defined as follows is a **partial order** on $\mathbb{P}$: For any $x, y \in \mathbb{P}$,*

$$x \preceq y \iff \text{there exists } \ell \in \mathbb{P} \text{ such that } I(\ell, x) = y . \tag{1}$$

Proof. Let us recall that a relation is said to be a partial order if it is reflexive, anti-symmetric and transitive.

(i) $\preceq$ **is reflexive:** From (**NP**) we see that $I(\mathbf{1}, x) = x$, for all $x \in \mathbb{P}$ which implies that $x \preceq x$.

(ii) $\preceq$ **is anti-symmetric:** Let $x, y \in \mathbb{P}$ such that $x \preceq y$ and $y \preceq x$. Then there exist $\ell_1, \ell_2 \in \mathbb{P}$ such that $I(\ell_1, x) = y$ and $I(\ell_2, y) = x$. Thus, we have

$$I(\ell_1, x) = y \implies I(\ell_2, I(\ell_1, x)) = I(\ell_2, y) = x .$$

Now, from (**QN**) we have $I(\ell_1, x) = x$ and hence $x = y$.

(iii) $\preceq$ **is transitive:** Let $x \preceq y$ and $y \preceq z$. Then there exist $\ell_1, \ell_2 \in \mathbb{P}$ such that $I(\ell_1, x) = y$ and $I(\ell_2, y) = z$. Now,

$$\begin{aligned}
I(\ell_2, y) = z &\implies I(\ell_2, I(\ell_1, x)) = z \\
&\implies I(C(\ell_1, \ell_2), x) = z \qquad \lceil \text{ by } (\textbf{LI}) \\
&\implies I(\ell, x) = z , \text{ where } \ell = C(\ell_1, \ell_2) , \\
&\implies x \preceq z .
\end{aligned}$$

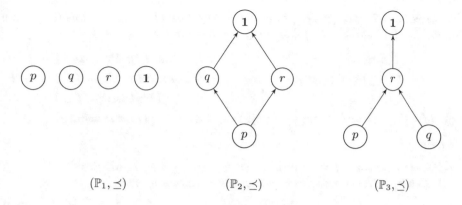

$$(\mathbb{P}_1, \preceq) \qquad\qquad (\mathbb{P}_2, \preceq) \qquad\qquad (\mathbb{P}_3, \preceq)$$

Fig. 1. The Hasse Diagrams of the Posets obtained from **Examples** 2(i)–(iii).

Remark 5. *Figure 1 gives the Hasse diagrams of the posets obtained from the importation algebras presented in Example 2.*

(i) *In the poset $(\mathbb{P}_1, \preceq)$ obtained from Example 2(i) every element is both maximal and minimal and there are no maximum or minimum elements.*

(ii) *The poset $(\mathbb{P}_3, \preceq)$ in Example 2(iii) is bounded above but not below.* **1** *is the maximum element while p, q are the minimal elements.*

(iii) *Clearly, the poset in Example 2(ii) is bounded (both above and below) with* **1** *and p as the maximum and minimum elements. In fact, $(\mathbb{P}_2, \preceq)$ is a lattice.*

Proposition 6. *Let $(\mathbb{P}, I, C, \mathbf{1})$ be an Importation algebra.*

(i) $y \preceq I(x, y)$, *for all $x, y \in \mathbb{P}$.*

(ii) *If $I(x, x) = \mathbf{1}$ for all $x \in \mathbb{P}$ then* **1** *is the maximal element.*

(iii) **1** *is the maximum element if and only if for every $x \in \mathbb{P}$ there exists $\alpha \in \mathbb{P}$ such that $I(\alpha, x) = \mathbf{1}$.*

(iv) *For any $z \in \mathbb{P}$, $x, y \in U_z^I \iff C(x, y) \in U_z^I$.*

(v) *Let $\mathcal{A} = \{z \in \mathbb{P} \mid I(x, z) = z$ for all $x \in \mathbb{P}\}$. $\mathcal{A}$ is the set of all maximal elements of $\mathbb{P}$. Furthermore, no two elements of A are comparable w.r.t. $\preceq$.*

(vi) *If $\mathbb{P} = U_I$ then no two elements of $\mathbb{P}$ are comparable w.r.t. $\preceq$.*

The converse of Proposition 6(ii) is not true. Example 2(iii) shows that $I_3(p, p) \neq \mathbf{1}$ but **1** is the maximum element. Proposition 6(v) shows that the annihilators of I become the maximal elements of $\mathbb{P}$.

Let $(\mathbb{P}, \leq)$ be a poset. If $I : \mathbb{P} \times \mathbb{P} \to \mathbb{P}$, we denote by (I1) and (I2) the properties that I is decreasing (increasing) in the first (second) variable w.r.t. the order $\leq$ on $\mathbb{P}$, i.e., for any $x, y, z \in \mathbb{P}$,

$$x \leq y \implies I(y, z) \leq I(x, z) , \tag{I1}$$

$$x \leq y \implies I(z, x) \leq I(z, y) . \tag{I2}$$

Proposition 7. *Let* $(\mathbb{P}, I, C, \mathbf{1})$ *be an Importation algebra and thus* $(\mathbb{P}, \preceq)$ *is a poset. Let* I, C *satisfy the following properties for any* $x, y, z \in \mathbb{P}$:

$$I(x, x) = \mathbf{1}, \qquad\qquad \textit{(Identity Principle)} \qquad \text{(IP)}$$
$$C(x, y) = C(y, x), \qquad\qquad \textit{(Commutativity of C)} \qquad \text{(CS)}$$
$$C(\mathbf{1}, y) = y, \qquad\qquad \textit{(Neutrality of C)} \qquad \text{(CN)}$$
$$I(x, C(y, z)) = C(I(x, y), I(x, z)) . \qquad \textit{(Distributivity)} \qquad \text{(CD)}$$

Then

(i) I *is increasing in the second variable w.r.t.* $\preceq$, *i.e.,* I *satisfies (I2) w.r.t.* $\preceq$.
(ii) The pair (I, C) *satisfies a partial residuation property:*

$$C(x, z) \preceq y \Longrightarrow z \preceq I(x, y) . \qquad\qquad \text{(FRP)}$$

(iii) $C(x, y) \preceq I(x, y)$.

Further, if the pair (I, C) *satisfies (MP), as defined below,*

$$C(x, I(x, y)) \preceq y , \qquad\qquad \text{(MP)}$$

then

(iv) I *satisfies the following iterative functional equation:*

$$I(x, I(x, y)) = I(x, y) . \qquad\qquad \text{(IBL)}$$

(v) If I *is one-one in the first variable for some fixed* y *then* C *is idempotent.*

Once again, it can be shown that the converse statements of the above proposition need not be true. None of the above properties demand I to satisfy (I1) w.r.t. $\preceq$ - a study which is worthy of its own. Interestingly, this non-insistence of (I1) on I turns out to be a blessing since, as shown in Sect. 4.2, I which are not fuzzy implications can also give rise to importation algebras.

4   Importation Algebras When $\mathbb{P} = [0, 1]$

In this section, we investigate importation algebras when $\mathbb{P} = [0, 1]$, which is endowed with a natural order, which we denote by $\leq$. We begin by presenting a preliminary result that will be useful for further discussion.

Proposition 8. *Let* $I : [0, 1] \times [0, 1] \to [0, 1]$.

(i) If $I(\alpha, \beta) \geq \beta$ *for every* $\alpha, \beta \in [0, 1]$ *then* I *satisfies* **(QN)**.
(ii) If I *satisfies* **(NP)** *and (I1) w.r.t.* $\leq$ *then* I *satisfies* **(QN)**.
(iii) If I *satisfies* **(NP)**, *(I1) w.r.t.* $\leq$ *and* **(LI)** *with a* C *then* $x \preceq y \Longrightarrow x \leq y$.

Further, let $([0, 1], I, C, 1)$ *be an importation algebra.*

(iv) If $I(0, x) = 1$ *for every* $x \in [0, 1]$ *then* 1 *is the maximum element.*
(v) For any $\alpha \in [0, 1]$, *let* $N_I^\alpha : [0, 1] \to [0, 1]$ *be defined as* $N_I^\alpha(x) = I(x, \alpha)$.
If N_I^α *is onto for some* $\alpha \in [0, 1]$, *then* α *is the least element of* $([0, 1], \preceq)$,
i.e., $\alpha \preceq x$ *for every* $x \in [0, 1]$.

Table 3. Some Importation Algebras from R-, (S,N)-, f-, (U,N)- and Probabilistic (S,N)-implications. The last column indicates the Hasse diagram in Fig. 2 that corresponds to the poset thus obtained.

C	I	$([0,1], \preceq)$
$T_{\mathbf{P}} : xy$	$I_{\mathbf{GG}} : \begin{cases} 1, & \text{if } x \leq y, \\ \dfrac{y}{x}, & \text{otherwise,} \end{cases}$	Fig. 2(i)
$T_{\mathbf{P}} : xy$	$I_{\mathbf{YG}} : \begin{cases} 1, & \text{if } x = 0, \\ y^x, & \text{otherwise ,} \end{cases}$	Fig. 2(i)
$T_{\mathbf{M}} : \min(x,y)$	$I_{\mathbf{GD}} : \begin{cases} 1, & \text{if } x \leq y, \\ y, & \text{otherwise,} \end{cases}$	Fig. 2(ii)
$T_{\mathbf{M}} : \min(x,y)$	$I_{\mathbf{D}} : \begin{cases} 1, & \text{if } x = 0, \\ y, & \text{otherwise,} \end{cases}$	Fig. 2(ii)
$T_{\mathbf{D}} : \begin{cases} 0, & \text{if } (x,y) \in [0,1), \\ \min(x,y), & \text{otherwise,} \end{cases}$	$I_{\mathbf{WB}} : \begin{cases} y, & \text{if } x = 1, \\ 1, & \text{otherwise,} \end{cases}$	Fig. 2(ii)
$T_{\mathbf{LK}} : \max(x+y-1, 0)$	$I_{\mathbf{LK}} : \min(1, 1-x+y)$	Fig. 2(iii)
$C_{\mathbf{H}} : \begin{cases} 0, & \text{if } (x,y) = (0,0), \\ \dfrac{xy}{x+y-xy}, & \text{otherwise,} \end{cases}$	$I_{\mathbf{H}}^{ps} : \dfrac{1-x-y+2xy}{1-y+xy}$	Fig. 2(iii)
$T_{\mathbf{M}}^{\#} : \begin{cases} 0, & \text{if } x \in [0, 0.5] \\ & \& \; y \in [0,1), \\ \min(x,y), & \text{otherwise,} \end{cases}$	$I_{T_{\mathbf{M}}^{\#}, N_{\mathbf{C}}} : \begin{cases} 1, & \text{if } x \in [0, 0.5] \\ & \& \; y \in (0,1], \\ \max(1-x, y), & \text{otherwise,} \end{cases}$	Fig. 2(iv)
$U_{\mathbf{M}} : \begin{cases} \max(x,y), & \text{if } (x,y) \in [0.5,1]^2, \\ \min(x,y), & \text{otherwise,} \end{cases}$	$I_{U_{\mathbf{M}}, N_{\mathbf{C}}} : \begin{cases} \min(1-x, y), & \text{if } \max(1-x, y) \leq 0.5, \\ \max(1-x, y), & \text{otherwise,} \end{cases}$	Fig. 2(v)
$T_{\mathbf{nM}} : \begin{cases} 0, & \text{if } x+y \leq 1, \\ \min(x,y), & \text{otherwise,} \end{cases}$	$I_{\mathbf{FD}} : \begin{cases} 1, & \text{if } x \leq y, \\ \max(1-x, y) & \text{if } x > y, \end{cases}$	Fig. 2(vi)
$T_{\mathbf{D}} : \begin{cases} 0, & \text{if } (x,y) \in [0,1), \\ \min(x,y), & \text{otherwise,} \end{cases}$	$I_{\mathbf{DP}} : \begin{cases} y, & \text{if } x = 1, \\ 1-x, & \text{if } y = 0, \\ 1, & \text{if } x < 1 \text{ and } y > 0, \end{cases}$	Fig. 2(vii)

4.1   Importation Algebras from Fuzzy Implications

Clearly, (**LI**) is the usual (weak) law of importation when I is a fuzzy implication and C is a conjunction, be it a t-norm, t-subnorm, uninorm or a copula. (For definitions of these connectives, and in general to those referred to in this section, we refer the readers to [3,7].)

Similarly, if $\mathbf{1} = 1$ then (**NP**) is the usual left-neutrality property of a fuzzy implication I. Thus we begin by discussing the importation algebras obtainable from some known families of fuzzy implications.

Importation Algebras from R-implications

Let T be any left-continuous t-norm and I_T be the R-implication obtained from it. Since I_T satisfies (**NP**), we let $\mathbf{1} = 1 \in \mathbb{U}_{I_T}$ and hence $\mathbb{U}_{I_T} \neq \emptyset$. It is also well-known that $I_T(\alpha, \beta) \geq \beta$ for every $\alpha, \beta \in [0,1]$ (thus I_T satisfies (**QN**) by Proposition 8(i)) and the pair (I_T, T) satisfies the law of importation (**LI**), see [3], **Theorem 7.3.5**. Thus $\big([0,1], I_T, T, 1\big)$ is an importation algebra.

Importation Algebras from (S,N)-implications

Let S be a t-conorm and N any strict negation. Let $I = I_{S,N}$ be the (S,N)-implication obtained from the pair (S,N). It is well-known that $I_{S,N}$ (see [3],

Theorem 7.3.2) satisfies **(LI)** w.r.t. the t-norm T obtained as the (N, N^{-1})-dual of S. Clearly, $1 \in \mathbb{U}_{I_{S,N}}$ and hence $\mathbb{U}_{I_{S,N}} \neq \emptyset$ and $I_{S,N}$ satisfies **(NP)**. From Proposition 8(i), since $I_{S,N}(\alpha, \beta) = S(N(\alpha), \beta) \geq \beta$ for any $\alpha, \beta \in [0,1]$, we have that $I_{S,N}$ satisfies **(QN)**. Thus $\left([0,1], I_{S,N}, T, 1\right)$ is an importation algebra.

Importation Algebras from f, g-implications. For any f- or g-implication (denoted I_f or I_g), since it satisfies (NP) we have that $1 \in \mathbb{U}_I \neq \emptyset$. Since an f- or g-implication is non-increasing in the first variable, we have by Proposition 8(ii), that it satisfies **(QN)**. Once again, by [3], **Section 7.5**, it is known that these families satisfy **(LI)** with the product t-norm $T_{\mathbf{P}}$. Thus we obtain that $\left([0,1], I_f, T_{\mathbf{P}}, 1\right)$ and $\left([0,1], I_g, T_{\mathbf{P}}, 1\right)$ are importation algebras.

Example 1. Table 3 presents some examples of fuzzy implications I and the corresponding conjunctions C with which $\mathbb{P} = [0,1]$ becomes an importation algebra. Figures 2(i)–(vii) show the Hasse diagrams of the different posets obtained from such importation algebras. While $I_{\mathbf{GG}}, I_{\mathbf{GD}}$ are R-implications, $I_{\mathbf{D}}, I_{\mathbf{DP}}$ are (S,N)-implications, $I_{\mathbf{LK}}, I_{\mathbf{FD}}$ are both R- and (S,N)-implications and $I_{\mathbf{YG}}$ is both an f- and a g-implication.

The following observations are worth noting:

(i) While, in general, (U,N)- and Probabilistic (S,N)-implications may not yield importation algebras, we do have some positive examples in $I_{\mathbf{U_M, N_c}}$ and $I_{\mathbf{H}}^{ps}$.

(ii) In $\left([0,1], I_{T_{\mathbf{M}}^{\#}, N_{\mathbf{C}}}, T_{\mathbf{M}}^{\#}, 1\right)$, $T_{\mathbf{M}}^{\#}$ is not commutative and hence is not a t-norm.

(iii) In $\left([0,1], I_{\mathbf{U_M, N_c}}, U_{\mathbf{M}}, \mathbf{1} = \frac{1}{2}\right)$, the special element $\mathbf{1} = \frac{1}{2}$ and is also the least element in the obtained poset.

(iv) In $\left([0,1], I_{\mathbf{D}}, T_{\mathbf{M}}, \mathbf{1}\right)$, $\mathbb{U}_{I_{\mathbf{D}}} = (0,1]$ and hence any $x \in (0,1]$ can be chosen as the special element $\mathbf{1}$.

(v) In $\left([0,1], I_{\mathbf{WB}}, T_{\mathbf{D}}, 1\right)$, $T_{\mathbf{D}}$ is not left-continuous and does not lead to a residuated structure on $[0,1]$ but we still obtain an importation algebra.

(vi) The posets obtained from these importation algebras show a gamut of order-theoretic structures: From posets that are not bounded below - with finite or infinitely many minimal elements (as in Figs. 2(i) & (ii)) to bounded lattices (as in Figs. 2(iv), (vi) & (vii)) to complete and totally ordered lattices (as in Figs. 2(iii) & (v)). Since 1 is the only annihilator for a fuzzy implication I, we always obtain a poset that is bounded above with 1 as the maximum element.

4.2   Importation Algebras Where I Is Not a Fuzzy Implication

Quite interestingly, there is nothing in the definition of an importation algebra to suggest that I should be a fuzzy implication. Note that, to begin with, there was no assumption on the underlying set $\mathbb{P}$ with regards to the order, and hence no order-theoretic properties like monotonicity or boundedness can be imposed on either I or C.

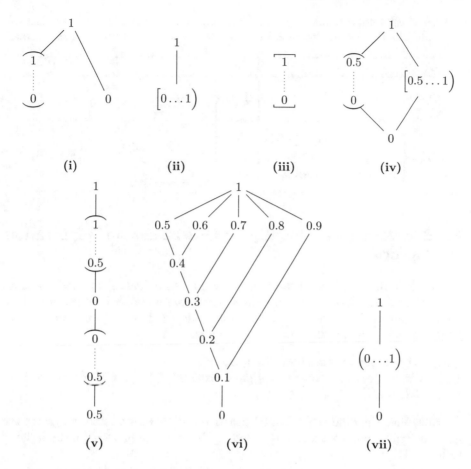

Fig. 2. Posets obtained from Importation Algebras given in Tables 3 and 4.

In Sect. 4.1, we have seen many families fuzzy implications which were obtained from other fuzzy logic connectives or unary functions. In the literature one can find many more such attempts to obtain fuzzy implications from other fuzzy logic connectives. However, not all of them readily give rise to fuzzy implications - typically the property (**I1**) fails - see for instance, the families of QL-operations, Probabilistic and Conditional implications based on Copulas. However, it can still satisfy (**NP**) and (**QN**) on $[0, 1]$ and also satisfy (**LI**) with a binary operation C thus leading to an importation algebra and hence a, possibly, different order on $[0, 1]$ than the usual one.

In Table 4 we present some functions I that are not fuzzy implications but still there exist C with which they form importation algebras.

Table 4. Examples of Importation Algebras from non fuzzy-implications. The last column indicates the Hasse diagram in Fig. 2 that corresponds to the poset thus obtained.

$C(x,y)$	$I(x,y)$	$([0,1], \preceq)$
$T_P : xy$	$I_{C_H}^p : \begin{cases} 1, & \text{if } x = 0, \\ \dfrac{y}{x+y-xy}, & \text{if } x > 0, \end{cases}$	Fig. 2(i)
$T_M^\# : \begin{cases} 0, & \text{if } x \in [0,0.5] \\ & \&\ y \in [0,1), \\ \min(x,y), & \text{otherwise}, \end{cases}$	$I_{T_M^\#} : \begin{cases} 1, & \text{if } x \in [0,0.5], \\ 1, & \text{if } x \in [0.5,1] \\ & \text{and } x \leq y, \\ y, & \text{otherwise}. \end{cases}$	Fig. 2(ii)

5   Recovering the Underlying Order from an Importation Algebra

As has been discussed so far, an importation algebra $(\mathbb{P}, I, C, \mathbf{1})$ can impose an order $\preceq$ on the underlying set $\mathbb{P}$, even if there were no order-theoretic structure on $\mathbb{P}$. However, if $\mathbb{P}$ were endowed with an order $\leq$, it would be interesting to study the following natural questions which arise:

(Q1) What is the relation between $\preceq$ and $\leq$?
(Q2) When does one recover the original order on $\mathbb{P}$, i.e., the conditions on I, C such that $(\mathbb{P}, \preceq) \equiv (\mathbb{P}, \leq)$?

Note that the different obtained posets on $[0,1]$, whose Hasse diagrams are given in Fig. 2, show that an importation algebra may not preserve the underlying order on $\mathbb{P}$.

Proposition 9. *Let $(\mathbb{P}, \leq, \mathbf{1})$ be a bounded lattice and $(\mathbb{P}, I, C, \mathbf{1})$ be an importation algebra. Let I be decreasing in the first variable w.r. to $\leq$. If $x \preceq y$ then $x \leq y$.*

Clearly, the posets in Fig. 2 show that the converse of Proposition 9 is not true when $\mathbb{P} = [0,1]$ and I is a fuzzy implication.

If $I : \mathbb{P} \times \mathbb{P} \to \mathbb{P}$ then for a $y \in \mathbb{P}$, we denote the second partial function by $I_y(t) = I(t,y)$. If $(\mathbb{P}, \leq)$ is a poset, for an $x \in \mathbb{P}$, we denote the upset of x by $\{x\}^\uparrow = \{y \in \mathbb{P} | x \leq y\}$.

Theorem 10. *Let $(\mathbb{P}, \leq, \mathbf{1})$ be a   poset (bounded above) and $(\mathbb{P}, I, C, \mathbf{1})$ be an importation algebra. The following are equivalent:*

(i) $x \preceq y \Longleftrightarrow x \leq y$ for all $x, y \in \mathbb{P}$.
(ii) $\mathcal{R}an(I_x) = \{x\}^\uparrow$ for each $x \in \mathbb{P}$.

Corollary 11. *Let $([0,1], I, C, \mathbf{1})$ be an importation algebra, with $\mathbf{1} = 1$ as the special element. If I is such that $I_x : [0,1] \to [x,1]$ is onto for each $x \in [0,1)$ then $([0,1], \preceq) = ([0,1], \leq)$.*

Note that the importation algebras $([0,1], I_{\mathbf{LK}}, T_{\mathbf{LK}}, \mathbf{1})$ and $([0,1], I_{\mathbf{H}}^{ps}, C_{\mathbf{H}}, \mathbf{1})$ from Table 3 allow us to recover the usual order from $[0,1]$.

Corollary 11, on the one hand shows that one can recover the usual order on $[0,1]$, even when a fuzzy implication I does not satisfy the ordering property, viz., $I(x,y) = 1 \iff x \le y$. On the other hand, it also shows that not only can one obtain a poset from an importation algebra but also a bounded complete totally ordered lattice.

Further, note that if I is a fuzzy implication, the onto-ness of the second partial function immediately implies that I is continuous in the first variable. However, I need not be continuous in both the variables.

6   Concluding Remarks

In this work, we have proposed a novel algebra called the Importation Algebra, which gives rise to an order on the underlying set, by employing an implicative type operation. Our study has also shown that there exist many families of fuzzy implications which give rise to importation algebras and, under some mild conditions, also allow us to recover the underlying order.

Similar to the order from an implication I, in the literature one would find works defining orders based on conjunctive-type operations, viz., t-norms [5], uninorms [4] and nullnorms [2]. However, all of them seem to make use of the underlying order of $[0,1]$ and the subsequent monotonicity of these operations. In fact, we can show that, on a $\mathbb{P} \ne \emptyset$, not necessarily a poset, if there exists a C such that $(\mathbb{P}, C)$ is a monoid and C satisfies the quasi-neutrality (**QN**) then one can obtain an order on $\mathbb{P}$ as in [5]: $x \sqsubseteq y \iff C(\ell, y) = x$. In fact, for any t-norm T, both $([0,1], T)$ is a monoid and T satisfies (**QN**) and hence one can obtain the order without resorting to the monotonicity or commutativity of T or the presumption of order on $[0,1]$. If the operation C in an importation algebra happens to enjoy these properties, then it would be interesting to study the relationships between $\sqsubseteq$ and $\preceq$, since they can bring in complementary information in further exploration.

As was shown in Sect. 4, not only do we get a poset from an importation algebra, we also can obtain complete totally ordered lattices. Thus it would be interesting to study the conditions on the underlying pair of operations (I, C) that would give us richer order-theoretic structures. Note that importation algebras $([0,1], I, C, 1)$ where I is a fuzzy implication always give rise to posets that are bounded above.

In the literature, many algebras have been proposed based on implication-like operations. For instance, the Implication algebra of Abbott [1], the Implicative algebra of Rasiowa & Sikorski [12], the Implicative semi-lattice of Nemitz [10], the Implicative lattice of Martinez [8,9] or the fuzzy implication algebra of Wu (Cf. [11]). It would be interesting to study the relationships between these and the proposed importation algebra. For instance, it can be easily shown that every regular fuzzy implication algebra is an importation algebra.

We intend to present the results of the above studies in our future works.

References

1. Abbott, J.: Semi-boolean algebra. Mat. Vesn. N. Ser. **4**, 177–198 (1967)
2. Aşıcı, E.: An order induced by nullnorms and its properties. Fuzzy Sets Syst. **325**, 35–46 (2017)
3. Baczyński, M., Jayaram, B.: An introduction to fuzzy implications. In: Fuzzy Implications, pp. 1–35. Springer (2008)
4. Ertuğrul, Ü., Kesicioğlu, M.N., Karacal, F.: Ordering based on uninorms. Inf. Sci. **330**, 315–327 (2016)
5. Karaçal, F., Kesicioğlu, M.N.: A t-partial order obtained from t-norms. Kybernetika **47**(2), 300–314 (2011)
6. Kesicioğlu, M.N., Mesiar, R.: Ordering based on implications. Inf. Sci. **276**, 377–386 (2014)
7. Klement, E.P., Mesiar, R., Pap, E.: Triangular Norms, Trends in Logic, vol. 8. Kluwer Academic Publishers, Dordrecht (2000)
8. Martínez, N.G.: A topological duality for some lattice ordered algebraic structures including *l*-groups. Algebra Universalis **31**(4), 516–541 (1994)
9. Martínez, N.G.: A simplified duality for implicative lattices and *l*-groups. Studia Logica **56**(1/2), 185–204 (1996)
10. Nemitz, W.: Implicative semi-lattices. Trans. Am. Math. Soc. **117**, 128–142 (1965)
11. Pei, D.: A survey of fuzzy implication algebras and their axiomatization. Int. J. Approx. Reason. **55**(8), 1643–1658 (2014)
12. Rasiowa, H., Sikorski, R.: The Mathematics of Metamathematics. Institut Mathematyczny, Polskiej Akademii Nauk: Monographie Mathematyczne. PWN-Polish Scientific Publishers (1970)

On Some Inner Dependence Relationships in Hierarchical Structure Under Hesitant Fuzzy Environment

Debashree Guha[1]([✉]) and Debasmita Banerjee[2]

[1] School of Medical Science and Technology, Indian Institute of Technology,
Kharagpur 721302, India
deb1711@gmail.com
[2] Department of Mathematics, Indian Institute of Technology,
Patna 800013, India
debasmitabanerjee12@gmail.com

Abstract. This paper presents an attempt to embed aggregation operators in large system for information aggregation. We develop a new aggregation operator to model inner dependence relation among the subcriteria of elementary level in multi-layer hierarchical structure under hesitant fuzzy environment.

Keywords: Hesitant fuzzy set · Hierarchy · Inner dependency ·
Aggregation operator

1 Introduction

Decision makers prefer to solve complex decision situation with large number of criteria, by decomposing it into smaller and manageable subtasks. Thus often criteria are structured into several levels and then decision makers provide their opinion for an alternative based on those criteria. This type of structure is labeled as a hierarchical system where partial results are computed first at some intermediate level of the hierarchy and later they are aggregated to get the comprehensive result. This structure is widely used for the case where entire problem management is extremely difficult. Researchers have developed mathematical models [15,16] to deal with such hierarchical structures. Further Multiple Criteria Hierarchy Process (MCHP) has been developed [1] and applied to solve many Multi-Criteria Decision Making (MCDM) problems like utility-based methods or, outranking methods for capturing interaction present in the criteria set or, to calculate imprecision or hesitation present in the weight space [2–4].

A hierarchical structure is a set of components containing criteria that relate together in a specific order to perform the overall evaluation. It can be depicted as a simple input-output propagation. In [5], Torra and Narukawa mention that models based on aggregation operators are capable to approximate functions at any arbitrary level of detail in the hierarchical system. These issues have been

© Springer Nature Switzerland AG 2019
R. Halaš et al. (Eds.): AGOP 2019, AISC 981, pp. 95–105, 2019.
https://doi.org/10.1007/978-3-030-19494-9_9

attempted in [17,18]. In the same spirit, we find the outcome of the multi-layer hierarchical system by modeling different types of interactions among criteria, sub-criteria of different hierarchy using appropriate aggregation technique. A hierarchical system can be observed as a simple linear chain of interactions where different types of interactions are possible among the criteria of different level, such as inner dependence and outer dependence relation in the sense of Saaty [12,13]. In this study we focus on modeling inner dependence relationship among the criteria of the hierarchical system in finding the decision output. The concept of inner dependency in hierarchical system is defined [12,13] as the functional dependence within the component of a set. With this view, we are considering a hierarchy where every subset of elementary sub-criteria set belonging to only one criterion of the level immediately above follow inner dependency relationship. Now, how to model this inner dependency relationship in an appropriate manner is needed to be focused on. In this study, inner dependence relationship is tackled by forming a partitioned structure [14] among the sub-criteria in the elementary level, while each class of the partition comprises of the elementary sub-criteria belonging to the same criteria of the immediately upper level. Thus, the members of each class posses a homogeneous relationship and no inter-relationship exists among inter-partition. By homogeneous relationship, we interpret that each member of a class has a relationship with the rest of the elements of that particular class. The expressed partitioned structure inter-relationship among the sub-criteria of the elementary level inspired us to develop an aggregation framework to capture the inner dependence relationship. Moreover in a MCDM problems, it is not always possible that all the experts agree to provide a fixed satisfaction value for an alternative on the basis of a criteria. To deal with such situation, in 2010 Torra [6] introduced a new notion in fuzzy set theory and labeled it as Hesitant Fuzzy Set (HFS).

In order to aggregate hesitant fuzzy information, several extensions and generalizations of different aggregation operators have been made. In [7] Xia and Xu developed and analyzed hesitant fuzzy weighted averaging and hesitant fuzzy weighted geometric operator under this environment. On the other hand, in [8] Yu et. al. developed hesitant fuzzy aggregation operator based on the Choquet integral. Further in order to capture the interrelationships among hesitant fuzzy arguments some extensions of the well-known operator Bonferroni mean was proposed in [9–11], as hesitant fuzzy Bonferroni mean, hesitant fuzzy geometric Bonferroni mean, hesitant fuzzy Choquet geometric Bonferroni mean. In the interest of considering a hierarchical system with inner dependence structure in the elementary sub-criteria set, in the present study we directly focus on input arguments and develop a hierarchical BM to combine the information of the different clusters. On that note the aim of this paper is to capture the inner dependence pattern in a hierarchical system by developing an aggregation operator in hesitant fuzzy environment which we refer to as Hesitant Hierarchical Inner Dependency Partial Aggregation operator ($HHIDPAgg$ operator).

The paper is organized as follows. Section 2 describes basic concept of hesitant fuzzy set. In Sect. 3 $HHIDPAgg$ operator is developed and its properties are analyzed. In Sect. 4 the application of the proposed operator is discussed. Conclusions and future scope of study is discussed in Sect. 5.

2   Hesitant Fuzzy Set and Its Basic Operations

As we aforementioned, Torra in his paper [6] first introduced the concept of HFS to model the situations in which different membership functions are possible. The definition is as follows:

Definition 1. [6] *Let X be a reference set, then HFS on X is a function h which returns a finite subset of $[0, 1]$ when applied to X.*

In other words, Let $M = \{\mu_1, \mu_2, ..., \mu_N\}$ be a set of N membership functions. Then, the HFS associated with M i.e., h_M is defined as, $h_M(x) = \cup_{\mu \in M}\{\mu(x)\}$ where $x \in X$. Later, Xia and Xu [7] presented the definition of HFS by including the mathematical symbol as given below:

Definition 2. [7] *Let X be a fixed set. A HFS is defined as $E = \{<x, h_E(x)> : x \in X\}$, where hesitant fuzzy element (HFE) $h_E(x)$ is a set of some values in [0, 1], denoting the possible membership degrees of the element $x \in X$ to the set E. For convenience, we call $h = h_E(x)$ a hesitant fuzzy element and $\mathcal{H}$ is the set of all (clearly finite) HFEs.*

As in his paper [6], Torra illustrated that the envelop of a HFE is an intuitionistic fuzzy value thus based on this relationship, the operational laws of on the HFEs h_1 and h_2 are defined as follows:

Definition 3. [7] *Let h_1 and h_2 be the two HFEs. then,*

  (I) $h_1 \oplus h_2 = \cup_{\gamma_1 \in h_1, \gamma_2 \in h_2}\{\gamma_1 + \gamma_2 - \gamma_1\gamma_2\}$
  (II) $h_1 \otimes h_2 = \cup_{\gamma_1 \in h_1, \gamma_2 \in h_2}\{\gamma_1\gamma_2\}$
  (III) $\lambda h_1 = \cup_{\gamma_1 \in h_1}\{1 - (1 - \gamma_1)^\lambda\}$
  (IV) $h_1{}^\lambda = \cup_{\gamma_1 \in h_1}\{\gamma_1{}^\lambda\}$

All the above mentioned operations on HFEs are suitable for HFSs. Next, in order to compare two HFEs one can calculate the score function as,

Definition 4. *For a HFE h, the score function is defined as $s = \frac{1}{\sharp h}\sum_{\gamma \in h}\gamma$, where $\sharp h$ denotes the total number of the elements in h. Thus for two HFEs h_1 and h_2, if $s(h_1) > s(h_2)$, then $h_1 > h_2$ and if $s(h_1) = s(h_2)$, then $h_1 = h_2$.*

3   Hierarchy of Criteria with Partitioned Structure

As mentioned in introduction, inner dependence relationship among the elementary sub-criteria can be tackled by forming a partition structure among the elementary sub-criteria set. As a consequence, the members of each class of the partition posses a homogeneous relationship with the same line of [14] and no inter-relationship exists among inter-partition. By homogenous relationship we interpret that each member of a class has relationship with the rest of the elements of that particular class. Evidently the ease of using a hierarchical system is that decision maker can obtain the results not only for the final goal as well as the results of the intermediate levels also. Thus for a multi-layer hierarchical structure of criteria where alternatives are evaluated under hesitant fuzzy environment, the inner dependence pattern is described mathematically as follows:

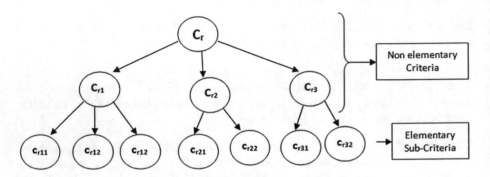

Fig. 1. Example of three layer hierarchical structure of criteria set starting from the root criterion C_r

3.1 Partial Result for Multi-layer Hierarchical Structure of Criteria in Hesitant Fuzzy Environment

- $A = \{A_1, A_2, ..., A_m\}$ is the finite set of m $(m \geq 2)$ alternatives.
- L is the total number of levels in the hierarchy of criteria.
- $\mathcal{C}$ is the set of all criteria comprises of elementary sub-criteria set and non-elementary criteria set including main criteria (The framework of three layer hierarchical structure of criteria set is shown in Fig. 1).
- $\mathcal{I}_\mathcal{C}$ is the set of indices of all criteria belong to $\mathcal{C}$ representing position of the criteria located at any level of the hierarchy.
- $\{c_l : l \in EL\}$ is the set of all elementary sub criteria where EL is the index set of all elementary sub-criteria and $EL \subset \mathcal{I}_\mathcal{C}$.
- $\{C_r : r \in \mathcal{I}_\mathcal{C} \backslash EL\}$ is the set of all non-elementary criteria including main criteria set.
- I_{C_r} is the set of indices of elementary sub-criteria descending from some non elementary criterion $C_r; (r \in \mathcal{I}_\mathcal{C} \backslash EL)$ using which an evaluation of criterion C_r can be provided and $I_{C_r} \subset EL$.
- Suppose P_r denotes the partition of the index set I_{C_r} where $r \in \mathcal{I}_\mathcal{C} \backslash EL$ where the partitions are pairwise disjoint and $\bigcup_r P_r = El$.
- Performance of the i-th alternative A_i, $i \in \{1, 2, ..., m\}$ with respect to the set of elementary sub-criteria $\{c_l : l \in EL\}$ is modeled by HFEs and can be written as $\{h_{il} : l \in EL\}$.

Then the partial aggregated value for an alternative A_i based on some sub-criteria $C_r \in \mathcal{C}$ where $r \in \mathcal{I_C} \backslash EL$ is as follows,

Definition 5. *For $p, q \geq 0$ with $p + q > 0$ and h_{il} ($l \in EL, i = 1, 2, ..., m$) be a collection of HFEs, the Hesitant Hierarchical Inner Dependency Partial Aggregation operator (HHIDPAgg operator) is a mapping $HHIDPAgg : \mathcal{H}^n \to \mathcal{H}$ such that*

$$HHIDPAgg^{p,q}(h_{i1}, ..., h_{in}) = \frac{1}{|r : r \in \mathcal{I_C} \backslash EL|} \bigoplus_{r \in \mathcal{I_C} \backslash EL} \left(\frac{1}{|P_r|} \bigoplus_{u \in P_r} \left(\frac{1}{|u|} \bigoplus_{l \in u} h_{il}^p \otimes \right.\right.$$

$$\left.\left. \left(\frac{1}{|u| - 1} \bigoplus_{k \in u \backslash \{l\}} h_{ik}^q \right) \right)^{\frac{1}{p+q}} \right). \qquad (1)$$

where $|P_r| =$ cardinality of the partition P_r and n is the total number of elementary sub-criteria descending from some non-elementary criteria or a set of non-elementary criteria for which we want to evaluate the partial result.

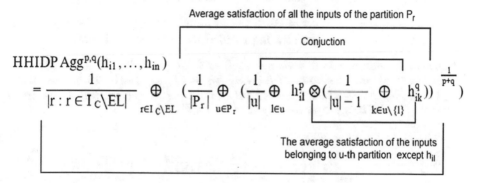

Fig. 2. Interpretation of the HHIDPAgg operator expressed by analyzing the components

Equation (1) depicts that we aggregate a set of input data evaluating alternative A_i, where there exists a partition structure inner dependency pattern amongst them. The expression $\frac{1}{|u|-1} \bigoplus_{k \in u \backslash \{l\}} h_{ik}^q$ indicates the average satisfaction of the inputs belonging to u-th partition for criteria $C_r; r \in \mathcal{I_C} \backslash EL$ except h_{il}. Now, $h_{il}^p \otimes \left(\frac{1}{|u|-1} \bigoplus_{k \in u \backslash \{l\}} h_{ik}^q \right)$ captures the conjunction of the satisfaction of input h_{il} with the average satisfaction of the rest of inputs of the u-th partition for criteria C_r. Then the expression $\frac{1}{|u|} \bigoplus_{l \in u} h_{il}^p \otimes \left(\frac{1}{|u|-1} \bigoplus_{k \in u \backslash \{l\}} h_{ik}^q \right)$ gives the satisfaction of the inter-related inputs of the partition and by $\frac{1}{|P_r|} \bigoplus_{u \in P_r} \left(\frac{1}{|u|} \bigoplus_{l \in u} h_{il}^p \otimes \left(\frac{1}{|u|-1} \bigoplus_{k \in u \backslash \{l\}} h_{ik}^q \right) \right)^{\frac{1}{p+q}}$ we obtain the average satisfaction of all the inputs of the partition P_r. Now to obtain the final aggregated value we take an average over all those intermediate level criteria value $C_r; r \in \mathcal{I}_C \backslash EL$ for which decision maker wants to get the result (Fig. 2).

Now, Depending on the nature of the input set, the proposed operator can be transformed into one of the following cases.

Case I. If the set of elementary sub-criteria are independent i.e. there exists no inter-relationship pattern in the sub criteria set, then the HHIDPAgg operator reduces to the aggregation function hesitant fuzzy power root arithmetic mean which is independent of q.

$$HIDPAgg^{p,q}(h_{i1},...,h_{in}) = \frac{1}{|r:r \in \mathcal{I}_C \backslash EL|} \left(\bigoplus_{r \in \mathcal{I}_C \backslash EL} \left(\frac{1}{|I_{C_r}|} \bigoplus_{l \in I_{C_r}} h_{il}^p \right)^{\frac{1}{p}} \right). \tag{2}$$

Case II. If all the elementary sub-criteria belong to the same class i.e. $|P_r| = 1$, then following the operations mentioned in Definition 3. on hesitant fuzzy sets, the HHIDPAgg operator will follow the structure of BM operator.

From the operational law of HFS, the formula for computing the aggregated value by the HHIDPAgg operator is provided by the theorem below:

Theorem 1. *For $p,q \geq 0$ with $p+q > 0$ and h_{il} ($l \in EL$) be a collection of HFEs where a_{il} are elements of h_{il} i.e., $h_{il} = \bigcup_{a_{il} \in h_{il}} \{a_{il}\}$, Then the aggregated value by the Hesitant Hierarchical Inner Dependency Partial Aggregation operator (HHIDPAgg operator) is also an HFE and*

$$HHIDPAgg^{p,q}(h_{i1},...,h_{in})$$

$$= \frac{1}{|r:r \in \mathcal{I}_C \backslash EL|} \bigoplus_{r \in \mathcal{I}_C \backslash EL} \left(\frac{1}{|P_r|} \bigoplus_{u \in P_r} \left(\frac{1}{|u|} \bigoplus_{l \in u} h_{il}^p \otimes \left(\frac{1}{|u|-1} \bigoplus_{k \in u \backslash \{l\}} h_{ik}^q \right) \right)^{\frac{1}{p+q}} \right)$$

$$= \bigcup_{a_{il} \in h_{il}, a_{ik} \in h_{ik}; k \neq l; \ l,k \in u; r \in \mathcal{I}_C \backslash EL} 1 - \left(\prod_{r \in \mathcal{I}_C \backslash EL} \left(\prod_{u \in |P_r|} \left(1 - \left(1 - \left(\prod_{l \in u} \right. \right. \right. \right.$$

$$\left. \left. \left. \left. \left(1 - a_{il}^p \left(1 - \left(\prod_{k \in u \backslash \{l\}} \left(1 - a_{ik}^q \right) \right)^{\frac{1}{|u|-1}} \right) \right) \right)^{\frac{1}{|u|}} \right)^{\frac{1}{p+q}} \right)^{\frac{1}{|P_r|}} \right)^{\frac{1}{|r:r \in \mathcal{I}_C \backslash EL|}} \tag{3}$$

where, $|r:r \in \mathcal{I}_C \backslash EL| =$ cardinality of the non elementary criteria set for which the partial result for multi-layer hierarchical structure is calculated.

Proof. The above equalities can be derived by applying the operations on hesitant fuzzy sets mentioned in Definition 3.

We now discuss the properties of the HHIDPAgg operator.

Property 1 (Idempotency). Let, $\{h_{il} : l \in EL\}$ be a collection of HFEs. If for every $l \in EL$, $h_{il} = \bigcup_{a_{il} \in h_{il}} \{a_{il}\} = \{a\}$. Then,

$$HHIDPAgg^{p,q}(h_{i1},...,h_{in}) = \{a\}.$$

Property 2 (Monotonicity). Let, $\{h_{il} : l \in EL\}$ and $\{h_{i'l} : l \in EL\}$ be two collections of HFEs representing the performances of i-th and i'-th alternative respectively. If for all $l \in EL$ and for $a_{il} \in h_{il}$, $a_{i'l} \in h_{i'l}$ we have $a_{il} \geq a_{i'l}$ then,

$$HHIDPAgg^{p,q}(h_{i1}, ..., h_{in}) \geq HHIDPAgg^{p,q}(h_{i'1}, ..., h_{i'n}).$$

Property 3 (Boundedness). i.e., Let $h^- = \bigcup_{a_{il} \in h_{il}} \min_{l \in EL}\{a_{il}\}$ and $h^+ = \bigcup_{a_{il} \in h_{il}}$ $\max_{l \in EL}\{a_{il}\}$ for all $l \in EL$. Then,

$$h^- \leq HHIDPAgg^{p,q}(h_{i1}, ..., h_{in}) \leq h^+.$$

In the following section, we will provide a brief approach for solving a MCDM problem under hesitant fuzzy environment considering a hierarchical structure of criteria set.

4   Decision Making Under Hesitant Fuzzy Environment

As aforementioned, if decision makers want to evaluate a finite set of m ($m \geq 2$) alternatives $\{A_1, A_2, ..., A_m\}$ based on a criteria set which are not at the same level but structured into several levels i.e., of a hierarchical structure, then by applying HHIDPAgg operator they can obtain a conclusion. With the same notations and assumptions as described in Sect. 3, we provide the decision making method using newly developed aggregation operator as follows:

Step 1. Utilize the developed aggregation operator i.e., HHIDPAgg operator to obtain the hesitant fuzzy elements for i-th alternative A_i (denote it by h_i) using the provided hesitant fuzzy values h_{il} against all or partial set of elementary sub-criteria where $l \in EL$.

Step 2. Compute the score values $s(h_i)$ using Definition 4. for each hesitant fuzzy elements for i-th alternative A_i where $i = 1, 2, ..., m$.

Step 3. Finally, get the optimal result by comparing all $s(h_i); i = 1, 2, ..., m$.

The motivation of using the developed operator is described in the next section using the following example.

4.1   Application

Here we want to focus on a problem where the ranking of an institute is a vital issue. The various aspects of an institute can be examined, based on multiple numbers of criteria which are structured into several levels. Here we are presenting a sample problem where the set of criteria are presented in a hierarchical structure. The parameters are organized into 2 broad categories C_1: Academic, C_2: Administration; that have been further grouped into a number of subcategories as shown in Fig. 3.

1. Learning Resources ($c_{1,1}$),
2. Research Productivity ($c_{1,2}$),
3. Percentage of Placement ($c_{1,3}$)
4. Quality of staff ($c_{2,1}$)
5. Accurate records ($c_{2,2}$)
6. Infrastructure management ($c_{2,3}$)

For sake of simplicity, we consider here a 2-layer hierarchical structure of the criteria set. Now it is clear from the structure that, the sub-criteria set has some inner dependency pattern.

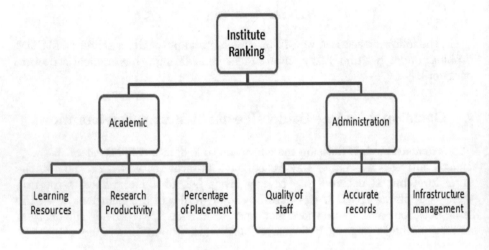

Fig. 3. Hierarchical structure of criteria set

The evaluations of institutes A_i ($i = 1, 2, 3$) based on elementary sub-criteria c_l where $l \in EL$ are summarized in the following hesitant decision matrix:

$$D =$$

	$c_{1,1}$	$c_{1,2}$	$c_{1,3}$	$c_{2,1}$	$c_{2,2}$	$c_{2,3}$
A_1	$\{0.2, 0.4\}$	$\{0.3, 0.9, 0.5\}$	$\{0.5, 0.4\}$	0.4	$\{0.7, 0.3\}$	0.8
A_2	$\{0.3, 0.5\}$	$\{0.8, 0.9\}$	0.2	$\{0.7, 0.5\}$	$\{0.4, 0.8\}$	0.3
A_3	0.9	$\{0.7, 0.4\}$	$\{0.2, 0.6\}$	$\{0.1, 0.3, 0.4\}$	$\{0.3, 0.2\}$	0.1

Utilize the newly developed HHIDPAgg operator to obtain the hesitant fuzzy elements for the institute $A_i(i = 1, 2, 3)$. Assuming $p = q = 1$, for first institute A_1 we obtain the hesitant fuzzy element as,

$h_1 = HHIDPAgg^{p,q}(\{0.2, 0.4\}, \{0.3, 0.9, 0.5\}, \{0.5, 0.4\}, 0.4, \{0.7, 0.3\}, 0.8)$

$= \{0.5537, 0.5834, 0.5232, 0.5657, 0.3457, 0.5436, 0.5687, 0.7843, 0.3428, 0.5692,$

$0.5432, 0.5534, 0.5698, 0.7652, 0.6128, 0.5459, 0.6042, 0.5921, 0.5328, 0.5421,$

$0.5692, 0.4962, 0.5371, 0.4358\}$

Similarly, for second institute A_2 we obtain the hesitant fuzzy element as,

$$h_2 = HHIDPAgg^{p,q}(\{0.3, 0.5\}, \{0.8, 0.9\}, 0.2, \{0.7, 0.5\}, \{0.4, 0.8\}, 0.3)$$
$$= \{0.4529, 0.5328, 0.4396, 0.4242, 0.4359, 0.5136, 0.4149, 0.4781, 0.4892, 0.3952,$$
$$0.4729, 0.4126, 0.3819, 0.4791, 0.4820, 0.4165\}$$

Similarly, for third institute A_3 we obtain the hesitant fuzzy element as,

$$h_3 = HHIDPAgg^{p,q}(0.9, \{0.7, 0.4\}, \{0.2, 0.6\}, \{0.1, 0.3, 0.4\}, \{0.3, 0.2\}, 0.1)$$
$$= \{0.3214, 0.3731, 0.4274, 0.5893, 0.4631, 0.4036, 0.5217, 0.4109, 0.3416, 0.4095,$$
$$0.4432, 0.5196, 0.5965, 0.4192, 0.5173, 0.4392, 0.4428, 0.5160, 0.5341, 0.4839,$$
$$0.4719, 0.4962, 0.4663, 0.3985\}$$

As the parameter p, q change we can get different results for each alternative. Next, compute the score values for each alternatives $s(A_i)$ by Definition 4 and get the results as $s(h_1) = 0.5533$, $s(h_2) = 0.4516$, $s(h_3) = 0.4586$. From which we can come to the conclusion about the ranking of alternatives as $A_1 \succ A_3 \succ A_2$. So decision maker can suggest institute A_1 as the best institute among the 3 institutes.

Now, as the ranking of the institute A_2 is low, it may be possible that the management committee of the institute A_2 wants to analyze the performances based on the two broad categories i.e., academic and administration. Then by using our proposed HHIDPAgg operator and score function, the following results are found as, for institute A_2, the evaluation for academic is, 0.5391 and administration is, 0.4807. So, the committee finds that the evaluation with respect to the administration is not up to the mark, then they can advise the institute to take care of this particular criteria only. Thus, we propose a general aggregation approach in this study which can not only evaluate the ultimate goal, as well as produces partial results by characterizing the situations with any possible partition. We hope that the implementation of our approach would help thousand of institutes to volunteer themselves based on the desired parameters and can improve their ranking to find a better place in world ranking subsequently.

5   Conclusions and Future Work

In this paper we have presented HHIDPAgg operator with a view to handle interdependency relationship among the elementary level sub-criteria of multi layer hierarchical system. We have analyzed its properties and illustrated with numerical example. Besides the Saaty's [12, 13] inner dependency relation among the elementary sub-criteria, we also took motivation mainly from the research article by Torra and Narukawa [5], where in they had suggested to select/develop an operator on the basis of the properties required in a particular problem. In this paper we have solved aggregation issue of hierarchical structure and found the partial results of the hierarchy by modeling inner dependency relationship based on partition structure interrelationship pattern. This is a reasonable star-ing point, as here we have not considered any additional information about the weights of the different input parameters. Thus selecting weights e.g., based on the orness measure of the operator is needed to be focused on. The idea of the paper can be further extended by including mechanisms for evaluation of the results when more data are needed to be included in the model. Then this issue can be solved by considering outer dependency relationship modeling among the inputs of different levels. In this context, we also would like to use the concept of heterogeneous relationship presented in a very recent research article by Das et al. [19]. This can be taken up in further extension of this paper.

Acknowledgement. We would like to thank Prof. Radko Mesiar for his insightful suggestions. The first author gratefully acknowledges the grant ECR/2016/001908 by SERB, India.

References

1. Corrente, S., Greco, S., Slowinski, R.: Multiple criteria hierarchy process in robust ordinal regression. Decis. Support Syst. **53**, 660–674 (2012)
2. Zhang, X.: Multi criteria Pythagorean fuzzy decision analysis: a hierarchical QUALIFLEX approach with the closeness index-based ranking methods. Inf. Sci. **330**, 104–124 (2016)
3. Corrente, S., Figueira, J.R., Greco, S., Slowinski, R.: A robust ranking method extending ELECTREIII to hierarchy of interacting criteria, imprecise weights and stochastic analysis. Omega **73**, 1–17 (2017)
4. Arcidiacono, S.G., Corrente, S., Greco, S.: GAIA-SMAA-PROMETHEE for a hier-archy of interacting criteria. Eur. J. Oper. Res. **270**, 1–19 (2018)
5. Torra, V., Narukawa, Y.: A view of averaging aggregation operators. IEEE Trans. Fuzzy Syst. **15**, 1063–1067 (2007)
6. Torra, V.: Hesitant fuzzy sets. Int. J. Intell. Syst. **25**, 529–539 (2010)
7. Xia, M.M., Xu, Z.S.: Hesitant fuzzy information aggregation in decision making. Int. J. Approx. Reason. **52**, 395–407 (2011)
8. Yu, D., Wu, Y., Zhou, W.: Multi-criteria decision making based on Choquet integral under hesitant fuzzy environment. J. Comput. Inf. Syst. **7**, 4506–4513 (2011)
9. Yu, D., Wu, Y., Zhou, W.: Generalized hesitant fuzzy Bonferroni mean and its application in multi-criteria group decision making. J. Inf. Comput. Sci. **9**, 267–274 (2012)
10. Zhu, B., Xu, Z.S., Xia, M.M.: Hesitant fuzzy geometric Bonferroni means. Inf. Sci. **205**, 72–85 (2012)
11. Zhu, B., Xu, Z.S.: Hesitant fuzzy Bonferroni means for multicriteria decision mak-ing. J. Oper. Res. Soc. **64**, 1831–1840 (2013)
12. Saaty, T.L., Takizawa, M.: Dependence and independence: from linear hierarchies to nonlinear networks. Eur. J. Oper. Res. **26**, 229–237 (1986)
13. Saaty, T.L.: How to handle dependence with the analytical hierarchy process. Math. Model. **9**, 369–376 (1987)
14. Dutta, B., Guha, D.: Partitioned Bonferroni mean based on linguistic 2-tuple for dealing with multi-attribute group decision making. Appl. Soft Comput. **37**, 166–179 (2015)

15. Klement, E.P., Mesiar, R., Pap, E.: On the order of triangular norms-comments on a triangular norm hierarchy by E. Cretu. Fuzzy Sets Syst. **131**, 409–413 (2002)
16. Calvo, T., Mesiarova, A., Valaskova, L.: Construction of aggregation operators-new composition method. Kybernetika **39**, 643–650 (2003)
17. Murofushi, T., Narukawa, Y.: A characterization of multi-step discrete Choquet integral. In: Proceedings of Abstracts, 6th International Conference on Fuzzy Sets Theory Its Application, p. 94 (2002)
18. Torra, V.: On some relationships between hierarchies of quasi-arithmetic means and neural networks. Int. J. Intell. Syst. **14**, 1089–1098 (1999)
19. Das, S., Guha, D., Mesiar, R.: Extended Bonferroni mean under intuitionistic fuzzy environment based on strict t-conorm. IEEE Trans. Syst. Man Cybern. Syst. **47**, 2083–2099 (2017)

The Relation Between F-partial Order and Distributivity Equation

Emel Aşıcı[1(✉)] and Radko Mesiar[2]

[1] Department of Software Engineering, Faculty of Technology,
Karadeniz Technical University, 61830 Trabzon, Turkey
emelkalin@hotmail.com
[2] Department of Mathematics and Descriptive Geometry,
Faculty of Civil Engineering, Slovak University of Technology,
Radlinského, 11, 81005 Bratislava, Slovakia
mesiar@math.sk

Abstract. The notations of the order induced by triangular norms, null-norms and uninorms have been studied widely. Nullnorms have been produced from triangular norms and triangular conorms with a zero element in the interior of the unit interval. They have been proved to be useful in several fields like expert systems, neural networks, fuzzy quantifiers. Also, the distributivity equation has been studied involving different classes of aggregation functions from triangular norms and triangular conorms to nullnorms. So, the study of the distributive property becomes very interesting, since nullnorms have been used several fields. In this paper we investigate distributivity equation for nullnorms on the unit interval $[0,1]$ and we give sufficient condition for two nullnorms to be equivalent.

1 Introduction

Aggregation operators [18] play an important role in theories of fuzzy sets and fuzzy logics. Schweizer and Sklar, in [27], introduced triangular norms and triangular conorms.

Also, t-operators were defined by Mas et al. in [23], and nullnorms by Calvo et al. in [8]. Though definitions of these two classes of aggregation functions are different, in [8] they were shown to be equivalent. Both of these operators, which also generalize concepts of t-norms and t-conorms, where an absorbing (or zero) element is from the whole unit interval, have been studied further. Nullnorms are special aggregation operators that have proven to be useful in many fields like fuzzy logic, expert systems, neural networks, utility theory and fuzzy system modeling. Nullnorms are interesting because their structure is a special combination of t-norms and t-conorms having a neutral element lying somewhere in the unit interval.

Karaçal, Ince and Mesiar [19] studied nullnorms on bounded lattices. They have shown the existence of nullnorms with zero element a for an arbitrary element $a \in L \setminus \{0,1\}$ with underlying t-norms and t-conorms on an arbitrary

© Springer Nature Switzerland AG 2019
R. Halaš et al. (Eds.): AGOP 2019, AISC 981, pp. 106–114, 2019.
https://doi.org/10.1007/978-3-030-19494-9_10

bounded lattice. Also, they have introduced the smallest and the greatest null-norm on a bounded lattice.

Karaçal and Kesicioğlu [20] introduced a partial order defined by means of t-norms.

Also, Aşıcı [2] defined an order induced by a nullnorm on a bounded lattice L and investigated some properties of such an order. In the literature, there are some other papers about nullnorms, uninorms, triangular norms and distributivity equation [1,3,4,6,7,9–16,25].

The distributivity equations have been studied from different points of view, by many authors during the last few years [8,24,26].

In this paper, we want to investigate the distributivity equation

$$F_1(x, F_2(y, z)) = F_2(F_1(x, y), F_1(x, z)) \ for \ all \ x, y, z$$

where the unknown functions F_1 and F_2 are nullnorms [24].

Also, we deeply investigate some properties of an order induced by nullnorms on the unit interval $[0, 1]$. The paper is organized as follows. We shortly recall some basic notions in Sect. 2. In Sect. 3, we investigate distributivity equation for nullnorms on the unit interval $[0, 1]$ and we give sufficient condition for two nullnorms are equivalent under the relation $\sim$. We give our concluding remarks in Sect. 4.

2   Preliminarics

In this section, some preliminaries concerning nullnorms (triangular norms, triangular conorms) are recalled.

Definition 1 [22]. *A triangular norm (t-norm for short) is a binary operation T on the unit interval $[0, 1]$, i.e., a function $T : [0, 1] \times [0, 1] \to [0, 1]$, such that for all $x, y, z \in [0, 1]$ the following four axioms are satisfied:*

(T1) $T(x, y) = T(y, x)$ (commutativity)
(T2) $T(x, T(y, z)) = T(T(x, y), z)$ (associativity)
(T3) $T(x, y) \leqslant T(x, z)$ whenever $y \leqslant z$ (monotonicity)
(T4) $T(x, 1) = x$ (boundary condition)

Example 1 [22]. The following are the four basic t-norms T_M, T_P, T_L and T_D given by, respectively,

$$T_M(x, y) = min(x, y)$$
$$T_P(x, y) = xy$$
$$T_L(x, y) = max(x + y - 1, 0)$$
$$T_D(x, y) = \begin{cases} 0, & if \ (x, y) \in [0, 1)^2 \\ min(x, y), & otherwise \end{cases}$$

Note: $T_D < T_L < T_P < T_M$

Example 2 [22]. The t-norm T^{nM} on $[0, 1]$ is defined as follows:

$$T^{nM}(x, y) = \begin{cases} 0, & x + y \leqslant 1 \\ min(x, y), & \text{otherwise} \end{cases}$$

$T^n M$ is called nilpotent minimum t-norm.

The t-norm $T^\star$ on $[0, 1]$ is defined as follows:

$$T^\star(x, y) = \begin{cases} 0 & (x, y) \in (0, k)^2, \\ min(x, y) & \text{otherwise.} \end{cases}$$

where $0 < k < 1$, k is parameter.

Definition 2 [22]. *A triangular conorm (t-conorm for short) is a binary opera-tion S on the unit interval $[0, 1]$, i.e., a function $S : [0, 1] \times [0, 1] \to [0, 1]$, which, for all $x, y, z \in [0, 1]$ satisfies (T1)–(T3) and*

(S4) $S(x, 0) = x$ (boundary condition)

Example 3 [22]. The following are the four basic t-conorms S_M, S_P, S_L and S_D given by, respectively,

$$S_M(x, y) = max(x, y)$$
$$S_P(x, y) = x + y - xy$$
$$S_L(x, y) = min(x + y, 1)$$
$$S_D(x, y) = \begin{cases} 1, & \text{if } (x, y) \in (0, 1)^2 \\ max(x, y), & \text{otherwise} \end{cases}$$

Note: $S_M < S_P < S_L < S_D$

Note that replacing the domain $[0, 1]$ in Definitions 1 and 2 by a bounded lattice (chain, poset) L, we obtain definitions of t-norms and t-conorms on L.

Extremal t-norms $T_\wedge$ and T_W on L are defined as follows, respectively:

$T_\wedge(x, y) = x \wedge y$

$$T_W(x, y) = \begin{cases} x & \text{if } y = 1, \\ y & \text{if } x = 1, \\ 0 & \text{otherwise.} \end{cases}$$

Then, for any t-norm T on L, it holds $T_W \leqslant T \leqslant T_\wedge$. Similarly, the extremal t-conorms $S_\vee$ and S_W can be defined, and $S_\vee \leqslant S \leqslant S_W$ for any t-conorm S on L.

Especially, if $L = [0, 1]$ is the standard real unit interval, then $T_W = T_D$ and $T_\wedge = T_M$, as well as $S_\vee = S_M$ and $S_W = S_D$.

Definition 3 [8]. *Operation $F : [0, 1]^2 \to [0, 1]$ is called nullnorm if it is com-mutative, associative, increasing, and there is an element $a \in [0, 1]$ such that*
$F(0, x) = x$ *for all* $x \leqslant a$,
$F(1, x) = x$ *for all* $x \geqslant a$.

It can be easily obtained that $F(x,a) = a$ for all $x \in [0,1]$. So, $a \in [0,1]$ is the zero (absorbing) element for F. Due to the monotonicity, we have even $F(x,y) = a$ whenever $\min(x,y) \leqslant a \leqslant \max(x,y)$.

Note: By the definition of nullnorm, the case $a = 0$ leads back to triangular norms, while the case $a = 1$ leads back to triangular conorms.

We use the D_a to represent the following set:

$$[0,a) \times (a,1] \cup (a,1] \times [0,a) \text{ for } a \in (0,1).$$

Note: Similarly the nullnorms are defined on a bounded lattice $(L, \leqslant, 0, 1)$.

Proposition 1 [8,23]. *Let $F : [0,1]^2 \to [0,1]$ be a nullnorm with zero element $F(1,0) = a \notin \{0,1\}$. Then,*

$$F(x,y) = \begin{cases} aS(\frac{x}{a}, \frac{y}{a}), & x,y \in [0,a] \\ a + (1-a)T(\frac{x-a}{1-a}, \frac{y-a}{1-a}), & x,y \in [a,1] \\ a, & \text{otherwise} \end{cases}$$

where S is a t-conorm and T is a t-norm.

A nullnorm F with zero element a, underlying t-conorm S and underlying t-norm T will be denoted by $F = <S,a,T>$.

Definition 4 [20]. *Let L be a bounded lattice and T be a t-norm on L. The order defined as follows is called a $T-$ partial order (triangular order) for t-norm T:*

$$x \preceq_T y :\Leftrightarrow T(\ell, y) = x \text{ for some } \ell \in L.$$

Definition 5 [2]. *Let $(L, \leqslant, 0, 1)$ be a bounded lattice and F be a nullnorm with zero element a on L. Define the following relation, for $x, y \in L$, as*

$$x \preceq_F y :\Leftrightarrow \begin{cases} \text{if } x, y \in [0,a] \text{ and there exist } k \in [0,a] \text{ such that } F(x,k) = y \text{ or,} \\ \text{if } x, y \in [a,1] \text{ and there exist } \ell \in [a,1] \text{ such that } F(y,\ell) = x \text{ or,} \\ \text{if } (x,y) \in L^* \text{ and } x \leqslant y. \end{cases}$$

$$(1)$$

where $I_a = \{x \in L \mid x \parallel a\}$ and $L^* = [0,a] \times [a,1] \cup [0,a] \times I_a \cup [a,1] \times I_a \cup [a,1] \times [0,a] \cup I_a \times [0,a] \cup I_a \times [a,1] \cup I_a \times I_a$.

Note: The partial order $\preceq_F$ in (1) is called F-partial order on L.

Proposition 2 [2]. *Let $(L, \leqslant, 0, 1)$ be a bounded lattice and F be a nullnorm on L. If $x \preceq_F y$ for any $x, y \in L$, then $x \leqslant y$.*

Proposition 3 [2]. *Let $(L, \leqslant, 0, 1)$ be a bounded lattice and F be a nullnorm with zero element a. Then, $(L, \preceq_F)$ is a bounded partially ordered set.*

Definition 6 [2]. *Define a relation $\sim$ on the class of all nullnorms on $[0,1]$ by $F_1 \sim F_2$ if and only if the F_1-partial order coincides with the F_2-partial order.*

Lemma 1 [2]. *The relation $\sim$ is an equivalence relation.*

Definition 7 [2]. *Let F be a nullnorm on $[0,1]$ and let K_F be defined by*

$$K_F = \{x \in (0,1) \mid \text{for some } y \in (0,1), [x < y \text{ and } x \not\preceq_F y]$$

$$\text{or } [y < x \text{ and } y \not\preceq_F x]\}.$$

Definition 8 [2]. *Define a relation β on the class of all nullnorms on $[0,1]$ by $F_1 \beta F_2$,*

$$F_1 \beta F_2 :\Leftrightarrow K_{F_1} = K_{F_2}.$$

Lemma 2 [2]. *The relation β given in Definition 8 is an equivalence relation.*

3   Distributivity Equation

In this section, we investigate the relationship between an order induced by nullnorms and distributivity equation for nullnorms on the unit interval $[0,1]$. We give here a sufficient condition for the coincidence of $\preceq_F$ orders, see Corollary 1.

Definition 9 [24]. *Let F_1 and F_2 be nullnorms on $[0,1]$. F_1 is distributive over F_2 if they satisfy the following condition:*

$$F_1(x, F_2(y,z)) = F_2(F_1(x,y), F_1(x,z)) \tag{2}$$

for all $x, y, z \in [0,1]$.

Proposition 4 [21]. *Let F_1 and F_2 be nullnorms on the unit interval $[0,1]$ with the same zero element a and F_1 be distributive over F_2. If S_1 and T_1 are divisible, then S_2 and T_2 are divisible.*

Remark 1. The converse of the above Proposition 4 may not be true. Here is an example illustrating a such case.

Example 4. Consider the nullnorms $F_D = <S_D, a, T_D>$ on $[0,1]$ with zero element a and $F_M = <S_M, a, T_M>$ on $[0,1]$ with zero element a defined as follows:

$$F_D(x,y) = \begin{cases} S_D(x,y), & x,y \in [0,a] \\ T_D(x,y), & x,y \in [a,1] \\ a, & \text{otherwise} \end{cases} \tag{3}$$

$$F_M(x,y) = \begin{cases} S_M(x,y), & x,y \in [0,a] \\ T_M(x,y), & x,y \in [a,1] \\ a, & \text{otherwise} \end{cases} \tag{4}$$

It is clear that F_D is distributive over F_M. Also, we know that S_M and T_M are divisible t-conorm and t-norm, respectively, but S_D and T_D are not divisible.

Theorem 1. *Let F_1 and F_2 be nullnorms on the unit interval $[0,1]$ with the same zero elements a. If F_1 is distributive over F_2, then $\preceq_{F_1} \subseteq \preceq_{F_2}$.*

Proof. Let F_1 be distributive over F_2, that is, $F_1(x, F_2(y, z)) = F_2(F_1(x, y), F_1(x, z))$ for all $x, y, z \in [0, 1]$. Firstly let $x, y \in [0, a]$ and $(x, y) \in \preceq_{F_1}$, i.e., $x \preceq_{F_1} y$. If $x = y$, then we have that $y \preceq_{F_2} y$. Let $x \neq y$. Then, there exists an element $k \in [0, a]$ such that $F_1(x, k) = y$. Since $x \neq y$, it must be $k \neq 0$ and

$$y = F_1(x, k) = F_1(x, F_2(k, 0)).$$

Since F_1 is distributive over F_2, then we have that

$$y = F_1(x, F_2(k, 0)) = F_2(F_1(x, k), F_1(x, 0)) = F_2(y, x).$$

So, $x \preceq_{F_2} y$, i.e., $(x, y) \in \preceq_{F_2}$.

Let $x, y \in [a, 1]$ and $(x, y) \in \preceq_{F_1}$, i.e., $x \preceq_{F_1} y$. If $x = y$, then we have similar case. Let $x \neq y$. Then, there exists an element $\ell \in [a, 1]$ such that $F_1(y, \ell) = x$. Since $x \neq y$, it is clear that $\ell \neq 1$ and

$$x = F_1(y, \ell) = F_1(y, F_2(\ell, 1)).$$

Since F_1 is distributive over F_2, then we have that

$$x = F_1(y, F_2(\ell, 1)) = F_2(F_1(y, \ell), F_1(y, 1)) = F_2(x, y).$$

So, $x \preceq_{F_2} y$, i.e., $(x, y) \in \preceq_{F_2}$.

Let $x, y \in D_a$. In this case, we have that $\preceq_{F_1} = \preceq_{F_2}$ by the definition of $\preceq_F$. $\qquad\blacksquare$

Example 5. We consider the nullnorm $F_D = <S_D, a, T_D>$ on $[0, 1]$ given by (3) and the nullnorm $F_M = <S_M, a, T_M>$ on $[0, 1]$ given by (4). It is clear that F_D is distributive over F_M. So, we have that $\preceq_{F_D} \subseteq \preceq_{F_M}$ by Theorem 1.

Corollary 1. *Let F_1 and F_2 be nullnorms on the unit interval $[0, 1]$ with the same zero elements a. If F_1 is distributive over F_2 and F_2 is distributive over F_1, then $F_1 \sim F_2$.*

Corollary 2. *Let F_1 and F_2 be nullnorms on the unit interval $[0, 1]$ with the same zero elements a. If F_1 is distributive over F_2 and F_2 is distributive over F_1, then $F_1 \beta F_2$.*

Remark 2. The converse of Corollary 2 may not be true. That is, if F_1 and F_2 are equivalent under the relation β, then it does not need to be F_1 is distributive over F_2. Please see the following example.

Example 6. Consider the unique idempotent nullnorm $F^{id} : [0, 1]^2 \to [0, 1]$ with zero element $\frac{1}{3}$ defined as follows (see [17]):

$$F^{id}(x, y) = \begin{cases} max(x, y), & x, y \in [0, \frac{1}{3}] \\ min(x, y), & x, y \in [\frac{1}{3}, 1] \\ \frac{1}{3}, & \text{otherwise} \end{cases} \qquad (5)$$

and consider the nullnorm $F : [0, 1]^2 \to [0, 1]$ with zero element $\frac{1}{3}$ defined as follows (see [18]):

$$F(x,y) = \begin{cases} max(x,y), & (x,y) \in [0, \frac{1}{3}]^2 \\ \frac{3xy-x-y+1}{2}, & (x,y) \in [\frac{1}{3}, 1]^2 \\ \frac{1}{3}, & \text{otherwise} \end{cases} \tag{6}$$

It is clear that $K_F = K_{F^{id}} = \emptyset$. So, F and F^{id} are equivalent under the relation β. But we claim that F^{id} is not distributive over F. Now, we will show this claim.

$F^{id}(\frac{2}{3}, F(\frac{1}{2}, \frac{3}{4})) = F^{id}(\frac{2}{3}, \frac{7}{16}) = \frac{7}{16}$ and $F(F^{id}(\frac{2}{3}, \frac{1}{2}), F^{id}(\frac{2}{3}, \frac{3}{4})) = F(\frac{1}{2}, \frac{2}{3}) = \frac{5}{12}$.

Since $\frac{7}{16} \neq \frac{5}{12}$, F^{id} is not distributive over F.

Definition 10 [5]. *Let F be a nullnorm on the unit interval $[0,1]$ with zero element a. The set $K_F^\star$ is defined by*

$$K_F^\star = \{x \in K_F \mid \text{for some } y, y' \in (0,1), [x < y \text{ but } x \npreceq_F y]$$

$$\text{and } [y' < x \text{ but } y' \npreceq_F x]\}.$$

Proposition 5. *Let F_1 and F_2 be nullnorms on the unit interval $[0,1]$ with the same zero elements a. If $\preceq_{F_1} \subseteq \preceq_{F_2}$, then $K_{F_2}^\star \subseteq K_{F_1}^\star$.*

Corollary 3. *Let F_1 and F_2 be nullnorms on the unit interval $[0,1]$ with the same zero elements a. If F_1 is distributive over F_2, then $K_{F_2}^\star \subseteq K_{F_1}^\star$.*

Remark 3. The converse of Corollary 3 may not be true. Here is an example illustrating such a case.

Example 7. Consider the greatest nullnorm $\overline{F_{\frac{1}{3}}} = <S_D, \frac{1}{3}, T_M>$ on $[0,1]$ with zero element $\frac{1}{3}$ defined as follows:

$$\overline{F_{\frac{1}{3}}}(x,y) = \begin{cases} min(x,y) & (x,y) \in [\frac{1}{3}, 1]^2, \\ \frac{1}{3} & (x,y) \in (0, \frac{1}{3}]^2 \cup D_{\frac{1}{3}}, \\ max(x,y) & \text{otherwise.} \end{cases} \tag{7}$$

and consider the nullnorm F on $[0,1]$ given by (6). It can be shown that $K_{\overline{F_{\frac{1}{3}}}}^\star = (0, \frac{1}{3})$ and $K_F^\star = \emptyset$. So, it is clear that $K_F^\star \subseteq K_{\overline{F_{\frac{1}{3}}}}^\star$. But we claim that $\overline{F_{\frac{1}{3}}}$ is not distributive over F. Now, we will show this claim.

$\overline{F_{\frac{1}{3}}}(\frac{1}{2}, F(\frac{1}{2}, \frac{2}{3})) = \overline{F_{\frac{1}{3}}}(\frac{1}{2}, \frac{5}{12}) = \frac{5}{12}$ and $F(\overline{F_{\frac{1}{3}}}(\frac{1}{2}, \frac{1}{2}), \overline{F_{\frac{1}{3}}}(\frac{1}{2}, \frac{2}{3})) = F(\frac{1}{2}, \frac{1}{2}) = \frac{3}{8}$.
Since $\frac{5}{12} \neq \frac{3}{8}$, $\overline{F_{\frac{1}{3}}}$ is not distributive over F.

4   Concluding Remarks

Nullnorms are useful tool in many different fields, for instance expert systems, neural networks and fuzzy logic. Also, they have been used as aggregators in fuzzy logic. Also, the distributivity equations have been studied from different points of view, by many authors [8,24]. On the other hand, F-partial order obtained from nullnorms was defined by [2] and in recent years, F-partial order denoted by $\preceq_F$ has been studied widely. In this study, mainly we have investigated the relation between the F-partial order and distributivity equation. Also, we have obtained some critical conclusions.

Acknowledgement. Second author was supported of grants VEGA 1/0006/19.

References

1. Aşıcı, E.: On the properties of the F-partial order and the equivalence of nullnorms. Fuzzy Sets Syst. **346**, 72–84 (2018)
2. Aşıcı, E.: An order induced by nullnorms and its properties. Fuzzy Sets Syst. **325**, 35–46 (2017)
3. Aşıcı, E.: Some notes on the F-partial order. In: Kacprzyk, J., Szmidt, E., Zadrożny, S., Atanassov, K., Krawczak, M. (eds.) Advances in Fuzzy Logic and Technology, IWIFSGN 2017, EUSFLAT 2017. Advances in Intelligent Systems and Computing, vol. 641, pp. 78–84. Springer, Cham (2018)
4. Aşıcı, E.: Some remarks on an order induced by uninorms. In: Kacprzyk, J., Szmidt, E., Zadrożny, S., Atanassov, K., Krawczak, M. (eds.) Advances in Fuzzy Logic and Technology, IWIFSGN 2017, EUSFLAT 2017. Advances in Intelligent Systems and Computing, vol. 641, pp. 69–77. Springer, Cham (2018)
5. Aşıcı, E.: An extension of the ordering based on nullnorms, Kybernetika, accepted
6. Birkhoff, G.: Lattice Theory, 3rd edn, Providence (1967)
7. Calvo, T.: On some solutions of the distributivity equation. Fuzzy Sets Syst. **104**, 85–96 (1999)
8. Calvo, T., De Baets, B., Fodor, J.: The functional equations of Frank and Alsina for uninorms and nullnorms. Fuzzy Sets Syst. **120**, 385–394 (2001)
9. Casasnovas, J., Mayor, G.: Discrete t-norms and operations on extended multisets. Fuzzy Sets Syst. **159**, 1165–1177 (2008)
10. De Baets, B., Mesiar, R.: Triangular norms on the real unit square. In: Proceedings of the 1999 EUSFLAT-ESTYLF Joint Conference, Palma de Mallorca, Spain, pp. 351–354 (1999)
11. Çaylı, G.D.: On the structure of uninorms on bounded lattices. Fuzzy Sets Syst. **357**, 2–26 (2019)
12. Çaylı, G.D., Karaçal, F., Mesiar, R.: On internal and locally internal uninorms on bounded lattices. Int. J. Gen. Syst. **48**(3), 235–259 (2019)
13. Çaylı, G.D.: A characterization of uninorms on bounded lattices by means of triangular norms and triangular conorms. Int. J. Gen. Syst. **47**, 772–793 (2018)
14. Çaylı, G.D., Drygaś, P.: Some properties of idempotent uninorms on a special class of bounded lattices. Inf. Sci. **422**, 352–363 (2018)
15. Drewniak, J., Drygaś, P., Rak, E.: Distributivity between uninorms and nullnorms. Fuzzy Sets Syst. **159**, 1646–1657 (2008)

16. Drygaś, P.: Distributive between semi-t-operators and semi-nullnorms. Fuzzy Sets Syst. **264**, 100–109 (2015)
17. Drygaś, P.: A characterization of idempotent nullnorms. Fuzzy Sets Syst. **145**, 455–461 (2004)
18. Grabisch, M., Marichal, J.-L., Mesiar, R., Pap, E.: Aggregation Functions. Cambridge University Press, Cambridge (2009)
19. Karaçal, F., Ince, M.A., Mesiar, R.: Nullnorms on bounded lattice. Inf. Sci. **325**, 227–236 (2015)
20. Karaçal, F., Kesicioğlu, M.N.: A T-partial order obtained from t-norms. Kybernetika **47**, 300–314 (2011)
21. Kesicioğlu, M.N.: On the property of T-distributivity. Fixed Point Theory Appl. **2013**, 32 (2013)
22. Klement, E.P., Mesiar, R., Pap, E.: Triangular Norms. Kluwer Academic Publishers, Dordrecht (2000)
23. Mas, M., Mayor, G., Torrens, J.: T-operators. Int. J. Uncertain. Fuzz. Knowl.-Based Syst. **7**, 31–50 (1999)
24. Mas, M., Mayor, G., Torrens, J.: The distributivity condition for uninorms and t-operators. Fuzzy Sets Syst. **128**, 209–225 (2002)
25. Mas, M., Mayor, G., Torrens, J.: The modularity condition for uninorms and t-operators. Fuzzy Sets Syst. **126**, 207–218 (2002)
26. Xie, A., Liu, H.: On the distributivity of uninorms over nullnorms. Fuzzy Sets Syst. **211**, 62–72 (2013)
27. Schweizer, B., Sklar, A.: Statistical metric spaces. Pacific J. Math. **10**, 313–334 (1960)

On Properties of Internal Uninorms
on Bounded Lattices

Gül Deniz Çaylı[(✉)]

Department of Mathematics, Faculty of Science, Karadeniz Technical University,
61080 Trabzon, Turkey
gdcayli@gmail.com

Abstract. In this paper, we study internal uninorms on bounded lattices. We show that an internal uninorm on a bounded lattice L for the chosen neutral element $e \in L \backslash \{0, 1\}$ does not always exist. We introduce two new construction methods for uninorms on a bounded lattice L based on the existence of a t-norm and a t-conorm, where some necessary and sufficient conditions on its neutral element $e \in L \backslash \{0, 1\}$ are required. These methods also show the existence of internal uninorms on some special bounded lattices.

Keywords: Bounded lattice · Internal uninorm · Neutral element ·
Uninorm · t-norm · t-conorm

1 Basic Notions and Preliminaries

Uninorms on the unit interval $[0, 1]$ are an important generalization of triangular norms (t-norms for short) and triangular conorms (t-conorms for short) introduced by Yager and Rybalov [27] and extensively studied by Fodor et al. [17]. These operators allow for the neutral element (sometimes called identity) lying anywhere in the unit interval, rather than at one or zero as in the case of t-norms and t-conorms, respectively. Uninorms play an important role not only in theoretical investigations but also in practical applications such as expert systems, neural networks, fuzzy logics, etc. For more studies concerning t-norms, t-conorms, uninorms and related operators on the unit interval, it can be referred to [3, 12, 13, 15, 16, 18, 20–22, 24].

The definition of uninorms was extended on bounded lattices and the existence of uninorms on bounded lattices was shown by Karaçal and Mesiar [19]. In the same paper, it was constructed the least and the greatest uninorms on a bounded lattice L, allowing to choose an arbitrary element $e \in L \backslash \{0, 1\}$ as the neutral one based on a t-norm T acting on the subinterval $[0, e]$ and a t-conorm S acting on the subinterval $[e, 1]$. In [8], it was demonstrated the existence of idempotent uninorms on a bounded lattice L for any element $e \in L \backslash \{0, 1\}$ playing the role of a neutral element. In addition, the least and the greatest idempotent uninorms on bounded lattices were obtained. It can be found some other studies related to uninorms on lattices in [6, 7, 9, 10, 14, 25, 26].

© Springer Nature Switzerland AG 2019
R. Halaš et al. (Eds.): AGOP 2019, AISC 981, pp. 115–128, 2019.
https://doi.org/10.1007/978-3-030-19494-9_11

The concept of internal uninorms on bounded lattices was introduced and some properties of such uninorms were investigated in [11]. In the meantime, it was showed that it is possible to construct internal uninorms on a bounded lattice L under some additional assumptions on the indicated neutral element $e \in L \setminus \{0, 1\}$. The main aim of this paper is to demonstrate that considering an arbitrary bounded lattice L, there need not always be an internal uninorm on L having a neutral element $e \in L \setminus \{0, 1\}$ and to propose two new construction methods for internal uninorms on L different from proposed in [11] with some constraints on the neutral element $e \in L \setminus \{0, 1\}$.

The paper is organized as follows. First, some basic results dealing with bounded lattices and uninorms on them are briefly discussed in Sect. 1. In Sect. 2, we show that on some bounded lattices L, an internal uninorm having the properly chosen neutral element e different from the top and bottom elements of L does not exist. In addition, we introduce two different methods for constructing uninorms on a bounded lattice L based on the existence of a t-norm on $[0, e]$ and a t-conorm on $[e, 1]$, where some necessary and sufficient conditions on the element $e \in L \setminus \{0, 1\}$ considered as the neutral one are required. Note that these conditions play an important role in our construction methods, as then our methods yield a uninorm in particular cases. We also provide some illustrative examples for well understanding the structure of these uninorms. As a by-product of our methods, we obtain two different internal uninorms on bounded lattices. Finally, some concluding remarks are added.

A bounded lattice $(L, \leq)$ is a lattice that has the top and bottom elements, which are written as 1 and 0, respectively, that is, there exist two elements 1, $0 \in L$ such that $0 \leq x \leq 1$, for all $x \in L$. Throughout this paper, unless otherwise is stated, we denote L as a bounded lattice with the top and bottom elements 1 and 0, respectively.

Given $a, b \in L$. If a and b are incomparable, we use the notation $a \parallel b$. If a and b are comparable, we use the notation $a \nparallel b$.

Definition 1 ([5]). *Let $a, b \in L$ such that $a \leq b$. A subinterval $[a, b]$ of L is defined by $[a, b] = \{x \in L \mid a \leq x \leq b\}$. Similarly, it is defined $]a, b] = \{x \in L \mid a < x \leq b\}$, $[a, b[= \{x \in L \mid a \leq x < b\}$ and $]a, b[= \{x \in L \mid a < x < b\}$.*

Definition 2 ([1,2,4,10]). *An operation $T: L^2 \to L$ $\left(S: L^2 \to L \right)$ is called a t-norm (t-conorm) on L if it is commutative, associative, increasing with respect to both variables and has the neutral element $1 \in L$ $(0 \in L)$ i.e., $T(1, x) = x$ $(S(0, x) = x)$ for all $x \in L$.*

Example 1. The smallest t-norm $T_W : L^2 \to L$ and the greatest t-norm $T_\wedge : L^2 \to L$, respectively, are defined by $T_W(x, y) = \begin{cases} x \wedge y & if\ 1 \in \{x, y\}, \\ 0 & otherwise \end{cases}$ and $T_\wedge(x, y) = x \wedge y$. The smallest t-conorm $S_\vee : L^2 \to L$ and the greatest t-conorm $S_W : L^2 \to L$, respectively, are defined by $S_\vee(x, y) = x \vee y$ and $S_W(x, y) = \begin{cases} x \vee y & if\ 0 \in \{x, y\}, \\ 0 & otherwise. \end{cases}$

Definition 3 ([9,19]). *An operation $U\colon L^2 \to L$ is called a uninorm on L (shortly a uninorm, if L is fixed) if it is commutative, associative, increasing with respect to both variables and there exists some element $e \in L$ called the neutral element i.e., $U(e,x) = x$ for all $x \in L$.*

Given $e \in L\backslash\{0,1\}$. Denote $A(e) = [0,e] \times \,]e,1] \cup \,]e,1] \times [0,e]$ and $I_e = \{x \in L \mid x \parallel e\}$.

Proposition 1 ([19]). *Let $e \in L\backslash\{0,1\}$ and U be a uninorm on L with the neutral element e. Then the following statements hold:*

(i) $T_e = U|[0,e]^2 : [0,e]^2 \to [0,e]$ is a t-norm on $[0,e]$.
(ii) $S_e = U|[e,1]^2 : [e,1]^2 \to [e,1]$ is a t-conorm on $[e,1]$.

2   Internal Uninorms on Bounded Lattices

In this section, we investigate the existence of internal uninorms on an arbitrary bounded lattice L having the indicated neutral element $e \in L\backslash\{0,1\}$. In Theorem 1, we propose that an internal uninorm on a bounded lattice L with the neutral element $e \in L\backslash\{0,1\}$ need not always exists. In order to show that there exist internal uninorms on some bounded lattices L, in Theorems 3 and 4, we introduce two new methods for building uninorms on L via t-norms acting on $[0,e]$ and t-conorms acting on $[e,1]$ under some constraints on the neutral element $e \in L\backslash\{0,1\}$. As a consequence of these methods, two internal uninorms defined on some bounded lattices are obtained in Corollaries 1 and 2. Note that the structure of internal uninorms on a bounded chain (i.e., on the real unit interval $[0,1]$ and on finite chains) was completely described in [23].

Definition 4 ([11]). *Let $e \in L\backslash\{0,1\}$ and U be a uninorm on L with the neutral element e. U is called an internal uninorm on L if it satisfies $U(x,y) \in \{x,y\}$ for all $x,y \in L$.*

Proposition 2 ([11]). *Let $e \in L\backslash\{0,1\}$ and U be an internal uninorm on L with the neutral element e. Then the subintervals $[0,e]$ and $[e,1]$ are chains.*

Proposition 3 ([11]). *Let $e \in L\backslash\{0,1\}$ and U be an internal uninorm on L with the neutral element e. Denote $U|[0,e]^2$ and $U|[e,1]^2$, respectively, by T_e and S_e. Then*

(i) $T_e(x,y) = x \wedge y$ for all $x,y \in [0,e]$.
(ii) $S_e(x,y) = x \vee y$ for all $x,y \in [e,1]$.

Proposition 4. *Let $e \in L\backslash\{0,1\}$ and U be an internal uninorm on L with the neutral element e. Then the following statements hold:*

(i) $U(x,y) \in I_e$ for $x,y \in I_e$.
(ii) $U(x,y) \in [0,e[$ or $U(x,y) \in \,]e,1]$ for $(x,y) \in A(e)$.

(iii) $U(x,y) \in I_e$ *or* $U(x,y) \in]e,1]$ *for* $(x,y) \in [e,1] \times I_e \cup I_e \times [e,1]$ *such that*
$x \nparallel y$.

(iv) $U(x,y) \in I_e$ *or* $U(x,y) \in [0,e[$ *for* $(x,y) \in [0,e] \times I_e \cup I_e \times [0,e]$ *such that*
$x \nparallel y$.

Note that when considering an internal uninorm U with the neutral element
$e \in L\backslash\{0,1\}$, $U(x,y) \in I_e$ for $(x,y) \in [e,1] \times I_e \cup I_e \times [e,1] \cup [0,e] \times I_e \cup I_e \times [0,e]$
such that $x \parallel y$ [11].

Theorem 1. *Let* $e \in L\backslash\{0,1\}$. *If there are some elements* $x,y,t,z \in L$ *such*
that $x \in]0,e[$, $y \in]e,1[$, $t,z \in I_e$, $x < z$, $t < y$, $y \parallel z$ *and* $x \parallel t$, *then there is no*
internal uninorm U *on* L *having a neutral element* e.

Proof. Let L be a lattice having some elements $x,y,t,z \in L$ such that $x \in$
$]0,e[$, $y \in]e,1[$, $t,z \in I_e$, $x < z$, $t < y$, $y \parallel z$ and $x \parallel t$, on which U is an
internal uninorm with the neutral element e. By the monotonicity of U, we have
$U(y,z) \geq U(e,z) = z$ and $U(x,t) \leq U(e,t) = t$. It holds $U(y,z) = z$ since
$U(y,z) \in \{y,z\}$ and $y \parallel z$. Similarly, it holds $U(x,t) = t$ since $U(x,t) \in \{x,t\}$
and $x \parallel t$. In addition, since $x < z$ and $t < y$, respectively, we have $U(x,y) \leq$
$U(z,y)$ and $U(x,t) \leq U(x,y)$ from the monotonicity of U. It is $U(x,y) \in \{x,y\}$.
Suppose that $U(x,y) = y$. Then, it is obtained $y \leq z$. This is a contradiction.
So, it cannot be $U(x,y) = y$. Suppose that $U(x,y) = x$. Then, it is obtained
$t \leq x$. This is a contradiction. So, it cannot be $U(x,y) = x$. Therefore, when
L has such determined elements, there is no internal uninorm on L with the
neutral element e.

Theorem 2. *Every uninorm defined on a bounded lattice* L *with the neutral*
element $e \in L\backslash\{0,1\}$ *is internal if and only if* $L = \{0,e,1\}$.

Proof. Let every uninorm defined on L with the neutral element $e \in L\backslash\{0,1\}$ be
internal. Suppose that there is an element $m \in]0,e[$. Considering the uninorm
U_t given in [19, Theorem 1], $U_t(x,y) = T(x,y)$ for all $(x,y) \in [0,e]^2$ such that
T is an arbitrary t-norm on $[0,e]$. So, it holds $U_t(m,m) = T(m,m) = 0$ for
the t-norm $T = T_W$ on $[0,e]$. Moreover, we have $U_t(m,m) = m$ since U_t is an
internal uninorm. Then, it holds $0 = m$. This is a contradiction. Therefore, there
is no element in $]0,e[$. Suppose that there is an element $n \in]e,1[$. Considering
the uninorm U_s given in [19, Theorem 1], $U_s(x,y) = S(x,y)$ for all $(x,y) \in [e,1]^2$
such that S is an arbitrary t-conorm on $[e,1]$. So, it holds $U_s(n,n) = S(n,n) = 1$
for the t-conorm $S = S_W$ on $[e,1]$. Moreover, we have $U_s(n,n) = n$ since U_s is an
internal uninorm. Then, it holds $1 = n$. This is a contradiction. Therefore, there
is no element in $]e,1[$. Suppose that there is an element $s \in I_e$. Considering the
uninorm U_t given in [19, Theorem 1], we have $U_t(s,s) = 1$. Moreover, we have
$U_t(s,s) = s$ since U_t is an internal uninorm. This is a contradiction. So, there
is no element in I_e. Therefore, if every uninorm defined on L with the neutral
element $e \in L\backslash\{0,1\}$ is internal, then $L = \{0,e,1\}$.

On the other hand if $L = \{0,e,1\}$, it is obvious that every uninorm on L
with the neutral element $e \in L\backslash\{0,1\}$ is internal.

Theorem 3. *Let $e \in L \backslash \{0, 1\}$ such that there be only one element in L incomparable with e, the subinterval $[0, e]$ be a chain and S_e be a t-conorm on the subinterval $[e, 1]$. The function U^S defined by*

$$
U^S(x, y) = \begin{cases}
S_e(x, y) & if \ (x, y) \in [e, 1]^2, \\
x \wedge y & \begin{array}{l} if \ (x, y) \in [0, e]^2 \ or \ x \in [0, e[, \ y \in I_e \ and \ x < y \\ or \ x \in I_e, \ y \in [0, e[\ and \ y < x, \end{array} \\
y & if \ x \in \,]0, e], \ y \in I_e \ and \ x \parallel y, \\
x & if \ x \in I_e, \ y \in \,]0, e] \ and \ x \parallel y, \\
x \vee y & otherwise
\end{cases}
\tag{1}
$$

is a uninorm on L with the neutral element e if and only if $a > b$ for all $a \in \,]e, 1]$, $b \in I_e$.

Example 2. Consider the lattice $L_1 = \{0, p, q, e, k, r, 1\}$ with Hasse diagram shown in Fig. 1. It is easy to see that this lattice satisfies all constraints in Theorem 3. Define the t-conorm $S_e : [e, 1]^2 \to [e, 1]$ as $S_e = S_W$. By applying the construction approach in Theorem 3, the uninorm $U : L_1^2 \to L_1$ with the neutral element e is defined by Table 1.

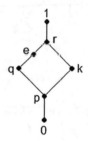

Fig. 1. The lattice L_1

Table 1. The uninorm U on L_1

U	0	p	q	e	k	r	1
0	0	0	0	0	0	r	1
p	0	p	p	p	p	r	1
q	0	p	q	q	k	r	1
e	0	p	q	e	k	r	1
k	0	p	k	k	k	r	1
r	r	r	r	r	r	1	1
1	1	1	1	1	1	1	1

Remark 1. Let $e \in L \backslash \{0, 1\}$. In Theorem 3 observe that the condition that the subinterval $[0, e]$ is a chain cannot be omitted, in general. Consider the lattice L having some elements $x, y \in]0, e[$, $z \in I_e$ such that $x \parallel z$, $y \parallel z$ and $x \wedge y = 0$. Then, by using the formula (1) in Theorem 3, we have that $U^S \left(U^S \left(x, y \right), z \right) = U^S \left(x \wedge y, z \right) = U^S \left(0, z \right) = 0$ and $U^S \left(x, U^S \left(y, z \right) \right) = U^S \left(x, z \right) = z$. Then the associativity of U^S is violated. So, the condition that the subinterval $[0, e]$ is a chain is a sufficient condition in Theorem 3. Then a natural question occurs: is this a necessary condition in Theorem 3? In the following example, we give a negative answer to this question.

Example 3. Consider the lattice $L_2 = \{0, m, p, q, r, e, k, 1\}$ with Hasse diagram shown in Fig. 2. Note that the subinterval $[0, e]$ is not a chain. By using the construction approach in Theorem 3, the function $U : L_2^2 \to L_2$ is defined by Table 2. It is possible to check that U is a uninorm on L_2 with the neutral element e.

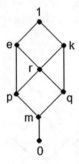

Fig. 2. The lattice L_2

Table 2. The uninorm U on L_2

U	0	m	p	q	r	e	k	1
0	0	0	0	0	0	0	0	1
m	0	m	m	m	m	m	m	1
p	0	m	p	m	p	p	p	1
q	0	m	m	q	q	q	q	1
r	0	m	p	q	r	r	r	1
e	0	m	p	q	r	e	k	1
k	0	m	p	q	r	k	k	1
1	1	1	1	1	1	1	1	1

Remark 2. Let $e \in L \backslash \{0, 1\}$. In Theorem 3 observe that the condition that there is only one element in L incomparable with e cannot be omitted, in general. Consider the lattice L having some elements $y, z \in I_e$ and $y \parallel z$. It can be $y \vee z > e$

or $y \vee z \parallel e$. In that case $y \vee z > e$, by using the formula (1) in Theorem 3, we have that $U^S \left(U^S \left(0, y\right), z\right) = U^S \left(0, z\right) = 0$ and $U^S \left(0, U^S \left(y, z\right)\right) = U^S \left(0, y \vee z\right) = y \vee z$. Since the associativity is violated, in that case $y \vee z > e$, the condition that there is only one element in L incomparable with e cannot be omitted. Now, we consider the case $y \vee z \parallel e$. If there is an element $x \in \,]0, e[$ such that $x < y$ and $x \parallel z$, then by using the formula (1) in Theorem 3, we have that $U^S \left(U^S \left(x, y\right), z\right) = U^S \left(x \wedge y, z\right) = U^S \left(x, z\right) = z$ and $U^S \left(x, U^S \left(y, z\right)\right) = U^S \left(x, y \vee z\right) = x$. Since the associativity is violated, in that case $y \vee z \parallel e$, the condition that there is only one element in L incomparable with e cannot be omitted. Therefore, we see that the condition that there is only one element in L incomparable with e is a sufficient condition in Theorem 3. Then a natural question occurs: is this a necessary condition in Theorem 3? In the following example, we give a negative answer to this question.

Example 4. Consider the lattice $L_3 = \{0, p, e, t, k, r, 1\}$ with Hasse diagram shown in Fig. 3. Note that there are two elements $k, t \in I_e$ and $k \neq t$. By using the construction approach in Theorem 3 and taking the t-conorm $S_e : [e, 1]^2 \to [e, 1]$ as $S_e = S_\vee$, the function $U : L_3^2 \to L_3$ is defined by Table 3. It is possible to check that U is a uninorm on L_3 with the neutral element e.

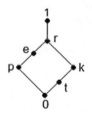

Fig. 3. The lattice L_3

Table 3. The uninorm U on L_3

U	0	p	e	t	k	r	1
0	0	0	0	0	0	r	1
p	0	p	p	t	k	r	1
e	0	p	e	t	k	r	1
t	0	t	t	t	k	r	1
k	0	k	k	k	k	r	1
r	r	r	r	r	r	r	1
1	1	1	1	1	1	1	1

Corollary 1. *Let $e \in L\backslash\{0,1\}$ such that there be only one element in L incomparable with e. In Theorem 3, take t-conorm $S_e : [e,1]^2 \to [e,1]$ as $S_e = S_\vee$. The function $U^\vee$ defined by*

$$U^\vee(x,y) = \begin{cases} x \wedge y & \begin{aligned} & \textit{if } (x,y) \in [0,e]^2 \textit{ or } x \in [0,e[,\ y \in I_e \textit{ and } x < y \\ & \textit{or } x \in I_e,\ y \in [0,e[\textit{ and } y < x, \end{aligned} \\ y & \textit{if } x \in]0,e],\ y \in I_e \textit{ and } x \parallel y, \\ x & \textit{if } x \in I_e,\ y \in]0,e] \textit{ and } x \parallel y, \\ x \vee y & \textit{otherwise} \end{cases} \quad (2)$$

is an internal uninorm on L with the neutral element e if and only if the subintervals $[0,e]$ and $[e,1]$ are chains and $a > b$ for all $a \in]e,1],\ b \in I_e$.

Theorem 4. *Let $e \in L\backslash\{0,1\}$ such that there be only one element in L incomparable with e, the subinterval $[e,1]$ be a chain and T_e be a t-norm on the subinterval $[0,e]$. The function U^T defined by*

$$U^T(x,y) = \begin{cases} T_e(x,y) & \textit{if } (x,y) \in [0,e]^2, \\ x \vee y & \begin{aligned} & \textit{if } (x,y) \in [e,1]^2 \textit{ or } x \in]e,1],\ y \in I_e \textit{ and } x > y \\ & \textit{or } x \in I_e,\ y \in]e,1] \textit{ and } y > x, \end{aligned} \\ y & \textit{if } x \in [e,1[,\ y \in I_e \textit{ and } x \parallel y, \\ x & \textit{if } x \in I_e,\ y \in [e,1[\textit{ and } x \parallel y, \\ x \wedge y & \textit{otherwise} \end{cases} \quad (3)$$

is a uninorm on L with the neutral element e if and only if $a < b$ for all $a \in [0,e[,\ b \in I_e$.

Example 5. Consider the lattice $L_4 = \{0,p,q,r,e,k,m,1\}$ with Hasse diagram shown in Fig. 4. It is easy to see that this lattice satisfies all constraints in Theorem 4. Define the t-norm $T_e : [0,e]^2 \to [0,e]$ as $T_e = T_W$. By applying the construction approach in Theorem 4, the uninorm $U : L_4^2 \to L_4$ with the neutral element e is defined by Table 4.

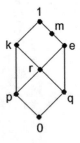

Fig. 4. The lattice L_4

Table 4. The uninorm U on L_4

U	0	p	q	r	e	k	m	1
0	0	0	0	0	0	0	0	0
p	0	0	0	0	p	p	p	p
q	0	0	0	0	q	q	q	q
r	0	0	0	0	r	r	r	r
e	0	p	q	r	e	k	m	1
k	0	p	q	r	k	k	k	1
m	0	p	q	r	m	k	m	1
1	0	p	q	r	1	1	1	1

Remark 3. Let $e \in L \backslash \{0, 1\}$. In Theorem 4 observe that the condition that the subinterval $[e, 1]$ is a chain cannot be omitted, in general. Consider the lattice L having some elements $x, y \in]e, 1[$, $z \in I_e$ such that $x > z$, $y \parallel z$ and $x \vee y = 1$. Then, by using the formula (3) in Theorem 4, we have that $U^T \left(U^T \left(x, y \right), z \right) = U^T \left(x \vee y, z \right) = U^T \left(1, z \right) = 1$ and $U^T \left(x, U^T \left(y, z \right) \right) = U^T \left(x, z \right) = x$. Then the associativity of U^T is violated. So, the condition that the subinterval $[e, 1]$ is a chain is a sufficient condition in Theorem 4. Then a natural question occurs: is this a necessary condition in Theorem 4? In the following example, we give a negative answer to this question.

Example 6. Consider the lattice $L_5 = \{0, e, k, s, n, m, 1\}$ with Hasse diagram shown in Fig. 5. Note that the subinterval $[e, 1]$ is not a chain. By using the construction approach in Theorem 4, the function $U : L_5^2 \to L_5$ is defined by Table 5. It is possible to check that U is a uninorm on L_5 with the neutral element e.

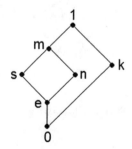

Fig. 5. The lattice L_5

Table 5. The uninorm U on L_5

U	0	e	k	s	n	m	1
0	0	0	0	0	0	0	0
e	0	e	k	s	n	m	1
k	0	k	k	k	k	k	1
s	0	s	k	s	m	m	1
n	0	n	k	m	n	m	1
m	0	m	k	m	m	m	1
1	0	1	1	1	1	1	1

Remark 4. Let $e \in L \backslash \{0,1\}$. In Theorem 4 observe that the condition that there is only one element in L incomparable with e cannot be omitted, in general. Consider the lattice L having some elements $y, z \in I_e$ and $y < z$. If there is an element $x \in]e, 1[$ such that $x > y$ and $x \parallel z$, then by using the formula (3) in Theorem 4, we have that $U^T(x, y) = x \vee y = x$ and $U^T(x, z) = z$. Then the monotonicity of U^T is violated. Therefore, we see that the condition that there is only one element in L incomparable with e is a sufficient condition in Theorem 4. Then a natural question occurs: is this a necessary condition in Theorem 4? In the following example, we give a negative answer to this question.

Example 7. Consider the lattice $L_6 = \{0, e, m, k, t, 1\}$ with Hasse diagram shown in Fig. 6. Note that there are some elements $k, t, m \in I_e$. By using the construction approach in Theorem 4, the function $U : L_6^2 \to L_6$ is defined by Table 6. It is possible to check that U is a uninorm on L_6 with the neutral element e.

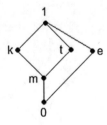

Fig. 6. The lattice L_6

Table 6. The uninorm U on L_6

U	0	e	m	k	t	1
0	0	0	0	0	0	0
e	0	e	m	k	t	1
m	0	m	m	m	m	1
k	0	k	m	k	m	1
t	0	t	m	m	t	1
1	0	1	1	1	1	1

Corollary 2. *Let $e \in L \backslash \{0,1\}$ such that there be only one element in L incomparable with e. In Theorem 4, take t-norm $T_e : [0,e]^2 \to [0,e]$ as $T_e = T_\wedge$. The function $U^\wedge$ defined by*

$$U^\wedge(x,y) = \begin{cases} x \vee y & \begin{aligned} &if \ (x,y) \in [e,1]^2 \ or \ x \in \,]e,1], \ y \in I_e \ and \ x > y \\ &or \ x \in I_e, \ y \in \,]e,1] \ and \ y > x, \end{aligned} \\ y & if \ x \in [e,1[, \ y \in I_e \ and \ x \parallel y, \\ x & if \ x \in I_e, \ y \in [e,1[\ and \ x \parallel y, \\ x \wedge y & otherwise \end{cases} \qquad (4)$$

is an internal uninorm on L with the neutral element e if and only if the subintervals $[0,e]$ and $[e,1]$ are chains and $a < b$ for all $a \in [0,e[\,, b \in I_e$.

Remark 5. Note that in Corollaries 1 and 2, respectively, the conditions $a > b$ for all $a > b$ for all $a \in \,]e,1]$, $b \in I_e$ and $c < d$ for all $c \in [0,e[, d \in I_e$ are both sufficient and necessary conditions for the formulas (2) and (4) yield internal uninorms on L with the neutral element $e \in L \backslash \{0,1\}$, respectively. Then one can ask whether there exists an internal uninorm with the neutral element $e \in L \backslash \{0,1\}$ on a bounded lattice L which does not satisfy the conditions $a > b$ for all $a \in \,]e,1]$, $b \in I_e$ and $c < d$ for all $c \in [0,e[, d \in I_e$. In the following, we provide an example of a lattice that does not satisfy these conditions, on which there exists an internal uninorm for a neutral element $e \in L \backslash \{0,1\}$.

Example 8. Consider the lattice $L_7 = \{0,n,z,e,m,y,t,1\}$ with Hasse diagram shown in Fig. 7. Note that $y \parallel m$ and $z \parallel m$ for $y \in \,]e,1[, z \in \,]0,e[, m \in I_e$. The function $U : L_7^2 \to L_7$ defined by Table 7 is an internal uninorm on L_7 with the neutral element e.

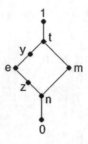

Fig. 7. The lattice L_7

Table 7. The internal uninorm U on L_7

U	0	n	z	e	m	y	t	1
0	0	0	0	0	0	0	0	0
n	0	n	n	n	n	n	t	1
z	0	n	z	z	m	z	t	1
e	0	n	z	e	m	y	t	1
m	0	n	m	m	m	m	t	1
y	0	n	z	y	m	y	t	1
t	0	t	t	t	t	t	t	1
1	0	1	1	1	1	1	1	1

3   Concluding Remarks

After the concept of uninorms on the unit interval $[0, 1]$ was introduced as a generalization of t-norms and t-conorms by Yager and Rybalov [27], uninorms related to algebraic structures on bounded lattices have attracted much attention from researchers. The definition of uninorms on bounded lattices was introduced by Karaçal and Mesiar in [19]. They showed that it is possible to construct the least and the greatest uninorms on a bounded lattice L with the neutral element $e \in L\backslash\{0,1\}$. And then in [8], idempotent uninorms on bounded lattices were studied and some construction methods for such idempotent uninorms having a neutral element were proposed. The concept of internal uninorms on bounded lattices was introduced in [11]. In the same paper, it was researched whether there is an internal uninorm on every bounded lattice L with a neutral element and proposed two methods for obtaining internal uninorms on L under some additional assumptions. However, the existence of internal uninorms on bounded lattices was demonstrated in [11], the characterization of such uninorms is still active research. In this paper, we study internal uninorms on bounded lattices and investigate their basic characteristics. We show that on some bounded lattices L, there exist an element $e \in L\backslash\{0,1\}$ such that none uninorm with the neutral element e is internal. It is also introduced two new construction methods for uninorms on a bounded lattice L with the neutral element $e \in L\backslash\{0,1\}$

by using a t-norm acting on $[0, e]$ and a t-conorm acting on $[e, 1]$ under some constraints. As a by-product, two new types of internal uninorms on bounded lattices are obtained. In addition, we add some illustrative examples to clearly understand the structure of the uninorms obtained by means of these methods.

References

1. Aşıcı, E.: An order induced by nullnorms and its properties. Fuzzy Sets Syst. **325**, 35–46 (2017)
2. Aşıcı, E.: On the properties of the F-partial order and the equivalence of nullnorms. Fuzzy Sets Syst. **346**, 72–84 (2018)
3. Aşıcı, E.: Some remarks on an order induced by Uninorms. In: Kacprzyk, J., et al. (eds.) Advances in Fuzzy Logic and Technology 2017, IWIFSGN 2017, EUSFLAT 2017. Advances in Intelligent Systems and Computing, vol. 641, pp. 69–77. Springer, Cham (2018)
4. Aşıcı, E.: Some notes on the F-partial order. In: Kacprzyk, J., et al. (eds) Advances in Fuzzy Logic and Technology 2017, IWIFSGN 2017, EUSFLAT 2017. Advances in Intelligent Systems and Computing, vol. 641, pp. 78–84. Springer, Cham (2018)
5. Birkhoff, G.: Lattice Theory. American Mathematical Society Colloquium Publishers, Providence (1967)
6. Bodjanova, S., Kalina, M.: Construction of uninorms on bounded lattices. In: IEEE 12th International Symposium on Intelligent Systems and Informatics, SISY 2014, Subotica, Serbia, pp. 61–66 (2014)
7. Bodjanova, S., Kalina, M.: Uninorms on bounded lattices-Recent development. In: Kacprzyk, J., et al. (eds.) Advances in Fuzzy Logic and Technology 2017, IWIFSGN 2017, EUSFLAT 2017. Advances in Intelligent Systems and Computing, vol. 641, pp. 224–234. Springer, Cham (2018)
8. Çaylı, G.D., Karaçal, F., Mesiar, R.: On a new class of uninorms on bounded lattices. Inf. Sci. **367–368**, 221–231 (2016)
9. Çaylı, G.D., Drygaś, P.: Some properties of idempotent uninorms on a special class of bounded lattices. Inf. Sci. **422**, 352–363 (2018)
10. Çaylı, G.D.: On the structure of uninorms on bounded lattices. Fuzzy Sets Syst. **357**, 2–26 (2019)
11. Çaylı, G.D., Karaçal, F., Mesiar, R.: On internal and locally internal uninorms on bounded lattices. Int. J. General Syst. **48**(3), 235–259 (2019)
12. De Baets, B.: Idempotent uninorms. Eur. J. Oper. Res. **118**, 631–642 (1999)
13. De Baets, B., Fodor, J., Ruiz-Aguilera, D., Torrens, J.: Idempotent uninorms on finite ordinal scales. Int. J. Uncertain Fuzziness Knowl.-Based Syst. **17**(1), 1–14 (2009)
14. Deschrijver, G.: Uninorms which are neither conjunctive nor disjunctive in interval-valued fuzzy set theory. Inf. Sci. **244**, 48–59 (2013)
15. Drewniak, J., Drygaś, P.: On a class of uninorms. Internat J. Uncertain Fuzziness Knowl.-Based Syst. **10**, 5–10 (2002)
16. Drygaś, P., Ruiz-Aguilera, D., Torrens, J.: A characterization of uninorms locally internal in A(e) with continuous underlying operators. Fuzzy Sets Syst. **287**, 137–153 (2016)
17. Fodor, J., Yager, R.R., Rybalov, A.: Structure of uninorms. Int. J. Uncertain Fuzziness Knowl.-Based Syst. **5**, 411–427 (1997)

18. Grabisch, M., Marichal, J.L., Mesiar, R., Pap, E.: Aggregation Functions. Cambridge University Press, Cambridge (2009)
19. Karaçal, F., Mesiar, R.: Uninorms on bounded lattices. Fuzzy Sets Syst. **261**, 33–43 (2015)
20. Klement, E.P., Mesiar, R., Pap, E.: Triangular Norms. Kluwer Acad. Publ, Dordrecht (2000)
21. Martin, J., Mayor, G., Torrens, J.: On locally internal monotonic operations. Fuzzy Sets Syst. **137**, 27–42 (2003)
22. Mesiarová-Zemánková, A.: The structure of n-contractive t-norms. Int. J. Gen. Syst. **34**(5), 625–637 (2005)
23. Mesiarová-Zemánková, A.: A note on decomposition of idempotent uninorms into an ordinal sum of singleton semigroups. Fuzzy Sets Syst. **299**, 140–145 (2016)
24. Su, Y., Zong, W., Liu, H.W., Zhang, F.: On migrativity property for uninorms. Inf. Sci. **300**, 114–123 (2015)
25. Takács, M.: Lattice ordered monoids and left continuous uninorms and t-norms. In: Castillo, O., et al. (eds.) Theoretical Advances and Applications of Fuzzy Logic and Soft Computing. Advances in Soft Computing, vol. 42, pp. 565–572. Springer, Heidelberg (2007)
26. Wang, Z.D., Fang, J.X.: Residual operators of left and right uninorms on a complete lattice. Fuzzy Sets Syst. **160**, 22–31 (2009)
27. Yager, R.R., Rybalov, A.: Uninorms aggregation operators. Fuzzy Sets Syst. **80**, 111–120 (1996)

Some Remarks on Generalized Hypothetical Syllogism and Yager's Implications

Piotr Helbin$^{(\boxtimes)}$ (iD), Katarzyna Miś(iD), and Michał Baczyński(iD)

Institute of Mathematics, University of Silesia in Katowice,
Bankowa 14, 40-007 Katowice, Poland
{piotr.helbin,kmis,michal.baczynski}@us.edu.pl

Abstract. In this paper we investigate some properties of generalized hypothetical syllogism (GHS). We focus on the class of Yager's implications and give some solutions of (GHS) among this family. Furthermore, we show some relations between the class of aggregated fuzzy implications and (GHS).

Keywords: Aggregation function · Fuzzy connectives ·
Fuzzy implication · T-norm · Generalized hypothetical syllogism ·
Yager's f- and g-generated implications

1 Introduction

The notion of a composition of fuzzy relations appeared many years ago, the first one was introduced by Zadeh [11] in 1973 in the following way

$$(R \circ S)(x,y) := \sup_{z \in Z} \min\{R(x,z), S(z,y)\}, \qquad x \in X, y \in Y,$$

where R, S are fuzzy relations and $R \circ S$ is a fuzzy relation as well. In this paper we focus our attention on the notion of generalized hypothetical syllogism. Firstly, it can come from one particular composition of fuzzy implications given as follows

$$(I \overset{T}{\circ} J)(x,y) := \sup_{z \in [0,1]} T(I(x,z), J(z,y)), \qquad x,y \in [0,1], \tag{1}$$

where, T is an arbitrary t-norm and I, J are fuzzy implications. Next, we can consider the following functional equation, which is a case of (1) when $I = J$,

$$I \overset{T}{\circ} I = I, \tag{2}$$

where I is a fuzzy implication (or, in general, any generalization of classical implication) and T is again a t-norm (or, in general, any generalization of classical conjunction). Note that (2) can be seen as I is a sup -T-idempotent fuzzy implication.

Secondly, the generalized hypothetical syllogism was presented by Klir and Yuan in [6] and simply defined in the following way.

R. Halaš et al. (Eds.): AGOP 2019, AISC 981, pp. 129–139, 2019.
https://doi.org/10.1007/978-3-030-19494-9_12

Definition 1.1. *Let T be a t-norm and I be a fuzzy implication. We say that the pair (T, I) satisfies the generalized hypothetical syllogism if*

$$\sup_{z \in [0,1]} T(I(x,z), I(z,y)) = I(x,y), \qquad x, y \in [0,1], \qquad \text{(GHS)}$$

i.e., when Eq. (2) is satisfied.

Since then (GHS) was widely studied. Recently, Vemuri [9] focused on the Eq. (GHS) and many families of fuzzy implications but he took into account only one t-norm—the minimum t-norm. Later, in [2] we investigated more general case—mostly left-continuous t-norms. However, we concentrated on R-implications. In this work we continue our research. This time we investigate Yager's implications. Furthermore, we recall how to obtain fuzzy implications using aggregation functions. We show a connection between this class of fuzzy implications and Eq. (GHS).

The paper is organized as follows. Section 2 contains some preliminaries, where we recall the basic concepts and definitions used later on. Section 3 contains the idea of an aggregation of fuzzy implications - they can be built as a result of composition of two given fuzzy implications through an aggregation function. Later, having some additional assumptions we show some properties of such class. In Sect. 4, we present the results concerning (GHS) for Yager's implications. In particular, we show that the Goguen implication (I_{GG}) is the only g-generated Yager's implication (when $g(1) < \infty$) such that (T, I_{GG}) satisfies (GHS) for some t-norm T.

2   Preliminaries

In this section we recall basic notations and facts used in the sequel.

Definition 2.1 ([1, Definition 1.1.1]). *A function $I \colon [0,1]^2 \to [0,1]$ is called a* **fuzzy implication** *if it satisfies the following conditions:*

(I1) *I is non-increasing with respect to the first variable,*
(I2) *I is non-decreasing with respect to the second variable,*
(I3) *$I(0,0) = I(1,1) = 1$ and $I(1,0) = 0$.*

Definition 2.2 (*see* [4]). *A function $A \colon [0,1]^2 \to [0,1]$ is called an* **aggregation function** *if it satisfies the following conditions:*

(A1) *$A(1,1) = 1$ and $A(0,0) = 0$,*
(A2) *$A(x_1, y_1) \le A(x_2, y_2)$, for all $x_1, x_2, y_1, y_2 \in [0,1]$ such that $x_1 \le x_2, y_1 \le y_2$.*

Definition 2.3 ([5, Definition 1.1]). *A function $T \colon [0,1]^2 \to [0,1]$ is called a* **triangular norm** *(shortly* **t-norm***) if it satisfies the following conditions, for all $x, y, z \in [0,1]$:*

(T1) $T(x, y) = T(y, x)$, *i.e.*, T *is commutative,*
(T2) $T(x, T(y, z)) = T(T(x, y), z)$, *i.e.*, T *is associative,*
(T3) T *is non-decreasing with respect to the second variable,*
(T4) $T(x, 1) = x$.

Example 2.4. The most important examples of t-norms are the following:

– the product t-norm

$$T_{\mathbf{P}}(x, y) = xy, \qquad x, y \in [0, 1],$$

– the Łukasiewicz t-norm

$$T_{\mathbf{LK}}(x, y) = \max(x + y - 1, 0), \qquad x, y \in [0, 1],$$

– the minimum t-norm

$$T_{\mathbf{M}}(x, y) = \min(x, y), \qquad x, y \in [0, 1],$$

– the drastic t-norm

$$T_{\mathbf{D}}(x, y) = \begin{cases} 0, & x, y \in (0, 1), \\ \min(x, y), & \text{elsewhere.} \end{cases}$$

Definition 2.5 ([5, Definition 11.3]). *A non-increasing function* $N \colon [0, 1] \to [0, 1]$ *is called a **fuzzy negation** if* $N(0) = 1$, $N(1) = 0$. *Moreover, a fuzzy negation* N *is called*

(i) **strict** *if it is strictly decreasing and continuous,*
(i) **strong** *if it is an involution, i.e.,* $N(N(x)) = x$ *for all* $x \in [0, 1]$.

Definition 2.6 ([1, Definition 1.4.15]). *Let* I *be a fuzzy implication. The function* N_I *defined by*
$$N_I(x) = I(x, 0), \qquad x \in [0, 1]$$
*is called the **natural negation** of* I.

One of the most important examples of fuzzy negations is the classical negation given by the formula $N_C(x) = 1 - x$, for $x \in [0, 1]$.

Definition 2.7. *We say that a fuzzy implication* I *satisfies*

(i) *the identity principle, if*

$$I(x, x) = 1, \qquad x \in [0, 1], \qquad \text{(IP)}$$

(ii) *the left neutrality property, if*

$$I(1, y) = y, \qquad y \in [0, 1], \qquad \text{(NP)}$$

(iii) the ordering property, if

$$x \le y \iff I(x,y) = 1, \qquad x,y \in [0,1]. \qquad (\text{OP})$$

Let us also recall definitions of several families of fuzzy implications.

Definition 2.8 ([1, Definition 2.5.1]). *A function* $I\colon [0,1]^2 \to [0,1]$ *is called an* **R-implication** *if there exists a t-norm* T *such that*

$$I(x,y) = \sup\{t \in [0,1] \mid T(x,t) \le y\}, \qquad x,y \in [0,1].$$

An R-implication generated by a t-norm T *will be denoted by* I_T.

Example 2.9. The most important examples of R-implications are the following:

– the Goguen implication (generated from $T_{\mathbf{P}}$)

$$I_{\mathbf{GG}}(x,y) = \begin{cases} 1, & x \le y, \\ \frac{y}{x}, & x > y, \end{cases} \qquad x,y \in [0,1],$$

– the Łukasiewicz implication (generated from $T_{\mathbf{LK}}$)

$$I_{\mathbf{LK}}(x,y) = \min(1 - x + y, 1), \qquad x,y \in [0,1],$$

– the Gödel implication (generated from $T_{\mathbf{M}}$)

$$I_{\mathbf{GD}}(x,y) = \begin{cases} 1, & x \le y, \\ y, & x > y, \end{cases} \qquad x,y \in [0,1].$$

One of the most important classes of fuzzy implications generated from unary functions are f- and g-generated Yager's implications.

Definition 2.10 ([10], **cf.** [1, Definition 3.1.1]). *Let* $f\colon [0,1] \to [0,\infty]$ *be a strictly decreasing and continuous function with* $f(1) = 0$. *The function* $I\colon [0,1]^2 \to [0,1]$ *defined by*

$$I(x,y) = f^{-1}(x \cdot f(y)), \qquad x,y \in [0,1],$$

with the understanding $0 \cdot \infty = 0$, *is called an* **f-generated implication**. *In such a case we will write* I_f *instead of* I.

Definition 2.11 ([10], **cf.** [1, Definition 3.2.1]). *Let* $g\colon [0,1] \to [0,\infty]$ *be a strictly decreasing and continuous function with* $g(0) = 0$. *A function* $I\colon [0,1]^2 \to [0,1]$ *defined by*

$$I(x,y) = g^{(-1)}\left(\frac{1}{x}g(y)\right), \qquad x,y \in [0,1],$$

with the understanding $0 \cdot \infty = \infty$ *and* $\frac{1}{0} = \infty$, *is called a* **g-generated implication**, *where the function* $g^{(-1)}$ *is the pseudo-inverse of* g *given by*

$$g^{(-1)}(x) = \begin{cases} g^{-1}(x), & x \in [0, g(1)], \\ 1, & x \in [g(1), \infty]. \end{cases}$$

Once again we will write I_g *instead of* I.

Let us denote different families of Yager's implications as follows:

- $\mathbb{I}_{F,\infty}$ - the family of all f-generated implication such as $f(0) = \infty$.
- $\mathbb{I}_{F,\aleph}$ - the family of all f-generated implication such as $f(0) < \infty$.
- $\mathbb{I}_F = \mathbb{I}_{F,\aleph} \cup \mathbb{I}_{F,\infty}$ - the family of all f-generated implication.
- $\mathbb{I}_{G,\infty}$ - the family of all g-generated implication such as $g(1) = \infty$.
- $\mathbb{I}_{G,\aleph}$ - the family of all g-generated implication such as $g(1) < \infty$.
- $\mathbb{I}_G = \mathbb{I}_{G,\aleph} \cup \mathbb{I}_{G,\infty}$ - the family of all g-generated implication.

Generators of f- and g-generated implications are unique up to positive multiplicative constant (see [1, Theorem 3.1.4, Theorem 3.2.5]). Thus, without loss of generality we will assume that $g(1) = 1$, when $g(1) < \infty$ and $f(0) = 1$, when $f(0) < \infty$.

3   Aggregation of Fuzzy Implications

As it is noted in [1,8] there exist many different methods to obtain new fuzzy implications. Some of them, like R-implications, S-implications or QL-implications are based on other fuzzy logic connectives. The other group of implication is generated from (additive) generators, like the above f-generated implications. However, there exist also methods to obtain implications from old ones: the N-reciprocation of a fuzzy implication, the medium contrapositivisation of a fuzzy implication, the convex combinations of fuzzy implications or the min and max operations:

$$(I \vee J)(x,y) := \max(I(x,y), J(x,y)), \qquad x,y \in [0,1], \qquad (3)$$
$$(I \wedge J)(x,y) := \min(I(x,y), J(x,y)), \qquad x,y \in [0,1]. \qquad (4)$$

Of course there are much more such different approaches. In this paper we will use general version of the last method. Let A be an aggregation function and I, J be fuzzy implications. Next, let us define a composition of I, J with the use of A by the following formula

$$A \circ (I, J)(x,y) := A(I(x,y), J(x,y)), \qquad x,y \in [0,1]. \qquad (5)$$

The composition $A \circ (I, J)$ will be denoted by $I_{I,J}^A$. The function $I_{I,J}^A : [0,1]^2 \to [0,1]$ defined as above is a fuzzy implication (see [3, Proposition 3.2]).

Example 3.1. First, let A be a geometric mean and I, J be fuzzy implications, then

$$I_{I,J}^A(x,y) = \sqrt{I(x,y) \cdot J(x,y)}, \qquad x,y \in [0,1]$$

is a fuzzy implication. Next, let $I = I_{\mathbf{GD}}, J = I_{\mathbf{LK}}$ and $A = T_{\mathbf{M}}$. Then $I_{I,J}^A = I_{\mathbf{GD}}$. However, if we take $A = T_{\mathbf{LK}}$, then we obtain

$$I_{I,J}^A(x,y) = \begin{cases} 1, & x \leq y, \\ \max\{0, 2y - x\}, & x > y, \end{cases} \qquad x,y \in [0,1].$$

Now, let us take $I = J = I_{\mathbf{GD}}$. Then in the case when A is a t-norm, $I_{\mathbf{GD}}$ is an idempotent fuzzy implication with respect to the composition (5) (i.e., $I_{I,J}^A = I_{\mathbf{GD}}$) if and only if $A = T_{\mathbf{M}}$ (see [5, Proposition 1.9]).

Note that this class of implications is connected with the (GHS) equation in a very natural way. However we need additional assumptions to describe some properties. Namely, it is not enough to take an aggregation function.

Theorem 3.2. *Let T be a continuous t-norm and let I, J be fuzzy implications such that pairs (T, I) and (T, J) satisfy (GHS). Then the pair $(T, I_{I,J}^T)$ satisfies (GHS).*

Proof. Let $x, y, z \in [0, 1]$. Then, from associativity and commutativity of T we have

$$T(T(I(x,z), J(x,z)), T(I(z,y), J(z,y)))$$
$$= T(T(T(I(x,z), J(x,z)), I(z,y)), J(z,y))$$
$$= T(T(I(x,z), T(J(x,z), I(z,y))), J(z,y))$$
$$= T(T(I(x,z), T(I(z,y), J(x,z))), J(z,y))$$
$$= T(T(T(I(x,z), I(z,y)), J(x,z)), J(z,y))$$
$$= T(T(I(x,z), I(z,y)), T(J(x,z), J(z,y))).$$

Using the above equation, the continuity of T and our assumptions for pairs (T, I) and (T, J), we obtain

$$\sup_{z \in [0,1]} T(I_{I,J}^T(x,z), I_{I,J}^T(z,y)) =$$

$$= \sup_{z \in [0,1]} T(T(I(x,z), I(z,y)), T(J(x,z), J(z,y)))$$

$$= T(\sup_{z \in [0,1]} T(I(x,z), I(z,y)), \sup_{z \in [0,1]} T(J(x,z), J(z,y)))$$

$$= T(I(x,y), J(x,y)).$$

$\square$

For some particular fuzzy implications the assumption of continuity of a t-norm is not necessary.

Proposition 3.3. *Let fuzzy implications I, J satisfy (OP). Then the pairs $(T_{\mathbf{D}}, I)$, $(T_{\mathbf{D}}, J)$ and $(T_{\mathbf{D}}, I_{I,J}^{T_{\mathbf{D}}})$ satisfy (GHS).*

Proof. From [2, Lemma 4.3] we know that if fuzzy implications I, J satisfy (OP), then the pairs $(T_{\mathbf{D}}, I)$ and $(T_{\mathbf{D}}, J)$ satisfy (GHS). Furthermore, $I_{I,J}^{T_{\mathbf{D}}}$ satisfy (OP) as well. Indeed,

$$1 = I_{I,J}^{T_{\mathbf{D}}}(x,y) \iff I(x,y) = 1 \wedge J(x,y) = 1 \iff x \leq y,$$

because $T_{\mathbf{D}}(x,y) = 1$ iff $x = 1$ and $y = 1$. Using again [2, Lemma 4.3] we have that the pair $(T_{\mathbf{D}}, I_{I,J}^{T_{\mathbf{D}}})$ satisfies (GHS). $\square$

Let us recall the following result which will be used in the next example.

Theorem 3.4 ([2, Theorem 4.12]). *Let T^* be a t-norm and T be a left-continuous t-norm. Then the following statements are equivalent:*

(i) The pair (T^*, I_T) satisfies (GHS).
(ii) $T^* \leq T$.

Example 3.5. From Theorem 3.4 we know that the pairs $(T_\mathbf{P}, I_\mathbf{GG})$, $(T_\mathbf{P}, I_\mathbf{GD})$ satisfy (GHS). Thus from Theorem 3.2 the pairs $(T_\mathbf{P}, I^{T_\mathbf{P}}_{I_\mathbf{GG}, I_\mathbf{GD}})$, $(T_\mathbf{P}, (I_\mathbf{GG})_n)$ also satisfy (GHS), where

$$I^{T_\mathbf{P}}_{I_\mathbf{GG}, I_\mathbf{GD}}(x, y) = \begin{cases} 1, & x \leq y, \\ \frac{y^2}{x}, & x > y, \end{cases}$$

$$(I_\mathbf{GG})_n(x, y) = \begin{cases} 1, & x \leq y, \\ \left(\frac{y}{x}\right)^n, & x > y, \end{cases}$$

for $n \geq 1$, $x, y \in [0, 1]$, $(I_\mathbf{GG})_n = I^{T^n_\mathbf{P}}_{I_\mathbf{GG}, I_\mathbf{GG}}$ and

$$I^{T^n_\mathbf{P}}_{I_\mathbf{GG}, I_\mathbf{GG}} = \begin{cases} I_\mathbf{GG}, & n = 0, \\ T_\mathbf{P}(I_\mathbf{GG}, I^{T^{n-1}_\mathbf{P}}_{I_\mathbf{GG}, I_\mathbf{GG}}), & n \geq 1. \end{cases}$$

4 Yager's Implications Satisfying (GHS)

In this section we present some results concerning Yager's fuzzy implications. It turns out there are not many fuzzy implications I from this family such that pairs (T, I) satisfy (GHS) for some t-norms T.

Lemma 4.1. *Let I be a fuzzy implication satisfying (NP) and $y, z \in [0, 1]$ be such that $z > y$ and $I(z, y) = 1$. Then, there does not exist any t-norm T such that (T, I) satisfies (GHS).*

Proof. Suppose that the pair (T, I) satisfies (GHS). Then,

$$y = I(1, y) = \sup_{t \in [0,1]} T(I(1, t), I(t, y)) \geq T(z, I(z, y)) = T(z, 1) = z,$$

which is a contradiction with the fact that $z > y$. $\qquad\square$

We will start with g-implications among which we have found a positive example regarding (GHS).

Theorem 4.2. *Let $I_g \in \mathbb{I}_{G, \aleph}$ and let $z \in (0, 1)$ be such that $g(z) > z$. Then there does not exist any t-norm T such that the pair (T, I_g) satisfies (GHS).*

Proof. Firstly note that every g-generated implication satisfies (NP) (see [1, Theorem 3.2.8]). Let $y \in [0, 1]$ be such that $g(y) > y$ and let $z = g(y)$. Then,

$$I_g(z, y) = g^{-1}\left(\min\left(1, \frac{g(y)}{z}\right)\right) = g^{-1}(1) = 1,$$

and from Lemma 4.1 we obtain the thesis. $\qquad\square$

Theorem 4.3. *Let* $I_g \in \mathbb{I}_{G,\aleph}$ *and let* $g(z) < z$ *for some* $z \in (0,1)$. *Then there does not exist any t-norm* T *such that* (T, I_g) *satisfies* (GHS).

Proof. Let us recall we can assume that $g(1) = 1$. Firstly, we show that this theorem is true for the minimum t-norm. From the definition of (GHS) and $\mathbb{I}_{G,\aleph}$, for every $x, y \in [0,1]$ we have

$$\sup_{z \in [0,1]} \min(I_g(x,z), I_g(z,y))$$

$$= \sup_{z \in [0,1]} \min\left(g^{-1}\left(\min\left(1, \frac{g(z)}{x}\right)\right), g^{-1}\left(\min\left(1, \frac{g(y)}{z}\right)\right)\right),$$

Let $z_0 \in (0,1)$ be the element such that $g(z_0) < z_0$, from continuity of g there exists $\varepsilon > 0$ such that

$$g(z) + 2\varepsilon < z, \qquad \text{for all } z \in [z_0 - \varepsilon, z_0 + \varepsilon]. \tag{6}$$

Indeed, let $h(z) = g(z) - z$. Thus, $h(z_0) < 0$ and from continuity of h there exist $\delta_1, \delta_2 > 0$ such that $h(z) < -\delta_2 < 0$ for all $z \in [z_0 - \delta_1, z_0 + \delta_1]$. Hence, it is enough to take $\varepsilon = \frac{\min(\delta_1, \delta_2)}{2}$. From Darboux's theorem there is $y \in [0,1]$ such that $g(y) = z_0 + \varepsilon$. Let $x = z_0$, then $\frac{g(y)}{x} > 1$ so $I_g(x, y) = 1$. Now, let us consider two cases.

- If $z \leq g(y)$, then $\frac{g(y)}{z} \geq 1$ and

$$\min(I_g(x,z), I_g(z,y)) = g^{-1}\left(\frac{g(z)}{x}\right) \leq g^{-1}\left(\frac{g(z_0 + \varepsilon)}{z_0}\right) \leq g^{-1}\left(\frac{z_0 - \varepsilon}{z_0}\right),$$

  but $g^{-1}\left(\frac{z_0 - \varepsilon}{z_0}\right) < 1$ from principle of g.
- Assume that $x < g(y) < z$ and let $f \colon [z_0 + \varepsilon, 1] \to [0,1]$ be defined by

$$f(z) = z \cdot g(z), \qquad z \in [z_0 + \varepsilon, 1].$$

Thus, f is strictly increasing and continuous. Moreover, $f(1) = 1$ and by the inequality (6) we obtain $f(z_0 + \varepsilon) < (z_0 + \varepsilon)(z_0 - \varepsilon)$. From Darboux's theorem there is exactly one $z_1 \in (z_0 + \varepsilon, 1]$ such that $f(z_1) = xg(y) = z_0(z_0 + \varepsilon)$. Therefore, $g(z_1) < z_0$. Indeed, suppose that $z_0 \leq g(z_1)$, then

$$z_1 \cdot z_0 \leq f(z_1) \cdot z_0(z_0 + \varepsilon),$$

but this is contradiction with fact that $z_0 + \varepsilon < z_1$. Hence, for $g(y) < z \leq z_1$ we obtain

$$\min(I_g(x,z), I_g(z,y)) = g^{-1}\left(\frac{g(z)}{x}\right) \leq g^{-1}\left(\frac{g(z_1)}{x}\right)$$

$$< g^{-1}\left(\frac{z_0 - 2\varepsilon}{z_0}\right) < 1.$$

Assume now that $z_1 \leq z$, then

$$\min(I_g(x,z), I_g(z,y)) = g^{-1}\left(\frac{g(y)}{z}\right)$$

$$\leq g^{-1}\left(\frac{z_0 + \varepsilon}{z_1}\right) < g^{-1}\left(\frac{z_0 - \varepsilon}{z_0 + \varepsilon}\right) < 1,$$

because $z_1 > z_0 + \varepsilon$. Since, $T \leq \min$ for every t-norm T, the above reasoning is true for every t-norm.

$\square$

Corollary 4.4. *The Goguen implication I_{GG} is the only fuzzy implication from the family of $\mathbb{I}_{G,\aleph}$ such that the pair (T, I_{GG}) satisfies (GHS) for some t-norm T. Moreover, the pair (T, I_{GG}) satisfies (GHS) if and only if $T \leq T_P$.*

Proof. Let T be a t-norm and g a generator of the Yager's implication I_g such that $g(1) = 1$. If $g(x) > x$ for some $x \in (0,1)$, then from Theorem 4.2 we know that the pair (T, I_g) does not satisfy (GHS). The same result we obtain for function g, when $g(x) < x$, for some $x \in (0,1)$ (Theorem 4.3). In the case when $g(x) = x$, for every $x \in (0,1)$, $I_g = I_{GG}$. Furthermore, I_{GG} is the R-implication generated from T_P. Hence, (T, I_{GG}) satisfies (GHS) if and only if $T \leq T_P$ (see Theorem 3.4). $\square$

In cases of families $\mathbb{I}_{G,\infty}$, $\mathbb{I}_{F,\infty}$, $\mathbb{I}_{F,\aleph}$ we have only partial results. First one is based on investigations of Vemuri [9].

Theorem 4.5. *Let $I_f \in \mathbb{I}_{F,\infty}$.*

(i) *If $T = T_M$, then (T, I_f) does not satisfy (GHS) (cf. [9]).*
(ii) *If T is a strict t-norm with a function f as an additive generator, i.e., $T(x,y) = f^{-1}(f(x) + f(y))$, for $x, y \in [0,1]$. Then the pair (T, I_f) does not satisfy (GHS).*

Proof. From [1, Theorem 7.2.22] we know that if either $T = T_M$, or T is the strict t-norm with f as the additive generator, then

$$I_f(x, T(y,z)) = T(I_f(x,y), I_f(x,z)), \qquad x, y \in [0,1].$$

Using the proof of [9, Theorem 3.15] we obtain (i) and (ii). $\square$

Theorem 4.6. *Let $I_g \in \mathbb{I}_{G,\infty}$.*

(i) *If $T = T_M$, then (T, I_g) does not satisfy (GHS).*
(ii) *Let T be the strict t-norm with a function $\frac{1}{g}$ as an additive generator, where $\frac{1}{0} = \infty$ and $\frac{1}{\infty} = 0$. Then, (T, I_g) does not satisfy (GHS).*

Proof. From [9, Theorem 3.18] we obtain (i). Let $f = \frac{1}{g}$, where $\frac{1}{0} = \infty$ and $\frac{1}{\infty} = 0$. Then, $I_g = I_f$ is an f-generated Yager's implication such that $f(0) = 1$. From Theorem 4.5 we obtain (ii). $\square$

138   P. Helbin et al.

Finally, let us recall a fact which will be used in the last example.

Lemma 4.7 ([9]). *Let I be a fuzzy implication and T be a t-norm. If the pair (T, I) satisfies (GHS), then $T(I(1, x), I(x, 0)) = 0$, for all $x \in [0, 1]$. If, in addition, I satisfies (NP), then*

$$T(x, N(x)) = 0, \qquad x \in [0, 1], \tag{7}$$

for any fuzzy negation $N \leq N_I$.

Now, let us consider a continuous t-norm T and a fuzzy implication $I \in \mathbb{I}_{\mathbb{F},\aleph}$. Every fuzzy implication from the family $\mathbb{I}_{\mathbb{F},\aleph}$ is continuous (see [7, Theorem 6]). Hence, if the pair (T, I) would satisfy (GHS), then T and N_I should have the following forms (see [1, Proposition 2.3.15]). For $x, y \in [0, 1]$

$$T(x, y) = \varphi^{-1} \max(\varphi(x) + \varphi(y) - 1, 0),$$
$$N_I(x) \leq \varphi^{-1}(1 - \varphi(x)),$$

where $\varphi \colon [0, 1] \to [0, 1]$ is an increasing bijection.

For the family of $\mathbb{I}_{\mathbb{F},\aleph}$ let us give just one example which shows necessary conditions from Lemma 4.7 for satisfying (GHS) are not sufficient in the case of this family.

Example 4.8. Consider the Reichenbach implication $I_{\mathbf{RC}}(x, y) = 1 - x + xy$, for $x, y \in [0, 1]$. In fact $I_{\mathbf{RC}} \in \mathbb{I}_{\mathbb{F},\aleph}$ and $N_{I_{\mathbf{RC}}} = N_{\mathbf{C}}$. The pair $(T_{\mathbf{LK}}, I_{\mathbf{RC}})$ does not satisfy (GHS) despite $T_{\mathbf{LK}}$ and $N_{\mathbf{C}}$ satisfy (7). Indeed, let $x, y \in (0, 1)$ be such that $x + y < 1$. Then,

$$\sup_{z \in [0,1]} T_{\mathbf{LK}}(I_{\mathbf{RC}}(x, z), I_{\mathbf{RC}}(z, y)) = \sup_{z \in [0,1]} \max(1 - x + xz + 1 - z + zy - 1, 0)$$
$$= \sup_{z \in [0,1]} \max(1 - x + z(x + y - 1), 0)$$
$$= \max(1 - x, 0) = 1 - x$$
$$< 1 - x + xy = I_{\mathbf{RC}}(x, y).$$

5   Conclusions

We have investigated the generalized hypothetical syllogism (GHS) and shown some of its properties. They were related with the class of aggregated fuzzy implications (cf. Eq. (5)) which were introduced in [3]. Further, we focused on Yager's implications. We have studied in details the relationships between the class $\mathbb{I}_{\mathbb{G},\aleph}$ and (GHS). However, results from Sect. 4 are not complete. In particular, remaining families ($\mathbb{I}_{\mathbb{G},\infty}$, $\mathbb{I}_{\mathbb{F},\infty}$, $\mathbb{I}_{\mathbb{F},\aleph}$) should be still investigated with reference to (GHS).

Acknowledgment. M. Baczyński and K. Miś acknowledge the support of the National Science Centre, Poland, under Grant No. 2015/19/B/ST6/03259.

References

1. Baczyński, M., Jayaram, B.: Fuzzy Implications, Studies in Fuzziness and Soft Computing, vol. 231. Springer, Heidelberg (2008)
2. Baczyński, M., Miś, K.: Selected properties of generalized hypothetical syllogism including the case of R-implications. In: Medina, J., Ojeda-Aciego, M., Verdegay, J.L., Pelta, D.A., Cabrera, I.P., Bouchon-Meunier, B., Yager, R.R. (eds.) Information Processing and Management of Uncertainty in Knowledge-Based Systems. Theory and Foundations. IPMU 2018. Communications in Computer and Information Science, vol. 853, pp. 673–684. Springer International Publishing, Cham (2018)
3. Calvo, T., Martín, J., Mayor, G.: Aggregation of implication functions. In: Pasi, G., Montero, J., Ciucci, D. (eds.) 8th conference of the European Society for Fuzzy Logic and Technology (EUSFLAT 2013), pp. 569–574. Atlantis Press (2013)
4. Grabisch, M., Marichal, J., Mesiar, R., Pap, E.: Aggregation Functions, Encyclopedia of Mathematics and Its Applications, vol. 127. Cambridge University Press, Cambridge (2009)
5. Klement, E., Mesiar, R., Pap, E.: Triangular Norms. Kluwer, Dordrecht (2000)
6. Klir, G., Yuan, B.: Fuzzy Sets and Fuzzy Logic: Theory and Applications. Prentice Hall, Upper Saddle River (1995)
7. Massanet, S., Torrens, J.: On the characterization of Yager's implications. Inf. Sci. **201**, 1–18 (2012)
8. Massanet, S., Torrens, J.: An overview of construction methods of fuzzy implications. In: Baczyński, M., Beliakov, G., Bustince Sola, H., Pradera, A. (eds.) Advances in Fuzzy Implication Functions, Studies in Fuzziness and Soft Computing, vol. 300, pp. 1–30. Springer, Heidelberg (2013)
9. Vemuri, N.R.: Investigations of fuzzy implications satisfying generalized hypothetical syllogism. Fuzzy Sets Syst. **323**, 117–137 (2018)
10. Yager, R.R.: On some new classes of implication operators and their role in approximate reasoning. Inf. Sci. **167**, 193–216 (2004)
11. Zadeh, L.: Outline of a new approach to the analysis of complex systems and decision processes. IEEE Trans. Syst. Man Cybern. **9**, 28–44 (1973)

Interval-Valued E_N-functions and Similarity Measures

Zdenko Takáč[1]([⊠]), Humberto Bustince[2], Javier Fernandez[2], Graçaliz Dimuro[3], Tiago Asmus[2], and Aitor Castillo[2]

[1] Slovak University of Technology, Bratislava, Slovakia
zdenko.takac@stuba.sk
[2] Universidad Publica de Navarra, Pamplona, Spain
{bustince,fcojavier.fernandez,tiago.dacruz,aitor.castillo}@unavarra.es
[3] Universidade Federal do Rio Grande, Rio Grande, Brazil
gracalizdimuro@furg.br

Abstract. In this work we introduce a definition of interval-valued similarity measures taking into account the width of the input intervals. We discuss a construction method based on the aggregation of interval-valued restricted equivalence functions.

Keywords: Similarity measure · E_N-function · Interval-valued fuzzy sets

1 Introduction

Although interval-valued fuzzy sets [10] are increasingly used in the fuzzy literature [1–6,12,13,17,20], in most of the cases, only the partial order between intervals is considered and furthermore, the widths of the intervals are not considered. This is in particular true for those applications which use similarity measures [7–9,18] for calculating the degree of resemblance between fuzzy sets [14,16,19,22].

In this work we propose a construction of similarity measures between interval-valued fuzzy sets using a total order and taking into account the widths of the involved intervals, as we consider that the latter is a measure of the uncertainty involved in the corresponding interval-valued fuzzy set.

The structure of the work is as follows. We start with some preliminaries. In Sect. 3, we introduce the concepts of interval-valued restricted equivalence functions and, in Subsect. 3.1, that of interval-valued aggregation functions preserving the widths of intervals. In Sect. 4, we discuss the definition of width-based interval-valued similarity measures and study different construction methods. We finish with some conclusions and references.

2 Preliminaries

Recall that an aggregation function is a non-decreasing function $M : [0,1]^n \to [0,1]$ with $M(0,\ldots,0) = 0$ and $M(1,\ldots,1) = 1$. An aggregation function $M :$

© Springer Nature Switzerland AG 2019
R. Halaš et al. (Eds.): AGOP 2019, AISC 981, pp. 140–150, 2019.
https://doi.org/10.1007/978-3-030-19494-9_13

$[0,1]^n \to [0,1]$ is called idempotent if $M(x,\ldots,x) = x$ for every $x \in [0,1]$, and it is called symmetric if $M(x_1,\ldots,x_n) = M(x_{\sigma(1)},\ldots,x_{\sigma(n)})$ for every $x_1,\ldots,x_n \in [0,1]$ and every permutation $\sigma : \{1\ldots,n\} \to \{1\ldots,n\}$.

$L([0,1])$ denotes the set of closed subintervals, $\{[\underline{X},\overline{X}] | 0 \le \underline{X} \le \overline{X} \le 1\}$. We use capital letters to denote elements in $L([0,1])$. The width of the interval $X = [\underline{X},\overline{X}] \in L([0,1])$ is $w(X) = \overline{X} - \underline{X}$. An interval function $f : (L([0,1]))^n \to L([0,1])$ is called width-preserving (or w-preserving, for simplicity) if, for any $X_1,\ldots,X_n \in L([0,1])$ such that $w(X_1) = \ldots = w(X_n)$, it holds that $w(f(X_1,\ldots,X_n)) = w(X_1)$.

We work on a finite universe $U = \{u_1,\ldots,u_n\}$. An interval-valued fuzzy set (IVFS) on the universe U is a mapping $A : U \to L([0,1])$. The class of all fuzzy sets in U is denoted by $FS(U)$ and the class of all interval-valued fuzzy sets in U by $IVFS(U)$.

Another key notion in this work is that of order relation.

Definition 1. *An order relation on $L([0,1])$ is a binary relation $\le$ on $L([0,1])$ such that, for all $X, Y, Z \in L([0,1])$,*

(i) $X \le X$, (reflexivity),
(ii) $X \le Y$ and $Y \le X$ imply $X = Y$, (antisymmetry),
(iii) $X \le Y$ and $Y \le Z$ imply $X \le Z$, (transitivity).

An order relation on $L([0,1])$ is called total or linear if any two elements of $L([0,1])$ are comparable, i.e., if for every $X, Y \in L([0,1])$, $X \le Y$ or $Y \le X$. An order relation on $L([0,1])$ which is not total is called partial.

$\precsim_L$ denotes the partial order relation on $L([0,1])$ induced by the usual partial order on $\mathbb{R}^2$:

$$[\underline{X},\overline{X}] \precsim_L [\underline{Y},\overline{Y}] \quad \text{if} \quad \underline{X} \le \underline{Y} \text{ and } \overline{X} \le \overline{Y}. \tag{1}$$

We denote by $\le_L$ any order on $L([0,1])$ (which can be partial or total) with minimal element $0_L = [0,0]$ (that is, $0_L \le_L X$ for all $X \in L([0,1])$) and maximal element $1_L = [1,1]$ (that is, $X \le_L 1_L$ for all $X \in L([0,1])$). To denote a total order on $L([0,1])$ with these minimal and maximal elements, we use the notation $\le_{TL}$.

Example 1. (i) A total order on $L([0,1])$ is, for example, the Xu and Yager's order (see [21]):

$$[\underline{X},\overline{X}] \le_{XY} [\underline{Y},\overline{Y}] \quad \text{if} \quad \begin{cases} \underline{X}+\overline{X} < \underline{Y}+\overline{Y} \text{ or} \\ \underline{X}+\overline{X} = \underline{Y}+\overline{Y} \text{ and } \overline{X}-\underline{X} \le \overline{Y}-\underline{Y}. \end{cases} \tag{2}$$

This definition of Xu and Yager's order was originally provided for Atanassov intuitionistic fuzzy pairs [21].

(ii) Another example of total order is provided by the lexicographical order with respect to the first variable, $\le_{lex1}$, and with respect to the second variable, $\le_{lex2}$. The first one is defined by :

$$[\underline{X},\overline{X}] \le_{lex1} [\underline{Y},\overline{Y}] \quad \text{if} \quad \begin{cases} \underline{X} < \underline{Y} \text{ or} \\ \underline{X} = \underline{Y} \text{ and } \overline{X} \le \overline{Y}, \end{cases}$$

whereas the second one follows just interchanging the role of the upper and the lower bounds.

Regarding total orders, a key notion is that of admissible order [11], that is, a total order $\leq_{TL}$ on $L([0,1])$ such that, if $X \precsim_L Y$ then $X \leq_{TL} Y$.

Admissible orders can be built using aggregation functions as follows.

Proposition 1. *([11]) Let $M_1, M_2 : [0,1]^2 \rightarrow [0,1]$ be two aggregation functions such that for all $X, Y \in L([0,1])$, the equalities $M_1(\underline{X}, \overline{X}) = M_1(\underline{Y}, \overline{Y})$ and $M_2(\underline{X}, \overline{X}) = M_2(\underline{Y}, \overline{Y})$ can only hold simultaneously if $X = Y$. The order $\leq_{M_1, M_2}$ on $L([0,1])$ given by*

$$X \leq_{M_1, M_2} Y \quad \text{if} \quad \begin{cases} M_1(\underline{X}, \overline{X}) < M_1(\underline{Y}, \overline{Y}) \ \text{or} \\ M_1(\underline{X}, \overline{X}) = M_1(\underline{Y}, \overline{Y}) \ \text{and} \ M_2(\underline{X}, \overline{X}) \leq M_2(\underline{Y}, \overline{Y}) \end{cases}$$

is an admissible order on $L([0,1])$.

Example 2. Take $\alpha \in [0,1]$. If we define the aggregation function

$$K_\alpha(x,y) = (1-\alpha)x + \alpha y$$

then, for $\alpha, \beta \in [0,1]$ with $\alpha \neq \beta$, we obtain an admissible order $\leq_{\alpha,\beta}$ taking $M_1(x,y) = K_\alpha(x,y)$ and $M_2(x,y) = K_\beta(x,y)$. See [11] for more details.

Another important notion in order to build similarity measures is that of restricted equivalence function [9].

Definition 2. *A function $R : [0,1]^2 \rightarrow [0,1]$ is called a restricted equivalence function (REF) if it satisfies:*

1. *$R(x,y) = 0$ if and only if $\{x,y\} = \{0,1\}$;*
2. *$R(x,y) = 1$ if and only if $x = y$;*
3. *$R(x,y) = R(y,x)$ for all $x, y \in [0,1]$;*
4. *If $x \leq y \leq z$, then $R(x,z) \leq R(x,y)$ and $R(x,z) \leq R(y,z)$ for all $x, y, z \in [0,1]$.*

Example 3. For any $p \in]0, \infty[$, the function $R^p(x,y) = 1 - |x-y|^p$ is a REF.

3   Width-Preserving Interval Valued Restricted Equivalence Functions

We start extending the notion of restricted equivalence function to the interval-valued setting in such a way that the width of the input intervals is preserved.

Definition 3. *Let $\leq_L$ be an order on $L([0,1])$. An interval-valued restricted equivalence function w.r.t. the order $\leq_L$ is a function $R_{IV} : L([0,1])^2 \rightarrow L([0,1])$ such that:*

1. $R_{IV}(X,Y) = 0_L$ if and only if $\{X,Y\} = \{0_L, 1_L\}$;
2. $R_{IV}(X,X) = [1 - w(X), 1]$ for all $X \in L([0,1])$;
3. $R_{IV}(X,Y) = R_{IV}(Y,X)$ for all $X,Y \in L([0,1])$;
4. If $X,Y,Z \in L([0,1])$ are such that $X \leq_L Y \leq_L Z$ and $w(X) = w(Y) = w(Z)$, then $R_{IV}(X,Z) \leq_L R_{IV}(X,Y)$ and $R_{IV}(X,Z) \leq_L R_{IV}(Y,Z)$.

Note that in this definition, as we consider that the width of the membership interval of an element in a given set is a measure of the lack of knowledge of the precise (real-valued) membership degree of that element, and it is assumed that the exact membership value is a value inside the membership interval, we take into account the fact that, if two elements have the same interval memberships, it may happen that their corresponding real-valued membership values are different.

We discuss now a construction method of IV REFs.

Lemma 1. *Let $X,Y \in L([0,1])$ be intervals such that $w(X) = w(Y)$. Then*

$$X \precsim_L Y \text{ if and only if } X \leq_{TL} Y$$

for any admissible order $\leq_{TL}$.

Proof. Note that two intervals with the same width are always comparable by the partial order $\precsim_L$.

Theorem 1. *Let $\alpha \in]0,1[$, let $M : [0,1]^2 \to [0,1]$ be an idempotent symmetric aggregation function and let $R : [0,1]^2 \to [0,1]$ be a restricted equivalence function. Then, the function $R_{IV} : L([0,1])^2 \to L([0,1])$ given by*

$$R_{IV}(X,Y) = \big[\max\big(0, R\left(K_\alpha(X), K_\alpha(Y)\right) - M(w(X), w(Y))\big),$$
$$\max\big(R\left(K_\alpha(X), K_\alpha(Y)\right), M(w(X), w(Y))\big)\big] \qquad (3)$$

is an IV restricted equivalence function w.r.t. any admissible order $\leq_{TL}$. Moreover, R_{IV} is w-preserving.

Proof. For simplicity we write $\mathcal{R}$ instead of $R\left(K_\alpha(X), K_\alpha(Y)\right)$, and $\mathcal{M}$ instead of $M(w(X), w(Y))$. Then (3) can be simplified:

$$R_{IV}(X,Y) = \big[\max\left(0, \mathcal{R} - \mathcal{M}\right), \max\left(\mathcal{R}, \mathcal{M}\right)\big] = \begin{cases} [\mathcal{R} - \mathcal{M}, \mathcal{R}], & \text{if } \mathcal{R} \geq \mathcal{M}, \\ [0, \mathcal{M}], & \text{otherwise.} \end{cases}$$
$$(4)$$

By (4) it is clear that R_{IV} is well-defined.

Observe that $R_{IV}(X,Y) = 0_L$ if and only if $\mathcal{R} = 0$ and $\mathcal{M} = 0$. The former holds if and only if $\{K_\alpha(X), K_\alpha(Y)\} = \{0,1\}$, which may happen if and only if $\{X,Y\} = \{0_L, 1_L\}$. So it follows that $w(X) = w(Y) = 0$ and we get the first condition in Definition 3.

The second and the third conditions in Definition 3 are straightforward.

The fourth condition in Definition 3 w.r.t. any admissible order follows from the monotonicity of R and Lemma 1, after observing that, if $X \leq_{TL} Y \leq_{TL} Z$ and $w(X) = w(Y) = w(Z)$, then $K_\alpha(X) \leq K_\alpha(Y) \leq K_\alpha(Z)$.

Finally, the fact that R_{IV} is w-preserving directly follows from Eq. (4) and idempotency of M.

Using the function K_α we can provide the following result.

Corollary 1. *Let $\alpha \in]0,1[$, let $M : [0,1]^2 \to [0,1]$ be an idempotent symmetric aggregation function such that $M(x,y) \leq \min\big((1-\alpha)x + \alpha y, \alpha x + (1-\alpha)y\big)$ for all $x, y \in [0,1]$, and let $R : [0,1]^2 \to [0,1]$ be a restricted equivalence function such that $R(x,y) \geq 1 - |x - y|$ for all $x, y \in [0,1]$. Then, the function $R_{IV} : L([0,1])^2 \to L([0,1])$ given by*

$$R_{IV}(X,Y) = \big[R\left(K_\alpha(X), K_\alpha(Y)\right) - M(w(X), w(Y)), R\left(K_\alpha(X), K_\alpha(Y)\right)\big] \quad (5)$$

is an IV restricted equivalence function w.r.t. any admissible order $\leq_{TL}$. Moreover, R_{IV} is w-preserving.

Proof. We only need to prove that $R\left(K_\alpha(X), K_\alpha(Y)\right) \geq M(w(X), w(Y))$ for all $X, Y \in L([0,1])$, since in that case Eq. (5) is a special case of Eq. (3). Due to the assumptions on M and R, it is enough to show that

$$1 - |K_\alpha(X) - K_\alpha(Y)| \geq \min\big((1-\alpha)w(X) + \alpha w(Y), \alpha w(X) + (1-\alpha)w(Y)\big). \quad (6)$$

Assume that $K_\alpha(X) \geq K_\alpha(Y)$. Then

$$1 - |K_\alpha(X) - K_\alpha(Y)| = 1 - (1-\alpha)\underline{X} - \alpha\overline{X} + (1-\alpha)\underline{Y} + \alpha\overline{Y}$$

and since

$$1 \geq \overline{X} - \underline{Y} = (1-\alpha)(\overline{X} - \underline{Y}) + \alpha(\overline{X} - \underline{Y}) = (1-\alpha)(\overline{X} - \underline{X} + \underline{X} - \underline{Y}) + \alpha(\overline{X} - \overline{Y} + \overline{Y} - \underline{Y}),$$

we have

$$1 - (1-\alpha)\underline{X} - \alpha\overline{X} + (1-\alpha)\underline{Y} + \alpha\overline{Y} \geq (1-\alpha)(\overline{X} - \underline{X}) + \alpha(\overline{Y} - \underline{Y}),$$

hence (6) is satisfied.

Now assume that $K_\alpha(X) < K_\alpha(Y)$. Then

$$1 - |K_\alpha(X) - K_\alpha(Y)| = 1 + (1-\alpha)\underline{X} + \alpha\overline{X} - (1-\alpha)\underline{Y} - \alpha\overline{Y}$$

and since

$$1 \geq \overline{Y} - \underline{X} = (1-\alpha)(\overline{Y} - \underline{X}) + \alpha(\overline{Y} - \underline{X}) = (1-\alpha)(\overline{Y} - \underline{Y} + \underline{Y} - \underline{X}) + \alpha(\overline{Y} - \overline{X} + \overline{X} - \underline{X}),$$

we have

$$1 + (1-\alpha)\underline{X} + \alpha\overline{X} - (1-\alpha)\underline{Y} - \alpha\overline{Y} \geq (1-\alpha)(\overline{Y} - \underline{Y}) + \alpha(\overline{X} - \underline{X}),$$

hence (6) is satisfied.

3.1   Width-Preserving IV Aggregation Functions

Since we intend to build IV similarity measures aggregating IV REFs, we also analyze those IV aggregation functions which preserve the width of the intervals.

Definition 4. *Let $n \geq 2$. An (n-dimensional) interval-valued (IV) aggregation function in $L([0,1])$ with respect to $\leq_L$ is a mapping $M_{IV} : (L([0,1]))^n \to L([0,1])$ which verifies:*

(i) $M_{IV}(0_L, \cdots, 0_L) = 0_L$.
(ii) $M_{IV}(1_L, \cdots, 1_L) = 1_L$.
(iii) M_{IV} *is a non-decreasing function with respect to* $\leq_L$.

We say that $M_{IV} : (L([0,1]))^n \to L([0,1])$ is a decomposable n-dimensional IV aggregation function associated with M_L and M_U, if there exist n-dimensional aggregation functions $M_L, M_U : [0,1]^n \to [0,1]$ such that $M_L \leq M_U$ pointwise and

$$M_{IV}(X_1, \ldots, X_n) = \left[M_L\left(\underline{X_1}, \ldots, \underline{X_n}\right), M_U\left(\overline{X_1}, \ldots, \overline{X_n}\right) \right] \tag{7}$$

for all $X_1, \ldots, X_n \in L([0,1])$.

Definition 5 ([1]). *Let $c \in [0,1]$ and $\alpha \in [0,1]$. We denote by $d_\alpha(c)$ the maximal possible width of an interval $Z \in L([0,1])$ such that $K_\alpha(Z) = c$. Moreover, for any $X \in L([0,1])$, let*

$$\lambda_\alpha(X) = \frac{w(X)}{d_\alpha(K_\alpha(X))}$$

where we set $\frac{0}{0} = 1$.

Proposition 2 ([1]). *For all $\alpha \in [0,1]$ and $X \in L([0,1])$ it holds that*

$$d_\alpha(K_\alpha(X)) = \min\left(\frac{K_\alpha(X)}{\alpha}, \frac{1 - K_\alpha(X)}{1 - \alpha} \right).$$

where we set $\frac{r}{0} = 1$ for all $r \in [0,1]$.

Now we propose a construction method of IV aggregation functions w.r.t. $<_{\alpha,\beta}$ which preserve the width of the input intervals. First of all, for an aggregation function $M : [0,1]^n \to [0,1]$, we consider the following two properties:

(P1) $M(cx_1, \ldots, cx_n) \geq cM(x_1, \ldots, x_n)$ for all $c \in [0,1]$, $x_1, \ldots, x_n \in [0,1]$.
(P2) $M(x_1, \ldots, x_n) \leq 1 - M(1 - x_1, \ldots, 1 - x_n)$ for all $x_1, \ldots, x_n \in [0,1]$.

Theorem 2. *Let $\alpha, \beta \in [0,1]$, $\beta \neq \alpha$. Let $M_1, M_2 : [0,1]^n \to [0,1]$ be aggregation functions such that M_1 is strictly increasing, $M_1(x_1, \ldots, x_n) \geq M_2(x_1, \ldots, x_n)$ for all $x_1, \ldots, x_n \in [0,1]$, M_1 or M_2 satisfies property (P1) and M_1 or M_2 satisfies property (P2). Then $M_{IV} : (L([0,1]))^n \to L([0,1])$ defined by:*

$$M_{IV}(X_1, \ldots, X_n) = Y, \quad where \quad \begin{cases} K_\alpha(Y) = M_1\left(K_\alpha(X_1), \ldots, K_\alpha(X_n)\right), \\ w(Y) = M_2\left(w(X_1), \ldots, w(X_n)\right), \end{cases}$$

for all $X_1, \ldots, X_n \in L([0,1])$, is an IV aggregation function with respect to $\leq_{\alpha,\beta}$. Moreover, if M_2 is idempotent, then M_{IV} is w-preserving.

Proof. We first show that M_{IV} is well defined. Observe that

$$Y = [\underline{Y}, \overline{Y}] = [K_\alpha(Y) - \alpha w(Y), K_\alpha(Y) + (1-\alpha)w(Y)].$$

Clearly, $\underline{Y} \leq \overline{Y}$, hence we only need to prove that

1. $\underline{Y} \geq 0$: For $\alpha = 0$ we have $\underline{Y} = M_1(\underline{X_1}, \dots, \underline{X_n}) \geq 0$ and for $\alpha \in]0, 1]$ we have

$$K_\alpha(Y) = M_1\left(K_\alpha(X_1), \dots, K_\alpha(X_n)\right) \geq \alpha M_2\left(\frac{K_\alpha(X_1)}{\alpha}, \dots, \frac{K_\alpha(X_n)}{\alpha}\right)$$

$$\geq \alpha M_2\left(w(X_1), \dots, w(X_n)\right) = \alpha w(Y)$$

where the first inequality follows from the fact that M_2 satisfies property (P1) and the second from the observation $K_\alpha(X) = (1-\alpha)\underline{X} + \alpha\overline{X} \geq \alpha(\overline{X} - \underline{X}) = \alpha w(X)$ for all $X \in L([0, 1])$.

2. $\overline{Y} \leq 1$: For $\alpha = 1$ we have $\overline{Y} = M_1(\overline{X_1}, \dots, \overline{X_n}) \leq 1$ and for $\alpha \in [0, 1[$ we have

$$K_\alpha(Y) + (1-\alpha)w(Y) = M_1\left(K_\alpha(X_1), \dots, K_\alpha(X_n)\right) + (1-\alpha)M_2\left(w(X_1), \dots, w(X_n)\right) \leq$$

$$\leq M_1\left(K_\alpha(X_1), \dots, K_\alpha(X_n)\right) + (1-\alpha)M_2\left(\frac{1 - K_\alpha(X_1)}{1-\alpha}, \dots, \frac{1 - K_\alpha(X_n)}{1-\alpha}\right) \leq$$

$$\leq M_1\left(K_\alpha(X_1), \dots, K_\alpha(X_n)\right) + M_2\left(1 - K_\alpha(X_1), \dots, 1 - K_\alpha(X_n)\right) \leq$$

$$\leq M_1\left(K_\alpha(X_1), \dots, K_\alpha(X_n)\right) + 1 - M_2\left(K_\alpha(X_1), \dots, K_\alpha(X_n)\right) = 1$$

where the first inequality follows from the observation $1 - K_\alpha(X) = 1 - (1 - \alpha)\underline{X} - \alpha\overline{X} \geq (1-\alpha)(\overline{X} - \underline{X}) = (1-\alpha)w(X)$ for all $X \in L([0, 1])$, and the second and third ones from the assumptions of the theorem.

Now we prove that M_{IV} is an IV aggregation function. (i) $M_{IV}(0_L, \cdots, 0_L) = Y$ where $K_\alpha(Y) = M_1(0, \dots, 0) = 0$ and $w(Y) = M_2(0, \dots, 0) = 0$, hence $Y = 0_L$. (ii) $M_{IV}(1_L, \cdots, 1_L) = Y$ where $K_\alpha(Y) = M_1(1, \dots, 1) = 1$ and $w(Y) = M_2(0, \dots, 0) = 0$, hence $Y = 1_L$. (iii) Let $X_i \leq_{\alpha, \beta} Y_i$ for all $i = 1, \dots, n$. Then $K_\alpha(X_i) \leq K_\alpha(Y_i)$ for all $i = 1, \dots, n$ and there are two cases:

1. There exists $j \in \{1, \dots, n\}$ such that $K_\alpha(X_j) < K_\alpha(Y_j)$. Then

$$M_1(K_\alpha(X_1), \dots, K_\alpha(X_n)) < M_1(K_\alpha(Y_1), \dots, K_\alpha(Y_n)),$$

since M_1 is strictly increasing, thus $M_{IV}(X_1, \dots, X_n) <_{\alpha, \beta} M_{IV}(Y_1, \dots, Y_n)$.

2. $K_\alpha(X_i) = K_\alpha(Y_i)$ for all $i = 1, \dots, n$. If $\beta > \alpha$, then $w(X_i) \leq w(Y_i)$ for all $i = 1, \dots, n$, hence $M_2(w(X_1), \dots, w(X_n)) \leq M_2(w(Y_1), \dots, w(Y_n))$. So $M_{IV}(X_1, \dots, X_n) \leq_{\alpha, \beta} M_{IV}(Y_1, \dots, Y_n)$.
   If $\beta < \alpha$, then $w(X_i) \geq w(Y_i)$ for all $i = 1, \dots, n$, thus $M_2(w(X_1), \dots, w(X_n)) \geq M_2(w(Y_1), \dots, w(Y_n))$, consequently $M_{IV}(X_1, \dots, X_n) \leq_{\alpha, \beta} M_{IV}(Y_1, \dots, Y_n)$.
   Finally, it is easy to check that M_{IV} is w-preserving from the idempotency of M_2.

Lemma 2. *Let $(v_1, \ldots, v_n) \in]0, 1]^n$ be a weighting vector with $v_1 + \ldots + v_n = 1$. Under the assumptions of Theorem 2, if $M_1(x_1, \ldots, x_n) = M_2(x_1, \ldots, x_n) = v_1 x_1 + \ldots + v_n x_n$ for all $x_1, \ldots, x_n \in [0, 1]$, then M_{IV} is the decomposable IV aggregation function associated with M_L and M_U where $M_L = M_U = M_1$.*

Proof. Let $X_1, \ldots, X_n \in L([0, 1])$ and $M_{IV}(X_1, \ldots, X_n) = Y$. From Theorem 2:

$$w(Y) = \sum_{i=1}^{n} v_i w(X_i) \quad \text{and} \quad K_\alpha(Y) = \sum_{i=1}^{n} v_i K_\alpha(X_i).$$

With $M_L = M_U = M_1$:

$$M_U \left(\overline{X_1}, \ldots, \overline{X_n}\right) - M_L \left(\underline{X_1}, \ldots, \underline{X_n}\right) = \sum_{i=1}^{n} v_i \overline{X_i} - \sum_{i=1}^{n} v_i \underline{X_i} = \sum_{i=1}^{n} v_i w(X_i) = w(Y)$$

and

$$(1-\alpha)M_L \left(\underline{X_1}, \ldots, \underline{X_n}\right) + \alpha M_U \left(\overline{X_1}, \ldots, \overline{X_n}\right) = (1-\alpha)\sum_{i=1}^{n} v_i \underline{X_i} + \alpha \sum_{i=1}^{n} v_i \overline{X_i} = \sum_{i=1}^{n} v_i K_\alpha(X_i)$$

which is equal to $K_\alpha(Y)$. So, from Eq. (7), M_{IV} is decomposable and associated with M_L and M_U.

Lemma 3. *Let $M_{IV} : (L([0, 1]))^n \to L([0, 1])$ be defined as in Theorem 2.*

(i) If
 - *$M_1(x_1, \ldots, x_n) = 0$ if and only if $x_1 = \ldots = x_n = 0$ and*
 - *$M_2(x_1, \ldots, x_n) = 0$ if and only if $x_1 = \ldots = x_n = 0$,*
 then $M_{IV}(X_1, \ldots, X_n) = 0_L$ if and only if $X_1 = \ldots = X_n = 0_L$. Moreover, if $\alpha \neq 0$, then the restriction on M_2 can be skipped.
(ii) If
 - *$M_1(x_1, \ldots, x_n) = 1$ if and only if $x_1 = \ldots = x_n = 1$ and*
 - *$M_2(x_1, \ldots, x_n) = 0$ if and only if $x_1 = \ldots = x_n = 0$,*
 then $M_{IV}(X_1, \ldots, X_n) = 1_L$ if and only if $X_1 = \ldots = X_n = 1_L$. Moreover, if $\alpha \neq 1$, then the restriction on M_2 can be skipped.
(iii) M_{IV} is idempotent if and only if M_1 and M_2 are idempotent.

Proof. The proof is straightforward.

Example 4. A function $M_{IV} : (L([0, 1]))^n \to L([0, 1])$ defined as in Theorem 2, is a w-preserving IV aggregation function with respect to $\leq_{\alpha,\beta}$, if, for instance M_1 and M_2 are the arithmetic mean.

4 Width-Based Interval Valued Similarity Measures

Finally, we study IV similarity measures that preserve the width of the input intervals. We start proposing a definition.

Definition 6. *Let $\leq_L$ be an order on $L([0,1])$ and $M : [0,1]^n \to [0,1]$ be an aggregation function. A width-based interval-valued similarity measure on $IVFS(U)$ w.r.t. $\leq_L$ associated with M is a mapping $S_M : IVFS(U) \times IVFS(U) \to L([0,1])$ such that, for all $A, B, A', B' \in IVFS(U)$,*

(SM1) $S_M(A,B) = S_M(B,A)$;
(SM2) $S_M(A,A) = \Big[1 - M\big(w(A(u_1)), \dots, w(A(u_n))\big), 1\Big]$;
(SM3) $S_M(A,B) = 0_L$ if and only if $\{A(u_i), B(u_i)\} = \{0_L, 1_L\}$ for all $i \in \{1, \dots, n\}$;
(SM4) If $A \subseteq A' \subseteq B' \subseteq B$ w.r.t. $\leq_L$ and $w(A(u_i)) = w(A'(u_i)) = w(B'(u_i)) = w(B(u_i))$ for all $i \in \{1, \dots, n\}$, then $S_M(A,B) \leq_L S_M(A',B')$, where, for $A, B \in IVFS(U)$, $A \subseteq B$ w.r.t. $\leq_L$ if $A(u_i) \leq_L B(u_i)$ for every $u_i \in U$.

Recall that an aggregation function $M : [0,1]^n \to [0,1]$ is called self-dual with respect to the standard negation if $M(x_1, \dots, x_n) = 1 - M(1 - x_1, \dots, 1 - x_n)$ for all $x_1, \dots, x_n \in [0,1]$. With this notion at hand, we propose the following construction method.

Theorem 3. *Let $M_{IV} : (L([0,1]))^n \to L([0,1])$ be a decomposable IV aggregation function w.r.t. $\leq_L$ associated with M_L and M_U where M_L is self-dual, and let $M_{IV}(X_1, \dots, X_n) = 0_L$ if and only if $X_1 = \dots = X_n = 0_L$. Let $R_{IV} : L([0,1])^2 \to L([0,1])$ be an IV restricted equivalence function w.r.t. $\leq_L$. Then the function $S_{M_L} : IVFS(U) \times IVFS(U) \to L([0,1])$ defined by:*

$$S_{M_L}(A,B) = M_{IV}\big(R_{IV}(A(u_1), B(u_1)), \dots, R_{IV}(A(u_n), B(u_n))\big)$$

for all $A, B \in IVFS(U)$ is a width-based IV similarity measure on $IVFS(U)$ w.r.t. $\leq_L$ associated with M_L.

Proof. It is straightforward.

Corollary 2. *Let $\alpha, \beta \in\ ]0,1[$ where $\beta \neq \alpha$. Let $(v_1, \dots, v_n) \in\ ]0,1]^n$ be a weighting vector such that $v_1 + \dots + v_n = 1$ and let $M_{IV} : (L([0,1]))^n \to L([0,1])$ be the IV aggregation function w.r.t. $\leq_{\alpha,\beta}$ defined as in Theorem 2 where $M_1(x_1, \dots, x_n) = M_2(x_1, \dots, x_n) = v_1 x_1 + \dots + v_n x_n$ for all $x_1, \dots, x_n \in [0,1]$. Let $R_{IV} : (L([0,1]))^2 \to L([0,1])$ be an IV REF defined as in Theorem 1. Then the function $S_M : IVFS(U) \times IVFS(U) \to L([0,1])$ defined by:*

$$S_M(A,B) = M_{IV}\big(R_{IV}(A(u_1), B(u_1)), \dots, R_{IV}(A(u_n), B(u_n))\big),$$

for all $A, B \in IVFS(U)$, is a width-based IV similarity measure on $IVFS(U)$ w.r.t. $\leq_L$ associated with M_1. Moreover, S_M satisfies the following for all $A, B \in IVFS(U)$:

$$w(S_M(A,B)) = w(A(u_1)) \tag{8}$$

whenever $w(A(u_1)) = w(B(u_1)) = \dots = w(A(u_n)) = w(B(u_n))$.

Proof. From Lemma 2, M_{IV} is the decomposable IV aggregation function associated with M_L, M_U where $M_L = M_U = M_1$. From Lemma 3, we have that $M_{IV}(X_1, \ldots, X_n) = 0_L$ if and only if $X_1 = \ldots = X_n = 0_L$. Since a weighted arithmetic mean is self-dual and idempotent, from Theorem 3 it follows that S_M is a width-based IV similarity measure associated with M_1. Finally, as M_{IV} and R_{IV} are w-preserving, the result follows.

5   Conclusions

In this work we have proposed a definition of interval-valued similarity measures which preserve the width of the input intervals and we have proposed a construction method in terms of interval-valued aggregation functions and restricted equivalence functions which also preserve the width of the intervals.

In future works we intend to analyze the application of these notions in problems such as stereo vision [15] or classification.

Acknowledgement. This work was partially supported by project TIN2016-77356-P (MINECO, UE/AEI, FEDER) of the Spanish Government and by Project VEGA 1/0614/18.

References

1. Asiain, M.J., Bustince, H., Mesiar, R., Kolesárová, A., Takáč, Z.: Negations with respect to admissible orders in the interval-valued fuzzy set theory. IEEE Trans. Fuzzy Syst. **26**, 556–568 (2018)
2. Barrenechea, E., Bustince, H., De Baets, B., Lopez-Molina, C.: Construction of interval-valued fuzzy relations with application to the generation of fuzzy edge images. IEEE Trans. Fuzzy Syst. **19**(5), 819–830 (2011)
3. Barrenechea, E., Fernandez, J., Pagola, M., Chiclana, F., Bustince, H.: Construction of interval-valued fuzzy preference relations from ignorance functions and fuzzy preference relations. Appl. Decis. Making Knowl.-Based Syst. **58**, 33–44 (2014)
4. Bentkowska, U., Bustince, H., Jurio, A., Pagola, M., Pekala, B.: Decision making with an interval-valued fuzzy preference relation and admissible orders. Appl. Soft Comput. **35**, 792–801 (2015)
5. Burillo, P., Bustince, H.: Construction theorems for intuitionistic fuzzy sets. Fuzzy Sets Syst. **84**, 271–281 (1996)
6. Bustince, H.: Indicator of inclusion grade for interval-valued fuzzy sets. Application to approximate reasoning based on interval-valued fuzzy sets. Int. J. Approx. Reason. **23**(3), 137–209 (2000)
7. Bustince, H., Barrenechea, E., Pagola, M.: Relationship between restricted dissimilarity functions, restricted equivalence functions and normal E_N-functions: image thresholding invariant. Pattern Recogn. Lett. **29**(4), 525–536 (2008)
8. Bustince, H., Barrenechea, E., Pagola, M.: Image thresholding using restricted equivalence functions and maximizing the measure of similarity. Fuzzy Sets Syst. **128**(5), 496–516 (2007)
9. Bustince, H., Barrenechea, E., Pagola, M.: Restricted equivalence functions. Fuzzy Sets Syst. **157**(17), 2333–2346 (2006)

10. Bustince, H., Barrenechea, E., Pagola, M., Fernández, J., Xu, Z., Bedregal, B., Montero, J., Hagras, H., Herrera, F., De Baets, B.: A historical account of types of fuzzy sets and their relationship. IEEE Trans. Fuzzy Syst. **24**(1), 179–194 (2016)
11. Bustince, H., Fernandez, J., Kolesárová, A., Mesiar, R.: Generation of linear orders for intervals by means of aggregation functions. Fuzzy Sets Syst. **220**, 69–77 (2013)
12. Choi, H.M., Mun, G.S., Ahn, J.Y.: A medical diagnosis based on interval-valued fuzzy sets. Biomed. Eng. Appl. Basis Commun. **24**(4), 349–354 (2012)
13. Couto, P., Jurio, A., Varejao, A., Pagola, M., Bustince, H., Melo-Pinto, P.: An IVFS-based image segmentation methodology for rat gait analysis. Soft Comput. **15**(10), 1937–1944 (2011)
14. Deng, G., Song, L., Jiang, Y., Fu, J.: Monotonic similarity measures of interval-valued fuzzy sets and their applications. Int. J. Uncertainty Fuzziness Knowl. Based Syst. **25**(4), 515–544 (2017)
15. Galar, M., Fernandez, J., Beliakov, G., Bustince, H.: Interval-valued fuzzy sets applied to stereo matching of color images. IEEE Trans. Image Process. **20**(7), 1949–61 (2011)
16. Heidarzade, A.: A new similarity measure for interval type-2 fuzzy sets: application in fuzzy risk analysis. Int. J. Appl. Decis. Sci. **9**(4), 400–412 (2016)
17. Jurio, A., Pagola, M., Mesiar, R., Beliakov, G., Bustince, H.: Image magnification using interval information. IEEE Trans. Image Process. **20**(11), 3112–3123 (2011)
18. Liu, X.: Entropy, distance measure and similarity measure of fuzzy sets and their relations. Fuzzy Sets Syst. **52**, 305–318 (1992)
19. Lu, Z., Ye, J.: Logarithmic similarity measure between interval-valued fuzzy sets and its fault diagnosis method. Information (Switzerland) **9**, 36 (2018)
20. Sanz, J.A., Fernandez, A., Bustince, H., Herrera, F.: IVTURS: a linguistic fuzzy rule-based classification system based on a new interval-valued fuzzy reasoning method with tuning and rule selection. IEEE Trans. Fuzzy Syst. **21**(3), 399–411 (2013)
21. Xu, Z.S., Yager, R.R.: Some geometric aggregation operators based on intuitionistic fuzzy sets. Int. J. Gen. Syst. **35**, 417–433 (2006)
22. Ye, J.: Multicriteria decision-making method based on cosine similarity measures between interval-valued fuzzy sets with risk preference. Econ. Comput. Econ. Cybern. Stud. Res. **50**(4), 205–215 (2016)

On Some Generalizations
of the Choquet Integral

Humberto Bustince[1], Javier Fernandez[1], Ľubomíra Horanská[2(✉)],
Radko Mesiar[3], and Andrea Stupňanová[3]

[1] Department of Statistics, Computer Science and Mathematics,
Institute of Smart Cities, Universidad Pública de Navarra,
Campus Arrosadía s/n, P.O. Box 31006, Pamplona, Spain
{bustince,fcojavier.fernandez}@unavarra.es
[2] Institute of Information Engineering, Automation and Mathematics,
Faculty of Chemical and Food Technology,
Slovak University of Technology in Bratislava,
Radlinského 9, 812 37 Bratislava, Slovak Republic
lubomira.horanska@stuba.sk
[3] Department of Mathematics and Descriptive Geometry,
Faculty of Civil Engineering, Slovak University of Technology in Bratislava,
Radlinského 11, 810 05 Bratislava 1, Slovak Republic
{radko.mesiar,andrea.stupnanova}@stuba.sk

Abstract. In the present paper we survey several generalizations of the
discrete Choquet integrals and we propose and study a new one. Our
proposal is based on the Lovász extension formula, in which we replace
the product operator by some binary function F obtaining a new n-ary
function $\mathfrak{I}_m^F$. We characterize all functions F yielding, for all capacities
m, aggregation functions $\mathfrak{I}_m^F$ with a priori given diagonal section.

Keywords: Aggregation function · Choquet integral · Capacity ·
Möbius transform

1 Introduction

The Choquet integral, as a useful tool in many applications, have been generalized in many ways. In our work we focus on those generalizations, which
can be obtained by replacing the standard arithmetical operations (summation,
subtraction, multiplication and minimum), occurring in basic formulas for the
discrete Choquet integral, by some more general function.

The structure of the paper is following: In Sect. 2, we recall the basic notions
and definitions including three equivalent expressions for the discrete Choquet
integral. In Sect. 3, we recall three generalizations of the discrete Choquet integral based on the above mentioned expressions and also crucial results concerning
those generalizations. In Sect. 4, we introduce a new generalization based on the
third of equivalent expressions. We illustrate the proposed construction by several examples and also our main results are confined in this section. Finally,
some concluding remarks are provided.

© Springer Nature Switzerland AG 2019
R. Halaš et al. (Eds.): AGOP 2019, AISC 981, pp. 151–159, 2019.
https://doi.org/10.1007/978-3-030-19494-9_14

2 Preliminaries

Let $n \in \mathbb{N}$ and $N = \{1, \cdots, n\}$.

Definition 1. *A function $H \colon [0,1]^n \to [0,1]$ is said to be an (n-ary) aggregation function if it is an order homomorphism, i.e., H is monotone and satisfies the boundary conditions $H(0, \ldots, 0) = 0$ and $H(1, \ldots, 1) = 1$.*

We denote the class of all n-ary aggregations functions by $\mathcal{A}_{(n)}$.

Definition 2. *A set function $m \colon 2^N \to [0,1]$ is said to be a capacity if it is an order homomorphism, i.e., $m(C) \leq m(D)$ whenever $C \subseteq D$ and satisfies boundary conditions $m(\emptyset) = 0$, $m(N) = 1$.*

We denote the class of all capacities on 2^N by $\mathcal{M}_{(n)}$.

Definition 3. *The set function $M_m \colon 2^N \to \mathbb{R}$, defined by*

$$M_m(I) = \sum_{K \subseteq I} (-1)^{|I \setminus K|} m(K)$$

for all $I \subseteq N$, is called Möbius transform corresponding to a capacity m.

Though the Choquet integral [3] was introduced for a general monotone measure space (Ω, Σ, m) as an extension of the Lebesgue integral, we will consider a capacity space $(N, 2^N, m)$ only.

Definition 4. *Let $m \colon 2^N \to [0,1]$ be a capacity and $\mathbf{x} = (x_1, \ldots, x_n) \in [0,1]^n$. Then the Choquet integral of $\mathbf{x}$ with respect to m is given by*

$$\mathbf{Ch}_m(\mathbf{x}) = \int_0^1 m(\{i \in N | x_i \geq t\}) \, \mathrm{d}t,$$

where the integral on the right-hand side is the Riemann integral.

We have three equivalent formulas for the discrete Choquet integral, see [2,4,5,9].

Proposition 1. *Let $m \colon 2^N \to [0,1]$ and $\mathbf{x} \in [0,1]^n$. Let $\sigma \colon N \to N$ be an arbitrary permutation such that $x_{\sigma(1)} \leq \cdots \leq x_{\sigma(n)}$. Then the discrete Choquet integral can be equivalently expressed as:*

(i)

$$\mathbf{Ch}_m(\mathbf{x}) = \sum_{i=1}^n (x_{\sigma(i)} - x_{\sigma(i-1)}) \cdot m(A_{\sigma(i)}), \tag{1}$$

where $A_{\sigma(i)} = \{\sigma(i), \ldots, \sigma(n)\}$ for $i = 1, \ldots, n$, and $x_{\sigma(0)} = 0$.

(ii)

$$\mathbf{Ch}_m(\mathbf{x}) = \sum_{i=1}^n x_{\sigma(i)} \cdot (m(A_{\sigma(i)}) - m(A_{\sigma(i+1)})), \tag{2}$$

where $A_{\sigma(n+1)} = \emptyset$.

(iii)

$$\mathbf{Ch}_m(\mathbf{x}) = \sum_{\emptyset \neq B \subseteq N} M_m(B) \cdot \left(\bigwedge_{i \in B} x_i \right), \tag{3}$$

where $\bigwedge$ is the minimum operator.

Observe that the Choquet integral, for any capacity $m \colon 2^N \to [0,1]$, is an idempotent n-ary aggregation function. It can be defined axiomatically as a comonotone additive aggregation function, see [14,15]. Note that all formulas for the discrete Choquet integral (compare (1), (2), (3)) are based on standard arithmetical operations (summation, subtraction, multiplication and minimum). There were several attempts to generalize these formulas to obtain new types of aggregation functions, or some weaker types of fusion functions, such as the pre-aggregation functions [10,11]. These proposals are summarized in the next section, while a new proposal, related to formula (3), is discussed and exemplified in Sect. 4.

3   An Overview of Generalizations of the Choquet Integral

As a first generalization we recall the proposal of Kolesárová et al. [8] generalizing formula (3). Its basic idea is in replacing the minimum operator $\bigwedge$ in (3) by some other aggregation function in the following way:

Let $m \in \mathcal{M}_{(n)}$ be a capacity, $A \in \mathcal{A}_{(n)}$ be an aggregation function. Define $F_{m,A} \colon [0,1]^n \to \mathbb{R}$ by

$$F_{m,A}(x_1, \ldots, x_n) = \sum_{I \subseteq N} M_m(I) \cdot A(\mathbf{x}_I), \tag{4}$$

where $(\mathbf{x}_I)_i = x_i$ whenever $i \in I$ and $(\mathbf{x}_I)_i = 1$ otherwise.

Note that taking the product operator in role of A in (4), we obtain the Owen extension of capacity m [16].

Theorem 1 ([8]). *Let $A \in \mathcal{A}_{(n)}$. Then $F_{m,A}$ is an aggregation function extending m for every $m \in \mathcal{M}_{(n)}$ if and only if A is an aggregation function with zero annihilator and for each $[\mathbf{a}, \mathbf{b}] \subseteq [0,1]^n$ such that $\{0,1\} \cap \{a_1, \ldots, a_n, b_1, \ldots, b_n\} \neq \emptyset$ the A-volume $V_A([\mathbf{a}, \mathbf{b}])$ is non-negative.*

Recall that the A-volume of the n-box $[\mathbf{a}, \mathbf{b}]$ is defined by

$$V_A([\mathbf{a}, \mathbf{b}]) = \sum (-1)^{\alpha(\mathbf{c})} A(\mathbf{c}),$$

where the sum is taken over all vertices $\mathbf{c}$ of $[\mathbf{a}, \mathbf{b}]$ and $\alpha(\mathbf{c})$ is the number of indices k such that $c_k = a_k$.

Observe also that any n-ary copula A [13] satisfies the constraints of Theorem 1.

Another types of generalization are based on replacing the product operator in formulas (1) and (2), respectively, by a function $F : [0,1]^2 \to [0,1]$ obtaining the following two functions:

$$C_m^F(\mathbf{x}) = \sum_{i=1}^{n} F(x_{(i)} - x_{(i-1)}, m(E_{(i)})), \tag{5}$$

$$C_F^m(\mathbf{x}) = \sum_{i=1}^{n} F(x_{(i)}, m(E_{(i)}) - m(E_{(i+1)})), \tag{6}$$

The functions C_m^F defined by (5) were deeply studied in [12], the functions C_F^m defined by (6) in [7]. The following characterization of all functions F yielding via (5) and (6), respectively, aggregation functions for all capacities $m \in \mathcal{M}_{(n)}$ was given there.

Theorem 2 ([12]). *Let $n = 2$. The function C_m^F is an aggregation function for all $m \in \mathcal{M}_{(2)}$ if and only if the following conditions are satisfied:*

(i) $F(0, y) = 0$ for all $y \in [0, 1]$.
(ii) $F(x, 1) = x$ for all $x \in [0, 1]$.
(iii) The function $F(\cdot, y) \colon [0, 1] \to [0, 1]$ is increasing and 1-Lipschitz for all $y \in [0, 1]$.

Theorem 3 ([12]). *Let $n > 2$. The function C_m^F is an aggregation function for all $m \in \mathcal{M}_{(n)}$ if and only if the following conditions are satisfied:*

(i) $F(0, y) = 0$ for all $y \in [0, 1]$.
(ii) $F(x, y) = x f(y)$, for some increasing function $f \colon [0, 1] \to [0, 1]$ satisfying $f(1) = 1$.

Theorem 4 ([7]). *Let $n = 2$. The function C_F^m is an aggregation function for all $m \in \mathcal{M}_{(2)}$ if and only if the following conditions are satisfied:*

(i) $F(x, u) + F(x, 1 - u) = 2F(x, 1/2)$, for all $x, u \in [0, 1]$.
(ii) $F(1, \frac{1}{2}) = \frac{1}{2}$.
(iii) F is an increasing function in the first variable satisfying $F(0, u) = 0$ for all $u \in [0, 1]$.

Theorem 5 ([7]). *Let $n > 2$. The function C_F^m is an aggregation function for all $m \in \mathcal{M}_{(n)}$ if and only if $F(x, u) = f(x)u$, for some increasing function $f \colon [0, 1] \to [0, 1]$ satisfying $f(0) = 0$ and $f(1) = 1$, i.e., $f \in \mathcal{A}_{(1)}$.*

4   A New Approach to Generalization of the Choquet Integral

Inspired by the above mentioned approaches to generalization of the Choquet integral, we focus on formula (3) replacing the product of $M_m(A)$ and $\bigwedge$ by some function $F \colon \mathbb{R} \times [0, 1] \to \mathbb{R}$.

Definition 5. *Let $n \geq 1$, $m \in \mathcal{M}_{(n)}$ and M_m be the Möbius transform of m. Let $F: \mathbb{R} \times [0,1] \to \mathbb{R}$ be a function bounded on $[0,1]^2$. We define the function $\mathfrak{I}_m^F: [0,1]^n \to \mathbb{R}$ as*

$$\mathfrak{I}_m^F(\mathbf{x}) = \sum_{\emptyset \neq B \subseteq N} F(M_m(B), \min_{i \in B}\{x_i\}). \tag{7}$$

We are interested in when $\mathfrak{I}_m^F$ is an aggregation function for all capacities $m \in \mathcal{M}_{(n)}$.

Note that for a fixed n the possible values of the Möbius transform cannot attain all real values. Denote $I_n = [a_n, b_n] \subset \mathbb{R}$ the range of the Möbius transform. Then, for a fixed n, we need to know the values of F on $I_n \times [0,1]$ only. The bounds a_n, b_n of the Möbius transform have recently been studied by Grabisch et al. in [6], wherein the following table can be found:

n	1	2	3	4	5	6	7	8	9	10	11	12
Upper bound b_n	1	1	1	3	6	10	15	35	70	126	210	462
Lower bound a_n	1(0)	-1	-2	-3	-4	-10	-20	-35	-56	-126	-252	-462

Note that, for a fixed $n \geq 2$, considering functions $F, G: \mathbb{R} \times [0,1] \to [0,1]$ such that

$$F(x,y) = G(x,y) + c\left(x - \frac{1}{2^n - 1}\right),$$

for some $c \in \mathbb{R}$, then, for any capacity $m \in \mathcal{M}_{(n)}$, it holds $\mathfrak{I}_m^F = \mathfrak{I}_m^G$.

Example 1. For any $f \in \mathcal{A}_{(1)}$ define $F: \mathbb{R} \times [0,1] \to [0,1]$ by $F(u,v) = uf(v)$. Then, for any $n \geq 1$ and any capacity $m \in \mathcal{M}_{(n)}$, $\mathfrak{I}_m^F$ is the Choquet integral of f-transformed inputs, i.e.,

$$\mathfrak{I}_m^F(x_1, \ldots, x_n) = \mathbf{Ch}_m(f(x_1), \ldots, f(x_n)).$$

Hence, $\mathfrak{I}_m^F$ is, for any capacity m, an aggregation function with diagonal section $\mathfrak{I}_m^F(x, \ldots, x) = f(x)$ for each $x \in [0,1]$. We conjecture that there are no other functions with the above properties, i.e., if $F(u,v) \neq uf(v)$, $f \in \mathcal{A}_{(1)}$, then there is a capacity m and $n \in \mathbb{N}$ such that $\mathfrak{I}_m^F$ is not an n-ary aggregation function.

Example 2. Let $F(u,v) = u^2 v$. For $n = 2$, let $m_{a,b}: 2^{\{1,2\}} \to [0,1]$ be a capacity such that $m_{a,b}(\{1\}) = a$ and $m_{a,b}(\{2\}) = b$, $a, b \in [0,1]$. Then

$$\mathfrak{I}_m^F(x_1, x_2) = a^2 x_1 + b^2 x_2 + (1 - a - b)^2 (x_1 \wedge x_2).$$

Clearly, $\mathfrak{I}_{m_{0,0}}^F = x_1 \wedge x_2$ and $\mathfrak{I}_{m_{1,1}}^F = x_1 \vee x_2$ define binary aggregation functions. On the other hand, for any $a, b \in]0,1[$ we have

$$\mathfrak{I}_{m_{a,b}}^F(1,1) = a^2 + b^2 + (1 - a - b)^2 \neq 1,$$

thus, $\mathfrak{I}_{m_{a,b}}^F$ is not an aggregation function.

Example 3. Let $F(u,v) = \begin{cases} \frac{1}{3}v^{2+u} & \text{if } u \in [-2, \infty[\\ 0 & \text{otherwise} \end{cases}$, with convention $0^0 = 0$.

Then, for $n = 2$ and any capacity $m \in \mathcal{M}_{(2)}$, $\mathfrak{I}_m^F \colon [0,1]^2 \to \mathbb{R}$ belongs to $\mathcal{A}_{(2)}$ with diagonal section

$$\mathfrak{I}_m^F(x,x) = \frac{1}{3}\left(x^{2+a} + x^{2+b} + x^{3-a-b}\right).$$

For $n > 2$ and for any capacity $m \in \mathcal{M}_{(n)}$, $\mathfrak{I}_m^F \notin \mathcal{A}_{(n)}$.

Though the full characterization of functions F such that, for a fixed $n > 1$ and any capacity $m \in \mathcal{M}_{(n)}$, it holds $\mathfrak{I}_m^F \in \mathcal{A}_{(n)}$, is still an open problem, we have the next interesting results. The first one gives a complete description of functions F yielding for any capacity $m \in \mathcal{M}_{(n)}$ an idempotent aggregation function $\mathfrak{I}_m^F$.

Theorem 6. *Let $n \geq 2$. Then the following are equivalent.*

(i) $\mathfrak{I}_m^F \colon [0,1]^n \to [0,1]$ *is an idempotent aggregation function for any capacity* $m \in \mathcal{M}_{(n)}$.

(ii) *For all $(u,v) \in I_n \times [0,1]$ it holds*

$$F(u,v) = u\,h(v) + \frac{v - h(v)}{2^n - 1}, \tag{8}$$

where $h \colon [0,1] \to \mathbb{R}$ is a function satisfying

$$-\frac{y - x}{2^n - 2} \leq h(y) - h(x) \leq y - x, \tag{9}$$

for all $(x,y) \in [0,1]^2$, such that $x < y$.

The proof of this fundamental result is based on the classical Cauchy functional equation and in all its details can be found in our forthcoming paper [1].

Example 4. Let $h \colon [0,1] \to \mathbb{R}$ be a constant function. Then, for any $n \geq 2$, F given by (8) does not depend on the first variable and is given by $F(u,v) = \frac{v}{2^n - 1}$. Then, for any capacity $m \in \mathcal{M}_{(n)}$ it holds

$$\mathfrak{I}_m^F(x_1,\dots,x_n) = \sum_{i=1}^{n} \frac{2^{n-i}}{2^n - 1} x_{\sigma(i)} = \mathrm{OWA}_n(x_1,\dots,x_n), \tag{10}$$

i.e., $\mathfrak{I}_m^F$ is an OWA operator [17] independently of m.

Example 5. For a fixed $n \geq 2$, let h considered in Theorem 6 be affine, i.e., $h(v) = \alpha + \beta v$. Due to variability constraints on h, necessarily $\beta \in [-\frac{1}{2^n - 2}, 1]$, and then, for any $m \in \mathcal{M}_{(n)}$, the related n-ary idempotent aggregation function $\mathfrak{I}_m^F$ for every $\mathbf{x} = (x_1,\dots,x_n) \in [0,1]^n$ is given by

$$\mathfrak{I}_m^F(\mathbf{x}) = \mathrm{sign}(\beta)\mathbf{Ch}_m(|\beta|\mathbf{x}) + (1 - \beta)\mathrm{OWA}_n(\mathbf{x}),$$

where OWA_n is given by (10).

Example 6. Let h be a 1-Lipschitz increasing function. Then it satisfies the variability constraints from Theorem 6 for any $n \geq 2$ and, for any $m \in \mathcal{M}_{(n)}$, the related n-ary idempotent aggregation function $\mathfrak{I}_m^F$ is given by

$$\mathfrak{I}_m^F(\mathbf{x}) = \mathbf{Ch}_m(h(x_1),\ldots,h(x_n)) + \frac{1}{2^n-1}\sum_{i=1}^n 2^{n-i}(x_{\sigma(i)} - h(x_{\sigma(i)}))$$

$$= \mathbf{Ch}_m(h(\mathbf{x})) + \mathrm{OWA}_n(\mathbf{x} - h(\mathbf{x})),$$

for every $\mathbf{x} = (x_1,\ldots,x_n) \in [0,1]^n$.

Recall that, for any n and $H \in \mathcal{A}_{(n)}$, the diagonal section $\delta_H \colon [0,1] \to [0,1]$ is given by $\delta_H(x) = H(x,\ldots,x)$ for $x \in [0,1]$. Obviously, $\delta_H \in \mathcal{A}_{(1)}$, and for any n and $\delta \in \mathcal{A}_{(1)}$ there is $H \in \mathcal{A}_{(n)}$ such that $\delta = \delta_H$. Our approach allows to generalize Theorem 6, where $\delta = \mathrm{id}_{[0,1]}$ is considered, for any $\delta \in \mathcal{A}_{(1)}$.

Theorem 7. *Let $n \geq 2$ and $\delta \in \mathcal{A}_{(1)}$. Then the following are equivalent.*

(i) $\mathfrak{I}_m^F \colon [0,1]^n \to [0,1]$ is an aggregation function with diagonal section δ for any capacity $m \in \mathcal{M}_{(n)}$.
(ii) For all $(u,v) \in I_n \times [0,1]$ it holds

$$F(u,v) = u\,h(v) + \frac{\delta(v) - h(v)}{2^n-1}, \tag{11}$$

where $h \colon [0,1] \to \mathbb{R}$ is a function satisfying

$$-\frac{\delta(y) - \delta(x)}{2^n-2} \leq h(y) - h(x) \leq \delta(y) - \delta(x), \tag{12}$$

for all $(x,y) \in [0,1]^2$, such that $x < y$.

Example 7. The smallest n-ary aggregation function $H_\star \in \mathcal{A}_{(n)}$ is given by

$$H(\mathbf{x}) = \begin{cases} 1 & \text{if } \mathbf{x} = (1,\ldots,1) \\ 0 & \text{otherwise} \end{cases} .$$

Obviously, $\delta_{H_\star} = \delta_\star = \mathbf{1}_{\{1\}}$ is the smallest diagonal section. Then $h \colon [0,1] \to \mathbb{R}$ satisfies the constraints of Theorem 7 for $n \geq 2$ fixed if and only if

$$h(v) = \begin{cases} \alpha & \text{if } v \in [0,1[\\ \alpha + \beta & \text{if } v = 1 \end{cases} ,$$

where $\alpha \in \mathbb{R}$ and $\beta \in \left[-\frac{1}{2^n-2}, 1\right]$. Then, for any $m \in \mathcal{M}_{(n)}$, the related n-ary idempotent aggregation function $\mathfrak{I}_m^F$ is given by

$$\mathfrak{I}_m^F(\mathbf{x}) = \beta \cdot m(\{i \mid x_i = 1\}) + \frac{1-\beta}{2^n-1}\left(2^{\mathrm{card}(\{i \mid x_i=1\})} - 1\right),$$

for every $\mathbf{x} = (x_1,\ldots,x_n) \in [0,1]^n$.

5   Concluding Remarks

We have introduced a new generalization of the discrete Choquet integral. Our approach is based on the Lovász extension, i.e., formula (3) evaluating the discrete Choquet integral by means of the Möbius transform M_m of the considered capacity m. We have characterized all functions F leading to idempotent aggregation function as well as all functions leading, for any capacity $m \in \mathcal{M}_{(n)}$, to n-ary aggregation function $\mathfrak{I}_m^F$ with an a priori given diagonal section δ. A complete characterization of functions F yielding for any capacity $m \in \mathcal{M}_{(n)}$ an aggregation function $\mathfrak{I}_m^F$ is a challenging open problem for next research.

Acknowledgments. This work was supported by the Slovak Research and Development Agency under the contract no. APVV-17-0066, grant VEGA 1/0682/16, grant VEGA 1/0614/18 and TIN2016-77356-P(AEI/FEDER,UE).

References

1. Bustince, H., Fernandez, J., Horanská, Ľ., Mesiar, R., Stupňanová, A.: A generalization of the Choquet integral defined in terms of the Möbius transform. In: Working Paper
2. Chateauneuf, A., Jaffray, J.Y.: Some characterizations of lower probabilities and other monotone capacities through the use of Möbius inversion. Math. Soc. Sci. **17**, 263–283 (1989)
3. Choquet, G.: Theory of capacities. Annales de l'Institut Fourier **5**, 131–295 (1953–1954)
4. Grabisch, M.: Set Functions, Games and Capacities in Decision Making. Springer, Cham (2016)
5. Grabisch, M., Marichal, J.-L., Mesiar, R., Pap, E.: Aggregation Functions. Cambridge University Press, Cambridge (2009)
6. Grabisch, M., Miranda, P.: Exact bounds of the Möbius inverse of monotone set functions. Discrete Appl. Math. **186**, 7–12 (2015)
7. Horanská, Ľ., Šipošová, A.: A generalization of the discrete Choquet and Sugeno integrals based on a fusion function. Inf. Sci. **451-452**, 83–99 (2018)
8. Kolesárová, A., Stupňanová, A., Beganová, J.: Aggregation-based extensions of fuzzy measures. Fuzzy Sets Syst. **194**, 1–14 (2012)
9. Lovász, L.: Submodular function and convexity. In: Mathematical Programming: The State of the Art, pp. 235–257. Springer, Berlin (1983)
10. Lucca, G., Sanz, J., Dimuro, G., Bedregal, B., Mesiar, R., Kolesárová, A., Bustince, H.: Pre-aggregation functions: construction and an application. IEEE Trans. Fuzzy Syst. **24**(2), 260–272 (2016)
11. Lucca, G., Sanz, J., Dimuro, G., Bedregal, B., Bustince, H., Mesiar, R.: C_F-integrals: a new family of pre-aggregation functions with application to fuzzy rule-based classification systems. Inf. Sci. **435**, 94–110 (2018)
12. Mesiar, R., Kolesárová, A., Bustince, H., Pereira Dimuro, G., Bedregal, B.: Fusion functions based discrete Choquet-like integrals. Eur. J. Oper. Res. **252**(2), 601–609 (2016)
13. Nelsen, R.B.: An Introduction to Copulas. Lecture Notes in Statistics, vol. 139. Springer, New York (1999)

14. Schmeidler, D.: Integral representation without aditivity. Proc. Am. Math. Soc. **97**(2), 255–261 (1986)
15. Schmeidler, D.: Subjective probability and expected utility without additivity. Econometrica **57**(3), 571–587 (1989)
16. Owen, G.: Multilinear extensions of games. In: Roth, A.E. (ed.) The Shapley Value. Essays in Honour of Lloyd S. Shapley, pp. 139–151. Cambridge University Press, Cambridge (1988)
17. Yager, R.R.: On ordered weighted averaging aggregation operators in multicriteria decisionmaking. IEEE Trans. Syst. Man Cybern. **18**(1), 183–190 (1988)

Penalty-Based Data Aggregation
in Real Normed Vector Spaces

Lucian Coroianu[1]([✉])[iD] and Marek Gagolewski[2,3][iD]

[1] Department of Mathematics and Informatics, University of Oradea,
1 Universitatii Street, 410087 Oradea, Romania,
lcoroianu@uoradea.ro
[2] Faculty of Mathematics and Information Science,
Warsaw University of Technology, ul. Koszykowa 75, 00-662 Warsaw, Poland
M.Gagolewski@mini.pw.edu.pl
[3] Systems Research Institute, Polish Academy of Sciences,
ul. Newelska 6, 01-447 Warsaw, Poland

Abstract. The problem of penalty-based data aggregation in generic real normed vector spaces is studied. Some existence and uniqueness results are indicated. Moreover, various properties of the aggregation functions are considered.

Keywords: Penalty-based aggregation · Prototype learning ·
Means, averages, and medians · Vector spaces · Fermat–Weber problem

1 Introduction

Aggregation theory is often concerned with the construction of means, see, e.g., [2,5,14,16,23]. Classically, there had been a focus on data on the real line, where monotonicity w.r.t. each coordinate used to play an important role. Due to this fact, a generalization of the notion of an aggregation function to bounded posets was considered, among others, in [10,17,19] . However, other data domains – where there is no natural notion of monotonicity – have recently started enjoying more and more attention of the aggregation community. These include the study of rankings [20], character strings [14], and different kinds of fuzzy sets [2], to name a few.

It is known that the construction of means can be performed via the notion of penalty minimizers [3,6,24]. The problem of aggregation of vectors in $\mathbb{R}^d$ for any $d \geq 1$ recently was given some treatment in [15]. Here we develop these results further, yet in a more generic context; namely, of real normed vector spaces.

Suppose that X is a vector space endowed with a norm $\|\cdot\|$ whose field of scalars is the field of real numbers, i.e., a real normed space. Let us denote with $\dim X$ the dimension of X, i.e., the cardinality of one of its possible bases. If $\dim X < \infty$, we shall say that X is finite-dimensional. For any finite $S \subseteq X$, with aff S we denote the *affine hull* of S, i.e., the set of all affine combinations

© Springer Nature Switzerland AG 2019
R. Halaš et al. (Eds.): AGOP 2019, AISC 981, pp. 160–171, 2019.
https://doi.org/10.1007/978-3-030-19494-9_15

of vectors in S, $\{\sum_{i=1}^{n} c_i x_i : x_i \in S, c_i \in \mathbb{R}, \sum_{i=1}^{n} c_i = 1\}$. Moreover, the *convex hull* of S, for which we additionally assume that $c_i \geq 0$, is denoted with $\mathrm{conv}\, S$.

Assume we are given a data sample $\mathbf{x}$ formed of $n \geq 2$ vectors $x_1, \ldots, x_n \in \Omega$ for some nontrivial Ω such that $\mathrm{conv}\,\{x_1, \ldots, x_n\} \subseteq \Omega \subseteq X$. Let D be some fixed norm on $\mathbb{R}^n$ whose restriction to $[0, \infty]^n$ is nondecreasing in each variable, i.e., for all $0 \leq d_i \leq d'_i$, $i = 1, \ldots, n$, it holds $\mathsf{D}(d_1, \ldots, d_n) \leq \mathsf{D}(d'_1, \ldots, d'_n)$. We are interested in studying the problem (see, e.g., [12]):

$$\min_{y \in \Omega} \mathsf{D}\left(\|y - x_1\|, \ldots, \|y - x_n\|\right). \tag{1}$$

Note that, due to the monotonicity constraint, the objective function is convex. Intuitively, we look for vectors in Ω that minimize a certain penalty based on the distance between the inputs and the output being computed. Denoting with $\mathsf{S}(\{x_1, \ldots, x_n\}) \subseteq \Omega$ the (possibly empty) minimum set of the above objective function, an output $y_0 \in \mathsf{S}(\{x_1, \ldots, x_n\})$ represents a kind of "consensus" value.

In particular, if D is some weighted L_p-norm ($1 \leq p \leq \infty$), our problem becomes equivalent to:

$$\min_{y \in \Omega} \sum_{i=1}^{n} w_i \|x_i - y\|^p, \tag{2}$$

for some $p \geq 1$ and $w_1, \ldots, w_n > 0$. In such a case we shall denote the corresponding norm D with $\|\cdot\|_{\mathbf{w},p}$ The case $p = 1$ is often referred to as the Fermat–Weber, minisum, or 1-median problem (minimizes the average distance to all points), $p = 2$ – weighted centroid, and $p = \infty$ – minimax or 1-center (minimizes the distance to the farthest-away point); these are rich in natural applications, e.g., in operational research (facility location problems), data analysis and mining, visualization, machine learning, etc.

Up to now, this problem has been considered in few works. In particular, in [13] a generalized Weber's problem and Weiszfeld's algorithm in reflexive Banach spaces is discussed. In [11] the Fermat–Weber problem ($p = 1$) in inner-product spaces is studied; the characterization of X such that the intersection of the set of solutions and the convex hull of input points is non-empty is indicated. The paper [12] has a particular focus on (among others) the "hull properties", which we may consider as different types of internality conditions.

We should also note that some particular space classes were considered in the context of a generalized version of the Fermat–Weber problem and/or the Weiszfeld algorithm: Riemannian manifolds of non-negative sectional curvature [1], L^p spaces of functions for $p \in [1, \infty]$ [21], Sobolev spaces [22], real intervals [7], and so forth.

This contribution is set out as follows. In Sect. 2 we formulate some results concerning the existence and uniqueness of the solution to the aggregation problem posed. In Sect. 3 we study the methods' properties from the aggregation theory perspective. In Sect. 4 we present some new results regarding the limiting case of (2) as $p \to \infty$. Most notably, for a fixed sample, the solution of the aggregation problem is bounded with respect to p and, what is more, we can construct a sequence $(p_k)_{k \geq 1}$, $p_k \to \infty$, such that the solutions of the aggregation problems associated with the elements of this sequence converge to the solution of an aggregation problem with respect to the infinity-norm. Lastly, Sect. 5 concludes the paper.

2   Existence and Uniqueness Results

At the very beginning we shall be interested in the results concerning the existence and possible uniqueness of the solutions of (1). Then we shall be ready to explore the properties of the minimizers of a given penalty function. Yet, before proceeding any further, let us note that we can from now on assume $\dim X \geq 2$.

Remark 1. Let $\dim X = 1$. First of all, with no harm in generality let us assume that $x_2 \neq x_1$. Moreover, we let $x_1 = \mathbf{0}$ and $\|x_2\| = 1$, as due to translation and scale equivariance (see Proposition 2) we can always solve our problem in $\Omega' = \left\{ \frac{1}{\|x_2 - x_1\|} (\omega - x_1) : \omega \in \Omega \right\}$ with inputs scaled and translated like $x_i \mapsto \frac{1}{\|x_2 - x_1\|} (x_i - x_1)$ and then the solution transformed back to the original Ω. For every i let $c_i \in \mathbb{R}$ be such that $x_i = c_i x_2$. Letting $y = c x_2$, $c \in \mathbb{R}$, we have $\mathsf{D}(\|x_1 - y\|, \ldots, \|x_n - y\|) = \mathsf{D}(\|c_1 x_2 - c x_2\|, \ldots, \|c_n x_2 - c x_2\|) = \mathsf{D}(|c_1 - c|, \ldots, |c_n - c|)$. We see that our problem is reduced to aggregation of data in the real line.

In order to find some useful sufficient conditions for the existence of the solutions and also on the uniqueness of the solution, let us introduce the n-ary Cartesian power space $X^n = X \times X \times \cdots \times X$ endowed with the norm $\|\cdot\|_\mathsf{D}$, where for any $(u_1, \ldots, u_n) \in X^n$ it holds $\|(u_1, \ldots, u_n)\|_\mathsf{D} = \mathsf{D}(\|u_1\|, \ldots, \|u_n\|)$. It is easy to prove that for any norm D indeed $(X^n, \|\cdot\|_\mathsf{D})$ is a normed space; see also [8], where in case where D is an L_p norm, this space is used as an illustrative example for even more general constructions of norms on the power space X^n.

Next, we need the set $\widetilde{\Omega} = \{(x, \ldots, x) : x \in \Omega\}$ and the vector $\widetilde{\mathbf{x}} \in X^n$, $\widetilde{\mathbf{x}} = (x_1, \ldots, x_n)$. Now, obviously y_0 is a solution of the problem (1) if and only if $\widetilde{\mathbf{y}}_0 = (y_0, \ldots, y_0)$ is a solution of:

$$\min_{\widetilde{\mathbf{y}} \in \widetilde{\Omega}} \|\widetilde{\mathbf{x}} - \widetilde{\mathbf{y}}\|_\mathsf{D}. \tag{3}$$

It means that problem (1) is reduced to the problem of finding the metrical projection $\widetilde{\mathbf{y}}_0$ of $\widetilde{\mathbf{x}}$ onto $\widetilde{\Omega}$ w.r.t. the metric $\|\cdot\|_\mathsf{D}$ on X^n, i.e., problem (3).

Let us point out those properties of the minimum set of our objective function that only depend on the properties of Ω. Recall that the whole space X is a closed set by definition.

Proposition 1. *In any normed space* $(X, \|\cdot\|)$ *let* $\mathsf{S}(\{x_1, \ldots, x_n\}) \subseteq \Omega$ *denote the minimum set of (1) given some* $x_1, \ldots, x_n \in \Omega$.

(i) If $\mathsf{S}(\{x_1, \ldots, x_n\})$ *is nonempty, then* $\mathsf{S}(\{x_1, \ldots, x_n\})$ *is bounded.*
(ii) If Ω *is closed and* $\mathsf{S}(\{x_1, \ldots, x_n\})$ *is nonempty, then it is closed.*
(iii) If Ω *is convex, then* $\mathsf{S}(\{x_1, \ldots, x_n\})$ *is convex.*

Proof. (i) Let $y_0 \in S(\{x_1,\ldots,x_n\})$ be arbitrarily chosen. Denoting $\tilde{\mathbf{x}} = (x_1,\ldots, x_n)$, $\tilde{\mathbf{y}}_0 = (y_0,\ldots,y_0)$, and $M = D(\|x_1 - y_0\|,\ldots,\|x_n - y_0\|)$, we get $\|\tilde{\mathbf{y}}_0\|_D \leq \|\tilde{\mathbf{y}}_0 - \tilde{\mathbf{x}}\|_D + \|\tilde{\mathbf{x}}\|_D = M + \|\tilde{\mathbf{x}}\|_D$. It means that $D(\|y_0\|,\ldots,\|y_0\|) \leq M + \|\tilde{\mathbf{x}}\|_D$. But D is a norm on $\mathbb{R}^n$ and obviously it is equivalent to the Chebyshev norm (or any other) on the same set. More precisely, there exists a constant K such that $\|y_0\| \leq K(M + \|\tilde{\mathbf{x}}\|_D)$. As y_0 is an arbitrary element of $S(\{x_1,\ldots,x_n\})$, it follows that $S(\{x_1,\ldots,x_n\})$ is bounded in $(X,\|\cdot\|)$.

(ii) For some $z \in S(\{x_1,\ldots,x_n\})$, let $M \in [0,\infty)$ be such that $D(\|z - x_1\|,\ldots,\|z - x_n\|) = M$. This implies $\|\tilde{\mathbf{z}} - \tilde{\mathbf{x}}\|_D = M$, where $\tilde{\mathbf{z}} = (z,\ldots,z)$. Thus, $y \in S(\{x_1,\ldots,x_n\})$ if and only if $y \in \Omega$ and $\|\tilde{\mathbf{y}} - \tilde{\mathbf{x}}\|_D = M$, where $\tilde{\mathbf{y}} = (y,\ldots,y)$. Now, let $(y_k)_{k \geq 1}$ be a convergent sequence in $S(\{x_1,\ldots,x_n\})$, such that $y_k \to y_0$ in the norm $\|\cdot\|$. It means that $\|\tilde{\mathbf{y}}_k - \tilde{\mathbf{x}}\|_D = M$, where $\tilde{\mathbf{y}}_k = (y_k,\ldots,y_k)$, $y_k \in \Omega$, for all $k \geq 1$. By the continuity of the norm $\|\cdot\|_D$, in the limit we get $\|\tilde{\mathbf{y}}_0 - \tilde{\mathbf{x}}\|_D = M$, where $\tilde{\mathbf{y}}_0 = (y_0,\ldots,y_0)$ and we also used that $\|\tilde{\mathbf{y}}_k - \tilde{\mathbf{y}}_0\|_D \to 0$. As Ω is closed, it also follows that $y_0 \in \Omega$, which combined with the previous property yields $y_0 \in S(\{x_1,\ldots,x_n\})$. Hence, $S(\{x_1,\ldots,x_n\})$ is closed.

(iii) Suppose that $y_0, z_0 \in S(\{x_1,\ldots,x_n\})$ and $\alpha \in [0,1]$. Let us denote $t_0 = \alpha y_0 + (1 - \alpha)z_0$. Again, let $M \in [0,\infty)$ be such that for any $\tilde{\mathbf{y}} = (y,\ldots,y)$, $y \in \Omega$, we have $\|\tilde{\mathbf{y}}_0 - \tilde{\mathbf{x}}\|_D = \|\tilde{\mathbf{z}}_0 - \tilde{\mathbf{x}}\|_D = M \leq \|\tilde{\mathbf{y}} - \tilde{\mathbf{x}}\|_D$. It is elementary to prove that $\|\tilde{\mathbf{t}}_0 - \tilde{\mathbf{x}}\|_D \leq \alpha \|\tilde{\mathbf{y}}_0 - \tilde{\mathbf{x}}\|_D + (1 - \alpha)\|\tilde{\mathbf{z}}_0 - \tilde{\mathbf{x}}\|_D = \|\tilde{\mathbf{y}}_0 - \tilde{\mathbf{x}}\|_D$. As Ω is convex, we get that $t_0 \in \Omega$, and combining this fact with the above inequality and the definition of $\tilde{\mathbf{y}}_0$, it follows that we must have $\|\tilde{\mathbf{y}}_0 - \tilde{\mathbf{x}}\|_D = \|\tilde{\mathbf{t}}_0 - \tilde{\mathbf{x}}\|_D$. Therefore, $t_0 \in S(\{x_1,\ldots,x_n\})$, and hence $S(\{x_1,\ldots,x_n\})$ is convex. ∎

For finite-dimensional closed search spaces, a solution of (1) always exists.

Theorem 1. *If Ω is a closed subset of a finite-dimensional affine subspace of the normed space $(X,\|\cdot\|)$, then the minimum set $S(x_1,\ldots,x_n)$ of (1) is nonempty for any $x_1,\ldots,x_n \in \Omega$.*

Proof (sketch). In this setting, it is easily seen that problem (3) becomes a best approximation problem with respect to a closed subset of a finite-dimensional affine subspace. This problem surely has at least one solution, see, e.g., [9, Theorem 3.2.3] which needs just a very simple extension from vector subspaces to affine subspaces. Thus, let $\tilde{\mathbf{y}}_0 = (y_0,\ldots,y_0)$ be a solution of problem (3). This immediately implies that y_0 is a solution of problem (1). ∎

On a side note, from Proposition 1 we can easily deduce that the minimum set is compact for each Ω fulfilling the assumptions of the above theorem.

Next we shall present some results concerning the existence and possible uniqueness of the solution in case where Ω is a closed and convex subset of normed spaces fulfilling some more specific properties, which we shall review in the sequel.

A normed space X is called *reflexive* if any bounded sequence $(x_n)_n$ in X, has a weakly convergent subsequence $(x_{k_n})_n$. It means that there exists $x_0 \in X$ such that for any continuous linear functional $\lambda : X \to \mathbb{R}$, we have $\lambda(x_{k_n}) \to \lambda(x_0)$.

A normed space is called *strictly convex* if for any $x, y \in X$, $x \neq y$, $\|x\| = \|y\|$, we have that $\|x + y\| < \|x\| + \|y\|$. This is equivalent with the property that for any $x, y \in X$, such that $\|x + y\| = \|x\| + \|y\|$, there exists $c \in [0, \infty)$ such that either $y = cx$ or $x = cy$. A stronger concept is that of uniformly convex normed space. A normed space X is called *uniformly convex* if for any $\varepsilon > 0$ there exists $\delta(\varepsilon) > 0$ such that $\frac{\|x+y\|}{2} \leq 1 - \delta(\varepsilon)$, whenever $\|x - y\| \geq \varepsilon$ and $\|x\| = \|y\| = 1$. For example, the space $L_q[0, 1]$ of $q \in (1, \infty)$-integrable functions on $[0, 1]$ is a Banach, uniformly convex normed space.

Then, it is well-known that every Hilbert space is a Banach (i.e., complete and normed) space which in addition is uniformly convex in the norm generated by the inner product. Any uniformly convex normed space is reflexive and any uniformly convex normed space is strictly convex. Every finite dimensional normed space is a Banach reflexive space. Therefore, in finite dimensional spaces, the concepts of uniformly convex normed space and strictly convex normed space are equivalent. It is well-known that in Banach reflexive spaces best approximation always exists with respect to closed convex sets. In addition, in strictly convex Banach reflexive spaces the best approximation is unique. The following theorem formalizes these statements.

Theorem 2. *If Ω is a closed and convex subset of a reflexive Banach space $(X, \|\cdot\|)$, then for any $x_1, x_2, \ldots, x_n \in \Omega$ problem (1) has at least one solution. In addition, if D is strictly convex and X is strictly convex with respect to $\|\cdot\|$, then the solution is unique.*

Proof. (a) All the topological properties of Ω are passed onto $\widetilde{\Omega}$. It means that $\widetilde{\Omega}$ is a closed and convex subset of the space $(X^n, \|\cdot\|_{\mathsf{D}})$. What is more, it is not hard to prove that $(X^n, \|\cdot\|_{\mathsf{D}})$ is a Banach space. Suppose that $(\mathbf{x}_k)_{k \geq 1}$ is a Cauchy sequence in $(X^n, \|\cdot\|_{\mathsf{D}})$, $\mathbf{x}_k = (x_{k1}, \ldots, x_{kn})$. So $\lim_{(m,p) \to (\infty, \infty)} \mathsf{D}(\|x_{m1} - x_{p1}\|, \ldots, \|x_{mn} - x_{pn}\|) = 0$. As D is equivalent to the Chebyshev norm on $\mathbb{R}^n$, there exists $K \geq 0$ s.t. $\mathsf{D}(\|x_{m1} - x_{p1}\|, \ldots, \|x_{mn} - x_{pn}\|) \leq K \max\{\|x_{mi} - x_{pi}\| : i = 1, \ldots, n\}$. This easily implies $\lim_{(m,p) \to (\infty, \infty)} \|x_{mi} - x_{pi}\| = 0$, $i = 1, \ldots, n$. As $(X, \|\cdot\|)$ is a Banach space, there exists $\mathbf{x}_0 = (x_1, \ldots, x_n)$ s.t. $\lim_{k \to \infty} \|x_{ki} - x_i\| = 0$, $i = 1, \ldots, n$. Again by the equivalence between D and the Chebyshev norm, we get that $\lim_{k \to \infty} \mathsf{D}(\|x_{k1} - x_1\|, \ldots, \|x_{kn} - x_n\|) = 0$. Hence, $(X^n, \|\cdot\|_{\mathsf{D}})$ is a Banach space as we have just proved that any Cauchy sequence in $(X^n, \|\cdot\|_{\mathsf{D}})$ is convergent. As problem (1) is reduced to problem (3), considering the first statement, it suffices to prove that $(X^n, \|\cdot\|_{\mathsf{D}})$ is a reflexive Banach space. Indeed, if $(X^n, \|\cdot\|_{\mathsf{D}})$ is a reflexive Banach space, then problem (3) is a best approximation problem in a reflexive Banach normed space with respect to a closed convex subset. As said, such a problem always has at least one solution. This implies that problem (1) has at least one solution. Thus, it remains to prove the reflexivity of $(X^n, \|\cdot\|_{\mathsf{D}})$. To this end, suppose that $(\mathbf{x}_k)_{k \geq 1}$, $\mathbf{x}_k = (x_{k1}, \ldots, x_{kn})$ is a bounded sequence in $(X^n, \|\cdot\|_{\mathsf{D}})$. Let us consider on X^n the product norm, $\|(x_1, \ldots, x_n)\|_1 = \sum_{i=1}^n \|x_i\|$. It is well-known that the product space of reflexive normed spaces is also a reflexive normed space with respect to the product norm. The definition of $\|\cdot\|_{\mathsf{D}}$ and the fact that obviously D and the Minkowski norm are equivalent on $\mathbb{R}^n$, all these imply that there exists a constant $M \geq 0$ such that $\|\mathbf{x}_k\|_{\mathsf{D}} \leq M \|\mathbf{x}_k\|_1$. It means that $(\mathbf{x}_k)_{k \geq 1}$ is bounded in $(X^n, \|\cdot\|_1)$ too. All these imply that in order to prove the reflexivity of X^n with respect to $\|\cdot\|_{\mathsf{D}}$, it suffices to prove that any linear functional $\lambda : X^n \to \mathbb{R}$, which is continuous with respect to $\|\cdot\|_{\mathsf{D}}$, is also continuous with respect to $\|\cdot\|_1$. Obviously, continuity of linear functionals is equivalent with the boundness property. Thus, λ is bounded with respect to $\|\cdot\|_{\mathsf{D}}$ and since $\|\cdot\|_{\mathsf{D}} \leq M \|\cdot\|_1$, we get that λ is bounded with respect to $\|\cdot\|_1$. From here we easily get that $(\mathbf{x}_k)_{k \geq 1}$ has a weakly convergent subsequence in $(X^n, \|\cdot\|_{\mathsf{D}})$. It means that indeed $(X^n, \|\cdot\|_{\mathsf{D}})$ is a reflexive Banach space. This completes the proof of the first statement of the theorem.

(b) Suppose now that X is strictly convex with respect to $\|\cdot\|$. One way to prove the uniqueness of the solution of problem (1) would be to show that $(X^n, \|\cdot\|_{\mathsf{D}})$ is a strictly convex normed space. First, we notice that if $x_1 = x_2 = \cdots = x_n = x$, then by the strict convexity of $\|\cdot\|$, it follows that problem (1) has a unique solution. Indeed, if y_1 and y_2 are solutions of problem (1), we get that $\mathsf{D}(\|x - y_1\|, \ldots, \|x - y_1\|) = \mathsf{D}(\|x - y_2\|, \ldots, \|x - y_2\|)$, which gives $\|x - y_1\| = \|x - y_2\|$. Now, if $y_1 \neq y_2$, taking into account that $(X, \|\cdot\|)$ is strictly convex, we get that $\|(x - y_1) + (x - y_2)\| < \|x - y_1\| + \|x - y_2\|$. As D is strictly increasing on $[0, \infty)^n$, it holds:

$$\|(\mathbf{x} - 0.5(\mathbf{y}_1 + \mathbf{y}_2)\|_{\mathsf{D}}$$
$$< \mathsf{D}(0.5\|x - y_1\| + 0.5\|x - y_2\|, \ldots, 0.5\|x - y_1\| + 0.5\|x - y_2\|)$$
$$\leq \mathsf{D}(0.5\|x - y_1\|, \ldots, 0.5\|x - y_1\|) + \mathsf{D}(0.5\|x - y_2\|, \ldots, 0.5\|x - y_2\|)$$
$$= \mathsf{D}(\|x - y_1\|, \ldots, \|x - y_1\|).$$

This yields $\|(\mathbf{x} - 0.5(\mathbf{y}_1 + \mathbf{y}_2)\|_{\mathsf{D}} < \|\mathbf{x} - \mathbf{y}_1\|_{\mathsf{D}}$, where $\mathbf{x} = (x, \ldots, x)$ and $\mathbf{y}_i = (y_i, \ldots, y_i)$, $i = 1, 2$. Since $0.5(y_1 + y_2) \in \Omega$, it follows that $0.5(y_1 + y_2)$ is a solution of problem (1) but the same is not true for y_1, a contradiction.

Therefore, in all that follows, we may suppose that the equality $x_1 = x_2 = \cdots = x_n$ does not hold. Let y_1 and y_2 be solutions of problem (1). For any $\mathbf{x} = (x_1, \ldots, x_n)$ we have $\|(\mathbf{x} - 0.5(\mathbf{y}_1 + \mathbf{y}_2)\|_{\mathsf{D}} \leq 0.5(\|\mathbf{x} - \mathbf{y}_1\|_{\mathsf{D}} + \|\mathbf{x} - \mathbf{y}_2\|_{\mathsf{D}})$, therefore, as $0.5(y_1 + y_2) \in \Omega$, we easily obtain that $\|(\mathbf{x} - 0.5(\mathbf{y}_1 + \mathbf{y}_2)\|_{\mathsf{D}} = \|\mathbf{x} - \mathbf{y}_1\|_{\mathsf{D}} = \|\mathbf{x} - \mathbf{y}_2\|_{\mathsf{D}}$. Moreover, by the monotonicity of D and then the triangle inequality, we obtain:

$$\|\mathbf{x} - 0.5(\mathbf{y}_1 + \mathbf{y}_2)\|_{\mathsf{D}} = \mathsf{D}(\|(x_1 - 0.5(y_1 + y_2)\|, \ldots, \|x_n - 0.5(y_1 + y_2)\|)$$
$$\leq \mathsf{D}(0.5(\|x_1 - y_1\| + \|x_1 - y_2\|), \ldots, 0.5(\|x_n - y_1\| + \|x_n - y_2\|))$$
$$\leq 0.5\mathsf{D}(\|x_1 - y_1\|, \ldots, \|x_n - y_1\|) + 0.5\mathsf{D}(\|x_1 - y_2\|, \ldots, \|x_n - y_2\|)$$
$$= \|\mathbf{x} - \mathbf{y}_1\|_{\mathsf{D}} = \|\mathbf{x} - \mathbf{y}_2\|_{\mathsf{D}}.$$

It means that we must have $\mathsf{D}(\|x_1 - 0.5(y_1 + y_2)\|, \ldots, \|x_n - 0.5(y_1 + y_2)\|) = \mathsf{D}(0.5(\|x_1 - y_1\| + \|x_1 - y_2\|), \ldots, 0.5(\|x_n - y_1\| + \|x_n - y_2\|))$, that is it holds $\mathsf{D}(\|(x_1 - y_1) + (x_1 - y_2)\|, \ldots, \|(x_n - y_1) + (x_n - y_2)\|) = \mathsf{D}(\|x_1 - y_1\| + \|x_1 - y_2\|, \ldots, \|x_n - y_1\| + \|x_n - y_2\|)$. By the monotonicity of D and the fact that in general we have $\|(x_i - y_1) + (x_i - y_2)\| \leq \|x_i - y_1\| + \|x_i - y_2\|$, $i = 1, \ldots, n$, it follows that we must have $\|(x_i - y_1) + (x_i - y_2)\| = \|x_i - y_1\| + \|x_i - y_2\|$, $i = 1, \ldots, n$. From the strict convexity of $\|\cdot\|$, there exist $c_i \in [0, \infty)$, $i = 1, \ldots, n$, such that:

$$x_i - y_1 = c_i(x_i - y_2) \text{ or } x_i - y_2 = c_i(x_i - y_1), i = 1, \ldots, n. \tag{4}$$

We must also have that $\mathsf{D}\,(\|x_1 - y_1\| + \|x_1 - y_2\|, \ldots, \|x_n - y_1\| + \|x_n - y_2\|) = \mathsf{D}\,(\|x_1 - y_1\|, \ldots, \|x_n - y_1\|) + \mathsf{D}\,(\|x_1 - y_2\|, \ldots, \|x_n - y_2\|)$. As D is strictly convex, there exists $c \in (0, \infty)$ such that $(\|x_1 - y_1\|, \ldots, \|x_n - y_1\|) = c(\|x_1 - y_2\|, \ldots, \|x_n - y_2\|)$. Note that $c > 0$ because in this setting both vectors are different from the null vector. Then, since $\mathsf{D}\,(\|x_1 - y_1\|, \ldots, \|x_n - y_1\|) = \mathsf{D}(\|x_1 - y_2\|, \ldots, \|x_n - y_2\|)$, we get that $c = 1$. It means that $\|x_i - y_1\| = \|x_i - y_2\|$, for all $i = 1, \ldots, n$. Therefore, in (4), we may suppose that $c_i = 1$ for all $i = 1, \ldots, n$. This implies $y_1 = y_2$, which completes the proof of the second statement of the theorem.   ∎

As an important side remark, note that from [8, Theorem 1] we get that $(X^n, \|\cdot\|_{\mathbf{w},p})$ (as in problem (2)) is a uniformly – and hence strictly – convex Banach space for $p > 1$.

However, in the special case of problem (2) with $p = 1$, the conclusion in the previous result with respect to uniqueness does not hold in general. The simplest example is to consider the case where $X = \Omega = \mathbb{R}$ and $w_1 = \cdots = w_n$, with the standard Euclidean norm. In this case, the minimum set of the objective function corresponds to the median set of the sample $x_1, \ldots, x_n$, i.e., a set which has more than just one element if n is even. More generally, for $p = 1$ and the Euclidean space $\mathbb{R}^d$ for $d \geq 1$, if points lie on some line, then the minimum might not unique, but it is always unique if they are not-colinear, see [18].

Remark 2. In the special case where Ω is a nonempty, closed and convex subset of a Hilbert space $(X, \langle \cdot, \cdot \rangle)$ and we are solving the problem (2) with $p = 2$, we can precisely determine the unique solution – the centroid, which is given by $y = \sum_{i=1}^{n} w_i x_i / \sum_{i=1}^{n} w_i$.

3   Noteworthy Aggregation Theory Properties

Assuming that set of solutions of (1) is not empty, let us denote it with $\mathsf{S}(x_1, \ldots, x_n) \subseteq \Omega$. First of all, considering the properties of a norm, we immediately obtain the following result.

Proposition 2. S *enjoys at least the following properties:*

- idempotence, *i.e.,* $\mathsf{S}(x, \ldots, x) = \{x\}$ *for all* $x \in \Omega$;
- translation equivariance, *i.e.,* $\mathsf{S}(x_1 + t, \ldots, x_n + t) = \mathsf{S}(x_1, \ldots, x_n) + t = \{y + t : y \in \mathsf{S}(x_1, \ldots, x_n)\}$ *for each* $x_1, \ldots, x_n, t \in \Omega$ *such that* $x_i + t \in \Omega$;

– scale equivariance *whenever* D *is homogeneous, i.e., if* $\mathsf{D}(sd_1,\ldots,sd_n) = s^k \mathsf{D}(d_1,\ldots,d_n)$ *for some* k *and all* d_i, *then* $\mathsf{S}(sx_1,\ldots,sx_n) = s\mathsf{S}(x_1,\ldots,x_n) = \{sy : y \in \mathsf{S}(x_1,\ldots,x_n)\}$ *for each* $x_1,\ldots,x_n \in \Omega$ *and* $s > 0$ *such that* $sx_i \in \Omega$;
– symmetry *whenever* D *is symmetric, i.e., if* $\mathsf{D}(d_1,\ldots,d_n) = \mathsf{D}(d_{\sigma(1)},\ldots,d_{\sigma(n)})$ *for any* d_i, *then* $\mathsf{S}(x_1,\ldots,x_n) = \mathsf{S}(x_{\sigma(1)},\ldots,x_{\sigma(n)})$ *for each* $x_1,\ldots,x_n \in \Omega$ *and any permutation* σ *of* $\{1,\ldots,n\}$.

In the sequel we shall consider two other natural properties of S, namely, concerning the internality and continuity.

3.1 Internalities

Let us review [11,12] some results concerning the *convex hull (CH)-internality*, $\mathsf{S}(x_1,\ldots,x_n) \cap \operatorname{conv}\{x_1,\ldots,x_n\} \neq \emptyset$, which generalizes the notion of internality in the case $X = \mathbb{R}$ (where a natural order exists, see, e.g., [16]), where we require $\min\{x_1,\ldots,x_n\} \leq y \leq \max\{x_1,\ldots,x_n\}$ for each $y \in \mathsf{S}(x_1,\ldots,x_n)$.

Theorem 3 ([12]). *Suppose that* $\dim X \geq 2$ *and fix* D. *Then for every* $x_1,\ldots, x_n \in \Omega$, $n \geq 3$, *and all* $w_1,\ldots,w_n$ *it holds* $\mathsf{S_w}(x_1,\ldots,x_n) \cap \operatorname{conv}\{x_1,\ldots,x_n\} \neq \emptyset$, *where* $\mathsf{S_w}(x_1,\ldots,x_n) \subseteq \Omega$ *is a set of minimizers of* $\mathsf{D}(w_1\|y-x_1\|,\ldots,w_n\|y-x_n\|)$.

Theorem 4 ([12]). *Suppose that* $\dim X \geq 3$ *and fix* D. *If for every* $x_1,\ldots,x_n \in \Omega$, $n \geq 3$, *and all* $w_1,\ldots,w_n$ *it holds* $\mathsf{S_w}(x_1,\ldots,x_n) \cap \operatorname{conv}\{x_1,\ldots,x_n\} \neq \emptyset$, *where* $\mathsf{S_w}(x_1,\ldots,x_n) \subseteq \Omega$ *is a set of minimizers of* $\mathsf{D}(w_1\|y-x_1\|,\ldots,w_n\|y-x_n\|)$, *then* X *is an inner product space.*

3.2 Stability of the Solution

Now let us we briefly analyze the behavior of the solution set of problem (2), where the sample $x_1,\ldots,x_n$ varies. In the case of uniformly convex Banach spaces, it is well-known that the metrical projection onto a closed and convex set is continuous (see, e.g., [4]). This easily implies the same property with respect to the solution function of problem (2). Thus, we can state the following continuity result which yields that small perturbations of a given sample lead to small perturbations of the aggregation result.

Theorem 5. *Suppose that* Ω *is a nonempty, closed and convex subset of a uniformly convex Banach space. For some* $p > 1$ *and any* $\mathbf{x} \in X^n$, $\mathbf{x} = (x_1,\ldots,x_n)$, *denote with* $\tilde{\mathsf{S}}(\mathbf{x})$ *the unique (see Theorem 2) solution of problem (2). Then for any* $\varepsilon > 0$ *there exists* $\delta > 0$ *such that for all* $\mathbf{x}, \mathbf{x}'$ *with* $\|\mathbf{x} - \mathbf{x}'\|_p < \delta$ *we have* $\|\tilde{\mathsf{S}}(\mathbf{x}) - \tilde{\mathsf{S}}(\mathbf{x}')\| < \varepsilon$.

Remark 3. In the special case where X is a Hilbert space and $p = 2$ (see Remark 2), we could easily prove a stronger, Lipschitz continuity property.

4   The Limiting Case $p = \infty$

In this section we will consider the special case of problem (2) where $w_1 = \cdots = w_n = 1$. Let $(X, \mathcal{M}, \mu)$ be now a measure space where $\mathcal{M}$ is a σ-algebra in X and $\mu : M \to [0, \infty)$ is a measure (see, e.g., [4, p. 89]). If $\mu(X) < \infty$, then $(X, \mathcal{M}, \mu)$ is called a finite measure space and if additionally $\mu(X) = 1$, it is called a probability space.

The well-known L_1 norm on X is given by $\|f\|_1 = \int_X |f| \, d\mu$, for any measurable function $f : X \to \mathbb{R}$ for which the integral exists. The space of L_1-integrable functions is denoted by $L_1(X)$. Then, for some $p > 1$, we say that f is L_p-integrable, if $|f|^p \in L_1(X)$. We denote with $L_p(X)$ the space of all L_p-integrable functions on X. The L_p norm on X is given by $\|f\|_p = (\int_X |f|^p \, d\mu)^{1/p}$. Now consider the subspace $L_\infty(X)$ where $f \in L_\infty(X)$ is measurable and there exists a constant C such that $\|f(x)\| \leq C$ almost everywhere on X. On $L_\infty(X)$ we define the infinity norm $\|\cdot\|_\infty$, where $\|f\|_\infty = \inf\{C \in \mathbb{R} : |f(x)| \leq C \text{ almost everywhere on } X\}$. For example, if $X = [a, b]$ and f is continuous, then $\|f\|_\infty$ becomes the maximum norm of f, that is $\|f\|_\infty = \max\{|f(x)| : x \in [a, b]\}$. It is well-known that $L_\infty(X) \subseteq L_p(X)$ for any $p \geq 1$. What is more, the $\|\cdot\|_\infty$ norm is the limiting case of $\|\cdot\|_p$, as $p \to \infty$, which means that $\|f\|_\infty = \lim_{p \to \infty} \|f\|_p$, $f \in L_\infty(X)$.

Let us discuss some properties of the normed spaces $(L_p(X), \|\cdot\|_p)$, $1 \leq p \leq \infty$ (see [4] for a detailed presentation). First, we recall that $(L_p(X), \|\cdot\|_p)$ is a uniformly convex Banach space for any $p \in (1, \infty)$. Then, the spaces $(L_p(X), \|\cdot\|_p)$, $p \in \{1, \infty\}$ are Banach spaces but they are not reflexive (and hence not uniformly convex) except for the cases where X has a finite number of measurable subsets, which also means that $L_p(X)$ is a finite-dimensional space.

Suppose that $\Omega \subseteq L_\infty(X)$ and consider a sample of n functions in $L_\infty(X)$, $f_1, \ldots, f_n$. We consider the problem

$$\arg\min_{f \in \Omega} \left\| \max_{i=1..n} |f_i - f| \right\|_\infty. \tag{5}$$

Now, if Ω is a nonempty, closed subset of a finite dimensional affine linear subspace of $L_\infty(X)$, then from Theorem 1, we know that the above problem has at least one solution. Indeed, it might be shown that the above problem is a best approximation problem in a normed space with respect to a closed subset of a finite dimensional vector space. For some $p \geq 1$, let us also consider the problem (2) in the setting of the space $L_p(X)$ and the same sample as above, that is,

$$\arg\min_{f \in \Omega} \sum_{i=1}^n \|f_i - f\|_p^p. \tag{6}$$

The main result of this section claims that we the following limiting behavior connecting the two above problems.

Theorem 6. *Suppose that $(X, \mathcal{M}, \mu)$ is a finite measure space and suppose that Ω is a nonempty and closed subset of a finite dimensional affine linear subspace*

$\mathcal{L}$, of $L_\infty(X)$. Furthermore, consider problems (5) and (6), respectively, with respect to the sample $f_1, \ldots, f_n$. Then, we have:

(i) both problems (5) and (6), respectively, have solutions. In addition, if Ω is convex, then for any $p > 1$, problem (6) has a unique solution;

(ii) if f_p^* is a solution of problem (6), for some $p \geq 1$, then the set $(f_p^*)_{p \geq 1}$ is bounded in $L_\infty(X)$. Then, for any convergent sequence $(f_{p_k}^*)_{k \geq 1}$ in the norm $\|\cdot\|_\infty$, extracted from $(f_p^*)_{p \geq 1}$, such that $p_k \to \infty$ and $f_{p_k}^* \to f^*$, as $k \to \infty$ (thus, f^* is a special type of a limit point for $(f_p^*)_{p \geq 1}$), we have that f^* is a solution of (5). Moreover, such a construction of f^* is always possible.

Proof. (i) Without any loss in generality suppose that $\mathcal{L}$ is a linear subspace of $L_\infty(X)$ (we can use translations in the case of affine spaces). It is well-known that in a finite dimensional vector space all norm-generated topologies are equivalent. Since Ω is closed in $L_\infty(X)$, it follows that Ω is closed in $L_p(X)$, for any $p \geq 1$. From Theorem 1, we get that problems (5) and (6), respectively, have solutions. If $p > 1$ then we get the uniqueness of the solution of problem (6), just by applying Theorem 2.

(ii) Again, without any loss of generality we may assume that $\mathcal{L}$ is a linear subspace of $L_\infty(X)$. We notice that $(f_p^*)_{p \geq 1}$ is a subset of the finite dimensional vector space $\mathcal{L}$. Since all norms are equivalent on $\mathcal{L}$, it suffices to prove that $(\|f_p^*\|_1)_{p \geq 1}$ is bounded. We need the set $\widetilde{\Omega} \subseteq L_\infty^n(X)$, $\widetilde{\Omega} = \{(f, \ldots, f) : f \in \Omega\}$ and the norm $\|\cdot\|_{n,p}$ on $L_\infty^n(X)$, where $\|(f_1, \ldots, f_n)\|_{n,p} = \left(\int_X \sum_{i=1}^n |f_i(x)|^p \, d\mu \right)^{1/p}$. Then, we need the vectors $\mathbf{F} \in \widetilde{\Omega}$ and $\mathbf{f} \in L_\infty^n(X)$, where $\mathbf{F} = (f_p^*, \ldots, f_p^*)$ and $\mathbf{f} = (f_1, \ldots, f_n)$. Then, noting that $\|f_p^*\|_1 \leq \|\mathbf{F}\|_{n,p}$ and then using an implication of the Hölder inequality (see also [4, Exercise 4.2, p. 118]) and the triangle inequality, we get

$$\|f_p^*\|_1 \leq \mu(X) \|f_p^*\|_p \leq \mu(X) \|\mathbf{F}\|_{n,p} \leq \mu(X) \left(\|\mathbf{F} - \mathbf{f}\|_{n,p} + \|\mathbf{f}\|_{n,p} \right). \qquad (7)$$

Now suppose that $\mathbf{g} = (g, \ldots, g)$ is arbitrary in $\widetilde{\Omega}$. From the definitions of $\mathbf{F}$ and $\mathbf{f}$, respectively, we get $\|\mathbf{F} - \mathbf{f}\|_{n,p} \leq \|\mathbf{g} - \mathbf{f}\|_{n,p}$. In general, for some $\mathbf{h} = (h_1, \ldots, h_n) \in L_\infty^n(X)$, it holds that $\left(\int_X (\max_{i=1..n} |h_i| (x))^p \, d\mu \right)^{1/p} \leq \left(\int_X \sum_{i=1}^n |h_i(x)|^p \, d\mu \right)^{1/p} = \|\mathbf{h}\|_{n,p} \leq \left(\int_X n (\max_{i=1..n} |h_i| (x))^p \, d\mu \right)^{1/p}$. First, since the extreme terms of the above inequalities have the same limit where $p \to \infty$, which is $\|\max_{i=1..n} |h_i|\|_\infty$, it follows

$$\|\mathbf{h}\|_{n,p} \to \left\| \max_{i=1..n} |h_i| \right\|_\infty, \text{ as } p \to \infty, \qquad (8)$$

a property which we shall find useful later on. In particular, we have $\|\mathbf{f} - \mathbf{g}\|_{n,p} \to \|\max_{i=1..n} |f_i - g|\|_\infty$, as $p \to \infty$. On the other hand, from the properties of the L_p norm and the above inequalities, we get $\|\mathbf{h}\|_{n,p} \leq n \cdot \mu(X) \|\max_{i=1..n} |h_i|\|_\infty$. Hence, $\|\mathbf{F} - \mathbf{f}\|_{n,p} \leq \|\mathbf{g} - \mathbf{f}\|_{n,p} \leq n \cdot \mu(X) \|\max_{i=1..n} \|g - f_i\|_\infty$, which, by relation (7), results in $\|f_p^*\|_1 \leq n \cdot \mu(X) \|\max_{i=1..n} |g - f_i\|_\infty + \|\mathbf{f}\|_{n,p} \leq$

$n \cdot \mu(X) \cdot (\|\max_{i=1..n} |g - f_i|\|_\infty + \|\max_{i=1..n} |f_i|\|_\infty)$. The expression $n \cdot \mu(X) \cdot (\|\max_{i=1..n} |g - f_i|\|_\infty + \|\max_{i=1..n} |f_i|\|_\infty)$ is constant with respect to p, and it means that $(\|f_p^*\|_1)_{p \geq 1}$ is a bounded set. As we said, this implies that the set $(f_p^*)_{p \geq 1}$ is bounded in $L_\infty(X)$.

Now suppose that $(f_{p_k}^*)_{k \geq 1}$ is extracted from $(f_p^*)_{p \geq 1}$ such that $p_k \to \infty$ and $f_{p_k}^* \to f^*$ in the $\|\cdot\|_\infty$ norm, as $k \to \infty$. As $(f_{p_k}^*)_{k \geq 1}$ is bounded, such a construction of f^* is indeed possible. For example, any limit point of the bounded sequence $(f_k^*)_{k \in \mathbb{N}_0}$ corresponds to this pattern. We have to prove that f^* is a solution of problem (5). First, the hypotheses imply that $f^* \in \Omega$ and what is more, for some arbitrary norm $\|\cdot\|$ defined on $L_\infty(X)$, we have $\|f_{p_k}^* - f^*\| \to 0$, as $k \to \infty$. This also means that $\|f_{p_k}^* - f\| \to \|f^* - f\|$, as $k \to \infty$, for any $f \in L_\infty(X)$. Next, we need the vector $\mathbf{F}^* = (f^*, \ldots, f^*)$ in $\widetilde{\Omega}$, and the sequence $(\mathbf{F}_{p_k}^*)_{k \geq 1}$ in $\widetilde{\Omega}$, where $\mathbf{F}_{p_k}^* = (f_{p_k}^*, \ldots, f_{p_k}^*)$. For some arbitrary norm $\|\cdot\|$ defined on $L_\infty^n(X)$, it is immediate that $\|\mathbf{F}_{p_k}^* - \mathbf{F}^*\| \to 0$, as $k \to \infty$. More generally, we have $\|\mathbf{F}_{p_k}^* - \mathbf{h}\| \to \|\mathbf{F}^* - \mathbf{h}\|$, for any $\mathbf{h}$ in $L_\infty^n(X)$. Now, consider an arbitrary $\mathbf{g} = (g, \ldots, g)$ in $\widetilde{\Omega}$, and an arbitrary $p_k \geq 1$. We have $\|\mathbf{f} - \mathbf{F}^*\|_{n,p_k} \leq \|\mathbf{f} - \mathbf{F}_{p_k}^*\|_{n,p_k} + \|\mathbf{F}_{p_k}^* - \mathbf{F}^*\|_{n,p_k}$. Suppose now that $\mathbf{g} = (g, \ldots, g)$ is arbitrary in $\widetilde{\Omega}$. The definition of $\mathbf{F}_{p_k}^*$ implies $\|\mathbf{f} - \mathbf{F}_{p_k}^*\|_{n,p_k} \leq \|\mathbf{f} - \mathbf{g}\|_{n,p_k}$. Thus, we have $\|\mathbf{f} - \mathbf{F}^*\|_{n,p_k} \leq \|\mathbf{f} - \mathbf{g}\|_{n,p_k} + \|\mathbf{F}_{p_k}^* - \mathbf{F}^*\|_{n,p_k}$. Then, we have $\|\mathbf{F}_{p_k}^* - \mathbf{F}^*\|_{n,p_k} \leq n \cdot \mu(X) \|f_{p_k}^* - f^*\|_\infty$, and since $\|f_{p_k}^* - f^*\|_\infty \to 0$, we get $\|\mathbf{F}_{p_k}^* - \mathbf{F}^*\|_{n,p_k} \to 0$, as $k \to \infty$. It means that we can pass to limit with $k \to \infty$ in the above inequality. Taking account of (8), this leads us to $\|\max_{i=1..n} |f_i - f^*|\|_\infty \leq \|\max_{i=1..n} |f_i - g|\|_\infty$ (here, it was important that $p_k \to \infty$, as $k \to \infty$). As g was chosen arbitrarily in Ω, we get that f^* is a solution of problem (5), and the proof is complete now. ∎

If we consider the general case of a not-necessarily-finite measure space, then the above reasoning does not work to prove that $(f_p^*)_{p \geq 1}$ is bounded in $L_\infty(X)$. But we can prove that $(f_p^* \cdot f_{\overline{p}}^*)_{p \geq 1}$ is bounded in $L_\infty(X)$, where for some $p \geq 1$, $\overline{p} \geq 1$ is such that $\frac{1}{p} + \frac{1}{\overline{p}} = 1$. Indeed, in this case from Hölder's inequality we get $\|f_p^* \cdot f_{\overline{p}}^*\|_1 \leq \|f_p^*\|_p \cdot \|f_{\overline{p}}^*\|_{\overline{p}} \leq \|\mathbf{F}\|_{n,p} \cdot \|\mathbf{F}\|_{n,\overline{p}}$ and from here on, the reasoning is similar to the one used in the proof of the previous theorem.

5   Conclusions

We have studied the problem of data aggregation in generic real normed vector spaces. Some existence and uniqueness results have been indicated. Moreover, we pointed out a few properties of such aggregation functions and studied the limiting case as the norm power $p \to \infty$.

Acknowledgments. The contribution of L. Coroianu was supported by a grant of Ministry of Research and Innovation, CNCS-UEFISCDI, project number PN-III-P1-1.1-PD-2016-1416, within PNCDI III. M. Gagolewski acknowledges the support by the Czech Science Foundation through the project No. 18-06915S.

References

1. Aftab, K., Hartley, R., Trumpf, J.: Generalized Weiszfeld algorithms for L_q optimization. IEEE Trans. Pattern Anal. Mach. Intell. **37**(4), 728–745 (2015)
2. Beliakov, G., Bustince, H., Calvo, T.: A Practical Guide to Averaging Functions. Springer, Heidelberg (2016)
3. Beliakov, G., James, S.: A penalty-based aggregation operator for non-convex intervals. Knowl.-Based Syst. **70**, 335–344 (2014)
4. Brezis, H.: Functional Analysis, Sobolev Spaces and Partial Differential Equations. Springer, New York (2010)
5. Bullen, P.: Handbook of Means and Their Inequalities. Springer, Dordrecht (2003)
6. Bustince, H., Beliakov, G., Dimuro, G.P., Bedregal, B., Mesiar, R.: On the definition of penalty functions in data aggregation. Fuzzy Sets Syst. **323**, 1–18 (2017)
7. Chavent, M., Saracco, J.: Central tendency and dispersion measures for intervals and hypercubes. Commun. Stat. - Theory Methods **37**, 1471–1482 (2008)
8. Clarkson, J.A.: Uniformly convex spaces. Trans. Am. Math. Soc. **40**, 396–414 (1936)
9. Coroianu, L.: Fuzzy approximation operators. Ph.D. thesis, Babes-Bolyai University, Cluj-Napoca, Romania (2013)
10. Demirci, M.: Aggregation operators on partially ordered sets and their categorical foundations. Kybernetika **42**, 261–277 (2006)
11. Durier, R.: The Fermat–Weber problem and inner-product spaces. J. Approx. Theory **78**(2), 161–173 (1994)
12. Durier, R.: Optimal locations and inner products. J. Math. Anal. Appl. **207**, 220–239 (1997)
13. Eckhardt, U.: Weber's problem and Weiszfeld's algorithm in general spaces. Math. Program. **18**, 186–196 (1980)
14. Gagolewski, M.: Data Fusion: Theory, Methods, and Applications. Institute of Computer Science, Polish Academy of Sciences, Warsaw (2015)
15. Gagolewski, M.: Penalty-based aggregation of multidimensional data. Fuzzy Sets Syst. **325**, 4–20 (2017)
16. Grabisch, M., Marichal, J.L., Mesiar, R., Pap, E.: Aggregation Functions. Cambridge University Press, Cambridge (2009)
17. Komorníková, M., Mesiar, R.: Aggregation functions on bounded partially ordered sets and their classification. Fuzzy Sets Syst. **175**, 48–56 (2011)
18. Milasevic, P., Ducharme, G.: Uniqueness of the spatial median. Ann. Stat. **15**(3), 1332–1333 (1987)
19. Ovchinnikov, S.: Means on ordered sets. Math. Soc. Sci. **32**, 39–56 (1996)
20. Pérez-Fernández, R., Rademaker, M., De Baets, B.: Monometrics and their role in the rationalisation of ranking rules. Inf. Fusion **34**, 16–27 (2017)
21. Puerto, J., Rodríguez-Chía, A.M.: Location of a moving service facility. Math. Methods Oper. Res. **49**(3), 373–393 (1999)
22. Puerto, J., Rodríguez-Chía, A.M.: New models for locating a moving service facility. Math. Methods Oper. Res. **63**(1), 31–51 (2006)
23. Torra, V. (ed.): Information Fusion in Data Mining. Studies in Fuzziness and Soft Computing, vol. 123. Springer, Heidelberg (2003)
24. Yager, R.R.: Toward a general theory of information aggregation. Inf. Sci. **68**(3), 191–206 (1993)

Gravitational Clustering Algorithm Generalization by Using an Aggregation of Masses in Newton Law

J. Armentia[1]([✉]), I. Rodríguez[2], J. Fumanal Idocin[2], Humberto Bustince[2], M. Minárová[3], and A. Jurio[2]

[1] Pamplona Planetarium, Sancho Ramírez, 31008 Pamplona, Spain
javier.armentia@gmail.com
[2] UPNA, Pamplona, Spain
bustince@unavarra.es
[3] SUT, 801 05 Bratislava, Slovakia
maria.minarova@stuba.sk

Abstract. The paper is devoted to the theoretical study of a specific clustering algorithm. Along the particular time steps of the algorithm the Newton gravitational law is used for attracting the particles each other. Our investigation is focused to a kind of generalization of Newton law by replacing the product of masses in the gravitational force equation by a more generalized aggregation of them. We study an overlap function employed, we establish a convergence criterion of such an algorithm. The controlling parameters that regulate the proximity of the particles and the convergence rate at the same time, are set up. The physical background of particle behaviour movement of which is governed by a generalized Newton gravitational force is provided along the investigation. The related geometrical interpretation is illustrated in figures.

Keywords: Gravitationl clustering · Overlap function ·
Particle system contracting

1 Introduction

Cluster analysis is a kind of a data set partitioning realized with regard to one or more attributes as to maximize the similarity within one cluster and minimize the similarity between individual clusters. Nowadays clustering is an important tool of the data mining.

Inspired by [3], the gravitational clustering algorithm utilizes the Newton gravitational force for attracting particles each others. Nevertheless, the acting of attracting force alternates with clustering itself: in each time step first the checking of the closeness of particles is done; resulting, or not, in joining some particles in one. The new particle's feature reflects the features (mass, position)

APVV 14-0013, APVV-17-0066.

of both (or more) former ones. When the clustering step is over, then an attracting force drives the actual set particles into their new positions. The magnitude of the attracting force is subjected to the parameter δ that is set prior to the algorithm launching. Parameter δ restricts the fastest particle shift within the actual step. The number of clusters, i.e. particles decreases gradually with lapsing time. Finally the unique particle remains enfolding all of them.

Herein, some theoretical treatment, regarding the physical background, is necessary, ensuring proper algorithm performance. And this theoretical investigation is described in the paper.

The paper is organized as follows. First some preliminaries is provided that introduced the entities, notions, properties that are of further use in the paper. As this theoretical investigation is devoted to a generalization of clustering algorithm as introduced by [4], this original algorithm is sketched in the next chapter. Then our design of a new - generalized algorithm is presented. In next chapters, we take the physical interpretation into account, point out to the source of possible problems and the way how to solve them. Two dimensional case is explored in details resulting in the convergence criterion establishment. The considerations are accompanied by figures.

2 Preliminaries

2.1 Aggregation Functions

Definition 1. *[2] Aggregation function is an n-dimensional function* $M :$ $[0,1]^n \rightarrow [0,1]$ *fulfilling the conditions:*

  - *M is non decreasing in each of its variable;*
  - $M(0,\ldots,0) = 0$ *and* $M(1,\ldots,1) = 1$.

As examples of aggregation function the product, averages, minimum, maximum, etc. can be taken. For $n = 2$ in Definition 1 we have so called bivariate aggregation function. Next we provide the following proprieties of bivariate functions and the definition of their special kind which will be of our interest within the further considerations.

Definition 2. *[2] Let* $M : [0,1]^n \rightarrow [0,1]$ *be an n-dimensional aggregation function*

  (i) $a \in [0,1]$ *is an annihilator of M if* $M(x_1,\ldots,x_n) = a$ *whenever* $a \in \{x_1,\ldots,x_n\}$.
  (ii) *If M does not have annihilator, then M is called strictly increasing as a real function of n variables on its domain* $[0,1]^n$ *if it is increasing in each of n variables. If M has an annihilator a, the strictly increase is done on* $([0,1]\backslash\{a\})^n$.
  (iii) *M has a zero divisors if there exist* $x_1,\ldots,x_n \in]0,1]$ *such that* $M(x_1,\ldots,x_n) = 0$.
  (iv) *M is idempotent if* $M(x,\ldots,x) = x$ *for all* $x \in [0,1]$.

Definition 3. *Let* $M : [0,1]^2 \rightarrow [0,1]$ *be an aggregation function of two variables.*

(i) M *is symmetric if* $M(x,y) = M(y,x)$ *for any* $x, y \in [0,1]$.
(ii) M *is associative if* $M(M(x,y), z) = M(x, M(y,z))$ *for any* $x, y, z \in [0,1]$.

Definition 4. *A t-norm (triangular norm) is a bivariate aggregation function* $T : [0,1]^2 \rightarrow [0,1]$ *which is symmetric, associative and* $T(1,x) = x$ *for all* $x \in [0,1]$.

In this section we bring the results of [1].

Definition 5. *[1] A function* $G_S : [0,1]^2 \rightarrow [0,1]$ *is an overlap function if*
(G_S1).- G_S *is symmetric;*
(G_S2).- $G_S(x,y) = 0$ *if and only if* $xy = 0$;
(G_S3).- $G_S(x,y) = 1$ *if and only if* $xy = 1$;
(G_S4).- G_S *is non-decreasing;*
(G_S5).- G_S *is continuous.*

Following bivariate functions are examples of overlap functions that are *t*-norms as well:

  – $G_S(x,y) = \min(x,y)$
  – $G_S(x,y) = xy$

Actually, each continuous t-norm without zero divisors ($T(x,y) = 0$ is hold only if $xy = 0$) is an overlap function. However, there exist overlap functions that are not t-norms, for example:

$$G_S(x,y) = \min(xy^k, x^k y) \quad with \quad k > 0 \tag{1}$$

which is not a t-norm if $k \neq 1$,

$$G_S(x,y) = (xy)^p \quad with \quad p > 0 \tag{2}$$

which is not a t-norm if $p \neq 1$,or

$$G_S(x,y) = \sin \frac{\pi}{2}(xy)^p \quad with \quad p > 0 \tag{3}$$

which is not a t-norm at all. The relation between overlap functions and t-norms can be summarized as follows.

Theorem 1. *[1] Let* G_S *be an associative overlap function. Then* G_S *is a t-norm.*

An overlap function can be constructed in accordance the following theorem.

Theorem 2. *[1] Function* $G_S : [0,1]^2 \rightarrow [0,1]$ *is an overlap function if and only if*

$$G_S(x,y) = \frac{f(x,y)}{f(x,y) + h(x,y)} \tag{4}$$

with $f, h : [0,1]^2 \rightarrow [0,1]$ *such that*

(1) f and h are symmetric;
(2) f is non-decreasing and h is non-increasing;
(3) f(x, y) = 0 if and only if xy = 0;
(4) h(x, y) = 0 if and only if xy = 1;
(5) f and h are continuous;

Taking $f(x, y) = \sqrt{xy}$ and $h(x, y) = \max(1 - x, 1 - y)$, we compose the overlap function

$$G_S(x, y) = \frac{\sqrt{xy}}{\sqrt{xy} + \max(1 - x, 1 - y)}. \tag{5}$$

2.2 Algorithm of Gravitational Clustering

Within the process of clustering the Newton gravitational law can be used as the governing law of attracting the particles. The scheme of the Gravitational Clustering as provided by [4] is as follows.

At the very beginning instant we have n particles $p_1, \ldots, p_n$. Their positions are know: $s_1, \ldots, s_n \in \mathcal{R}^n$.

1. Initially we assign a mass 1 to each particle p_i.
2. We take parameters ϵ and δ, real positive constants unchanged within the entire process.
   - We utilize δ for determining the actual time increment dt. So the time interval in which the most rapid particle moves by δ is $[l, l + dt]$.
   - If in a moment two particles find themselves in a distance less than ϵ, we unify them in one, the mass of resulting particle being equal to sum of masses of both of them and position done by their center of gravity.
3. Initial time $t = 0$ is set.
4. We repeat following steps (i)–(iv) until the unique particle remains.
   (i) In each time interval $[t, t + dt]$, for each particle i we compute its movement driving function:

   $$g(i, t, dt) = \frac{1}{2} G \sum_{j \neq i} \frac{m_i(t) m_j(t)}{m_i(t)} \frac{s_j(t) - s_i(t)}{|s_j(t) - s_i(t)|^3} dt^2 \tag{6}$$

   where G is a positive constant.
   (ii) The new position particle i is:

   $$s_i(t + dt) = s_i(t) + g(i, t, dt)$$

   (iii) We raise t to $t + dt$.
   (iv) If two particles i and j are in a distance less then ϵ, their unification is done as explained above.

At the very end just one particle remains.

If we denote the total time (the duration of entire process) by T, we will use the notation of the particular time steps as follows:

- Initial time is $t_n = 0$. Up to t_{n-1} time instant, there are n (all) particles remaining.
- between t_{n-1} and t_{n-2} time instants, there are $n-1$ particles remaining.
- ...
- between t_3 and t_2 there are 3 particles remaining.
- between t_2 and $t_1 = T$ there are 2 particles remaining.

The relative life duration R_k of the configuration stage with k clusters can be computed as

$$R_k = \frac{t_{k+1} - t_k}{T}$$

3   New Gravitational Clustering Algorithm

The algorithm proposed by [4] can be generalized by using more general expression for particle movement governing function instead of (6). We have done so followingly:

$$g(i, t, dt) = \frac{1}{2}G\sum_{j \neq i} \frac{m_i(t)^p m_j(t)^q}{m_i(t)} \frac{s_j(t) - s_i(t)}{|s_j(t) - s_i(t)|^3} dt^2 \qquad (7)$$

with $p, q > 0$.

And, as it is apparent from our experimental results, the best results are for $p = q = 0$, of unitary Markov model with

$$g(i, t, dt) = \frac{1}{2}G\sum_{j \neq i} \frac{1}{m_i(t)} \frac{s_j(t) - s_i(t)}{|s_j(t) - s_i(t)|^3} dt^2 \qquad (8)$$

3.1   Description and Actuating of the Proposed Gravitational Clustering Algorithm

In the next chapters we study theoretically the behaviour of gravitational cluster algorithm in case of $G_S(1, \frac{1}{m_i(t)})$ taken in Newton's law, where $G_S : [0,1]^2 \to [0,1]$ is an arbitrary overlap function instead of $\frac{1}{m_i(t)}$ in unitary Markov model (8).

Since an overlap function is given on the unit square $[0,1]^2$, a normalization is needed. It can be realised by using e.g.

$$G_S(\frac{m_i^{p-1}}{n^{p-1}}, \frac{m_j^q}{n^q})$$

with $p, q > 0$. However, we focus on the case p, q going to zero (the best referred case in original algorithm), it is hold that this function is always greater or equal then $G_S(\frac{1}{m_i}, 1)$. So, let us have an overlap function $G_S : [0,1]^2 \to [0,1]$. When a configuration of n particles is given: $p_1, \ldots, p_n$, and their positions $s_1, \ldots, s_n \in \mathcal{R}^n$, the new algorithm we propose, is:

1. Initially we assign a mass 1 to each particle p_i.
2. We take two real positive parameters ϵ and δ constant during whole process of clustering, $\delta < \epsilon$ if the clustering is prior to attracting.
   - Form δ we determine the actual time step longitude of the actual $dt = dt(t, \delta)$. Since δ is fixed, we have just $dt(t)$. Indeed, $[t, t + dt(t)]$ is the time period in which the most rapid particle moves by δ.
   - We check whether any two particles find themselves in a distance less than ϵ. If so, we unify them in one particle with the mass equal to the sum of masses of both of them and position in their center of gravity.
3. We set the initial time to zero.
4. Then we repeat steps (i)–(iv) until the unique particle remains.
   (i) In each time interval $[t, t + dt]$, for each particle i we compute its movement driving function:

$$\boldsymbol{g}_i(t) = \frac{1}{2} G \sum_{j \neq i} G_S(1, \frac{1}{m_i(t)}) \frac{\boldsymbol{s}_j(t) - \boldsymbol{s}_i(t)}{|\boldsymbol{s}_j(t) - \boldsymbol{s}_i(t)|^3} dt^2 \qquad (9)$$

   where G is a positive constant.
   (ii) First we find the fastest particle p_I

$$I = \arg(\max_i \{|\boldsymbol{g}_i|(t)\}),$$

   then we compute the length $dt(t)$ of current time step

$$|\boldsymbol{g}_I(t)| = \delta \Rightarrow dt(t) \qquad (10)$$

   and finally we assign the new position of each particle i:

$$\boldsymbol{s}_i(t + dt(t)) = \boldsymbol{s}_i(t) + \boldsymbol{g}_i(t) \qquad (11)$$

   (iii) We raise t to $t + dt(t)$.
   (iv) If two particles i and j are in a distance less then ϵ, their unification is done as explained above.

Finally we have just one particle. The duration of the entire process is denoted by T. The evolution of the process is recorded in the sequence $\{t_{n-k_1}, t_{n-k_2}, ..., t_{n-k_l} = t_1\}, 0 < k_1 < ... < k_q < n$; here we record all time moments when a change in the number of particles happened. The hierarchical indexing t_{n-k}, etc., reflects the fact that the number of particles can decrease by more than one within one level of configuration. The current number of particles can be referred by indexing. Then

- in time interval $[t_n = 0, t_{n-k_1}[$ the number of particles is n.
- in time interval $[t_{n-k_1}, t_{n-k_2}[$ the number of particles is $n - k_1$.
- ...
- in time interval $[t_{n-k_{q-1}} = t_2, t_{n-k_q} = t_1[$ just two particles remain.
- in $[t_1, \infty[$ all data are gathered in one cluster, one particle remains.

The relative life span of one level of clustering is defined as

$$R_k = \frac{t_{n-k_l} - t_{n-k_{l+1}}}{T}$$

And these 100% of the total time $T = R_1 = t_1 - t_n$ can be split at q parts with q values of relative particular times assigned.

3.2   Study of Convergence. Single Particle Movement. Governing Mapping Contraction

The movement of particles system is driven by Newton gravitational force and the Markov process is considered, i.e. no history influences are included. Therefore, in order to prove the convergence of the algorithm it is sufficient to ensure the contraction the system with lapsing time.

Physical interpretation gives us the imagination of this contraction of the system with time lapsing. In accordance with the gravitational force each single particle p_i is driven towards the center of gravity of all the other particles denoted by $C_{(-i)}$. However, generally this movement does not necessary accords with the movement to the global center of gravity, see Fig. 1b.

Accordingly, **the restriction of maximal shift value δ to a "sufficiently small" value prevents a possibility of escaping particles from their convex hull.** It means either particle p_i, when attracted by all other particles, tends to the global centre of gravity C (with its position vector $\boldsymbol{c} = (c_1, ..., c_D)$) as it proceeds towards $C_{(-i)}$ or, if not, then this particle is sufficiently far from the border of the convex hull and cannot over-cross the border of it within this step. Accordingly, the set $\{\boldsymbol{c}, \boldsymbol{s}_i(t), \boldsymbol{c}_{(-i)}(t), \delta_i = |g_i(t)|\}$ determines "oncoming diapason" for particle movement. As performed in Fig. 1 left, if the $2D$ driving vector of particle movement falls into the angle $(P_1 S_i P_2) = 2\alpha$, then the particle p_i approaches towards both $C_{(-i)}$ and C. In $3D$ a cone is the oncoming diapason, etc.

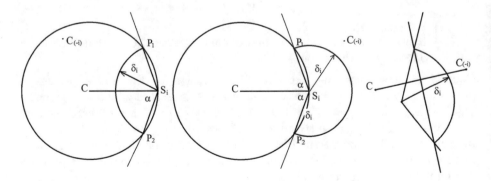

Fig. 1. Configuration of global and local centre of gravity when a particle is approaching (left), retreating the global centre of gravity C (right) within one time step for $\delta_i \leq 2|CS_i|$

As the constant G in (9) does not influence the convergence, we can put $\frac{1}{2}G = 1$. Then, rewritten (9) coordinate-wisely is

$$g_i^d(t) = \sum_{j \neq i} G_S(1, \frac{1}{m_i(t)}) \frac{s_j^d(t) - s_i^d(t)}{|\boldsymbol{s}_j(t) - \boldsymbol{s}_i(t)|^3} dt^2(t) \qquad (12)$$

As set by the algorithm, the fastest particle I moves by δ within one time interval. That determines the longitude of the particular $dt(t)$ for $G_S(x, y) = xy$:

$$\delta = |\boldsymbol{g}_I(t)| = G_S\left(1, \frac{1}{m_I(t)}\right)\left|\sum_{j\neq I} \frac{\boldsymbol{s}_j(t) - \boldsymbol{s}_I(t)}{|\boldsymbol{s}_j(t) - \boldsymbol{s}_I(t)|^3}\right| dt^2(t) \Rightarrow$$

$$\Rightarrow dt^2(t) = \frac{\delta m_I}{\left|\sum_{j\neq I} \frac{\boldsymbol{s}_j(t) - \boldsymbol{s}_I(t)}{|\boldsymbol{s}_j(t) - \boldsymbol{s}_I(t)|^3}\right|}(> 0) \tag{13}$$

2D Interpretation:

We deal with N particles, with position vectors in a time instant t: $\boldsymbol{s}_i(t) = (x_i(t), y_i(t))$, $i = 1, \ldots N$. First we create the convex hull of the space occupied by all particles and immerse the convex hull into a space invariant minimal covering rectangle $\langle mX(t), MX(t) \rangle \times \langle mY(t), MY(t) \rangle$, where

$$mX(t) = \min_i\{x_i(t)\}, mY(t) = \min_i\{y_i(t)\} \tag{14}$$

$$MX(t) = \max_i\{x_i(t)\}, MY(t) = \max_i\{y_i(t)\} \tag{15}$$

see in Fig. 2. Then the shrinking of the rectangle will proceed together with shrinking of the convex hull.

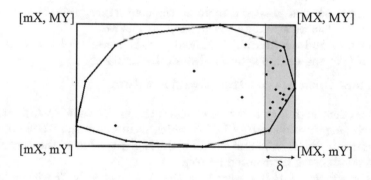

Fig. 2. Convex hull of the particles with its minimal covering rectangle and the right hand side (RHS) δ−strip

As reasoned above, we have to prevent the particles having their position in the δ margin to reach or over cross the boundary of the minimal covering rectangle. We do so coordinate by coordinate.

Let us focus our attention to the right hand side δ strip of the covering rectangle, i.e. the subdomain $\langle MX - \delta, MX) \times \langle mY, MY \rangle$, see Fig. 2. Each particle p_R in this strip has its first coordinate $s_R^1 \geq MX - \delta$. If we want to prevent this particle to over-cross the right hand side margin of the covering rectangle, we have to ensure $s_R^1 + g_R^1 < MX$, i.e.

$$g_R^1 < MX - s_R^1. \tag{16}$$

As the right hand side of (16) is greater than zero, by restricting g_R^1 as to make it non-positive, i.e. $g_R^1 \leq 0 < MX - s_R^1$, we put a stronger criterion for ensuring the convergence. Physically spoken, coordinate by coordinate we detect all particles occurring in the δ margin and suppress all their outward movement in sense of that coordinate. So we have established the criterion

$$G_S(1, \frac{1}{m_R(t)}) \sum_{j \neq R} \frac{s_j^1 - s_R^1}{|s_j - s_R|^3} dt^2 \leq 0 \iff \sum_{j \neq R} \frac{s_j^1 - s_R^1}{|s_j - s_R|^3} \leq 0 \iff$$

$$\iff \sum_{j \neq R} \frac{s_j^1 - s_R^1 + |s_j^1 - s_R^1|}{2|s_j - s_R|^3} + \sum_{j \neq R} \frac{s_j^1 - s_R^1 - |s_j^1 - s_R^1|}{2|s_j - s_R|^3} \leq 0 \iff$$

$$\iff \sum_{j \neq R} \frac{s_j^1 - s_R^1 + |s_j^1 - s_R^1|}{2|s_j - s_R|^3} \leq \sum_{j \neq R} \frac{s_R^1 - s_j^1 + |s_j^1 - s_R^1|}{2|s_j - s_R|^3} \tag{17}$$

The first sum of (17) involves all particles having its first coordinate from the interval $(MX - \delta, MX)$, i.e. those within the RHS δ strip, see Fig. 2. The second sum in (17) involves all other particles. Particles with the same first coordinate as s_R^1 will not have an influence here.

Herein (12) will be called the MX criterion of convergence (as far as the first coordinate concerns).

It is apparent that greater δ increase the convergence rate. But we have to keep in mind that we are supposed to keep (17) as well.

Let us now find the limiting (minimal possible) position for R that ensures validity of (17) and denote it by R^1. We do so iteratively.

Parameters Optimization Iterative Algorithm

(i) Let us deal with $d = 1$ first. Let us set the initial value of R for the RHS δ strip, e.g. to too small $\frac{MX - mX}{2}$ which means that the RHS δ strip span is starts in the middle and ends on the highest value of the coordinate d. Spatial step will be denoted as *step*.

The existence of R is ensured by the fact that $g^1 < 0$ within the RHS margin (RHS δ strip).

(ii) If (17) fulfilled then $R^d = R$, finish

(iii) $R = R + step$ if $R \geq MX$ then decrease *step* and proceed again with step (i)

The left hand side (LHS) δ strip can be treated likewise, yielding r^1. The criterion of convergence near lower border of the covering rectangle will be of the form

$$\sum_{j \neq r} \frac{s_j^1 - s_r^1 + |s_j^1 - s_r^1|}{2|s_j - s_r|^3} \geq \sum_{j \neq r} \frac{s_r^1 - s_j^1 + |s_j^1 - s_r^1|}{2|s_j - s_r|^3} \tag{18}$$

Then proceeding coordinate by coordinate, all values R^d and r^d can be computed and for all coordinates d the corresponding criteria are constituted.

Lemma 1. *Magnitude of the most rapid particle shift in space*

$$\delta = \min_{d}(\min\{r^d - mX, MX - R^d\}) \tag{19}$$

within one time step prevents the particles in the distance less or equal to δ from a margin of minimal covering hyper-prism from moving towards the margin.

Proof. The proof follows directly from the construction of δ.

Theorem 3. *Supposing $G_S(x,y)$ is an overlap function, the minimal space invariant covering rectangle $\langle mX, MX \rangle \times \langle mY, MY \rangle$ shrinks with time lapsing when the particles motion is driven by (12).*

Proof. 1. It is to be proven in $2D$ that

$$mX(t) < mX(t + dt(t)) \tag{20}$$
$$mY(t) < mY(t + dt(t)) \tag{21}$$
$$MX(t) > MX(t + dt(t)) \tag{22}$$
$$MY(t) > MY(t + dt(t)) \tag{23}$$

For $s_{mX} = (mX, Y)$ we have

$$g_{mX}^1(t) = G_S(1, \frac{1}{m_{mX}(t)}) \sum_{j \neq mX} \frac{s_j^1(t) - s_{mX}^1(t)}{|s_j(t) - s_{mX}(t)|} \frac{1}{|s_j(t) - s_{mX}(t)|^2} dt^2(t) > 0 \tag{24}$$

And

$$mX(t + dt(t)) = mX(t) + g_{mX}^1(t)$$

implies (20).
For $s_{mY} = (X, mY)$, $s_{MX} = (MX, Y)$, $s_{MY} = (X, MY)$, the proof of (21)–(23) is analogous. Moreover, it is apparent, that by proceeding coordinate by coordinate we can easily enhance 2 dimensional case to D dimensional.
2. As the Lemma (1) performs, the particles tending towards a margin of minimal covering hyper-prism, are situating sufficiently far from this border.

4   Conclusion

In this paper we have introduced the theoretical investigation on the new proposed gravitational algorithm. The generalization lies in the implementation of an overlap function in attracting force formula. The convergence treatment of the algorithm is done in accordance with the physical phenomena. For the proper functioning of the algorithm a set of convergence criteria is raised containing a couple of criteria for each dimension. The parameters controlling the convergence rate is optimized iteratively.

Regarding to the related computational experimentation running nowadays, it is natural to consider a more general function H instead of G_s in (9) in future.

References

1. Bustince, H., Fernandez, J., Mesiar, R., Montero, J., Orduna, R.: Overlap functions. Nonlinear Anal. **72**, 1488–1499 (2010)
2. Grabisch, M., Marichal, J.-L., Mesiar, R., Pap, E.: Aggregation Functions. Cambridge University Press, Cambridge (2009)
3. Rashedi, E., Nezamabadi-pour, H., Saryazdi, S.: GSA: a gravitational search algorithm. Inf. Sci. **179**, 2232–2248 (2009)
4. Wright, W.E.: Gravitational clustering. Pattern Recogn. **9**, 151–166 (1977)

Uninorms on Bounded Lattices with Given Underlying Operations

Slavka Bodjanova[1] and Martin Kalina[2($\boxtimes$)]

[1] Department of Mathematics, Texas A&M University-Kingsville,
MSC 172, Kingsville, TX 78363, USA
kfsb000@tamuk.edu
[2] Department of Mathematics, Faculty of Civil Engineering,
Slovak University of Technology in Bratislava,
Radlinského 11, 810 05 Bratislava, Slovakia
kalina@math.sk

Abstract. The main goal of this paper is to construct uninorms on a bounded lattice L which are neither conjunctive nor disjunctive (i.e., are of the third type), and have a given underlying t-norm and t-conorm. Three different cases will be studied. The first one will present some (quite general) conditions under which it is possible to construct a uninorm of the third type, regardless of the type of underlying t-norm and t-conorm. Then, two different cases of idempotent uninorm will be presented. Finally, a uninorm with a given underlying t norm and t-conorm will be presented.

Keywords: Bounded lattice · Idempotent uninorm · t-conorm ·
t-norm · Uninorm · Uninorm of the third type ·
Uninorm which is neither conjunctive nor disjunctive

1 Introduction

Uninorms on the unit interval are special types of aggregation functions since, due to their associativity they can be straightforwardly extended to n-ary operations for arbitrary $n \in \mathbb{N}$. They are important in various fields of applications, e.g., neuron nets, fuzzy decision making and fuzzy modelling. They are interesting also from a theoretical point of view. Important is, among others, the class of idempotent uninorms, studied, e.g., in [1]. Recently they have been studied on bounded lattices (see, e.g., [3,6–8,17]).

Uninorms were introduced by Yager and Rybalov [20]. Special types of associative, commutative and monotone operations with neutral elements had been already studied in [9,12,13]. Deschrijver [10,11] has shown that on the lattice $L^{[0,1]}$ of closed subintervals of the unit interval there exist uninorms which are neither conjunctive nor disjunctive (i.e., whose annihilator is different from both, $\mathbf{0}$ and $\mathbf{1}$). Particularly, he constructed uninorms having the neutral element $\mathbf{e} = [e, e]$, where $e \in {]0, 1[}$. In [17] the authors have shown that on arbitrary

© Springer Nature Switzerland AG 2019
R. Halaš et al. (Eds.): AGOP 2019, AISC 981, pp. 183–194, 2019.
https://doi.org/10.1007/978-3-030-19494-9_17

bounded lattice L it is possible to construct a uninorm regardless which element of L is chosen to be the neutral one. A different type of construction of uninorms on bounded lattices was presented in [3].

In [16] construction of a uninorm for arbitrary pair $(\mathbf{e}, \mathbf{a})$ of incomparable elements such that $\mathbf{e}$ is the neutral element and $\mathbf{a}$ the absorbing one, was presented. In [15] the author showed that on some special bounded lattices, one can construct operations which are both, proper uninorms and nullnorms, meaning that their neutral element, as well as annihilator are different from both, $\mathbf{0}$ and $\mathbf{1}$.

2   Basic Notations and Some Known Facts and Notions

Bounded lattices will be considered (for details on bounded lattices see the monograph [2]). L will denote the set of all elements of the lattice. If it will cause no confusion, by L will be denoted also the lattice itself. Every bounded lattice $(L, \wedge, \vee, \mathbf{0}, \mathbf{1})$ is equipped with a partial order $\leq$ given by

$$(\forall a, b \in L)(a \leq b \iff a \wedge b = a).$$

Then the notion of a closed interval can be introduced as follows

$$\text{for all } a \leq b \quad [a, b] = \{x \in L; a \leq x \leq b\}.$$

Further Notations

(1) Let $a \in L$ be a fixed element. Then $\|_a$ denotes the set of all elements of L that are incomparable with a.

(2) For all $x, y \in L$ we denote $x \underset{>}{\lessgtr} y$ if x and y are comparable.

(3) For all $x \neq \mathbf{0}$, by $[\mathbf{0}, x[$ we will denote the semi-open interval, i.e.,

$$[\mathbf{0}, x[= [\mathbf{0}, x] \setminus \{x\}.$$

(4) For all $x \neq \mathbf{1}$, by $]x, \mathbf{1}]$ we will denote the semi-open interval, i.e.,

$$]x, \mathbf{1}] = [x, \mathbf{1}] \setminus \{x\}.$$

Let us recall some basic properties of binary operations.

Definition 1. *Let L be a bounded lattice and $*$ be a binary commutative operation on L. Then*

(i) *element $c \in L$ is said to be* idempotent *if $c * c = c$,*
(ii) *element $e \in L$ is said to be* neutral *if $e * x = x$ for all $x \in L$,*
(iii) *element $a \in L$ is said to be* annihilator *if $a * x = a$ for all $x \in L$.*

Lemma 1. *Let $*$ be a commutative and associative operation on L. Further, let c be an idempotent element. Assume that there exist elements $x, y \in L$ such that $x * c = y$. Then also $y * c = y$.*

Proof. By the assumptions, $y = x * c = x * (c * c) = (x * c) * c = y * c$. $\square$

Schweizer and Sklar [19] introduced the notion of a triangular norm (t-norm for brevity).

Definition 2 ([19]). *An operation* $T : [0,1]^2 \to [0,1]$ *is a t-norm if it is associative, commutative, monotone, and* 1 *is its neutral element.*

T-norms and t-conorms are dual to each other. If $T : [0,1]^2 \to [0,1]$ is a t-norm, then

$$S(x,y) = 1 - T(1 - x, 1 - y)$$

is the dual t-conorm to T. For details on t-norms and t-conorms see, e.g., [18].

As a generalization of both t-norms and t-conorms Yager and Rybalov [20] proposed the notion of uninorm.

Definition 3 ([20]). *An operation* $U : [0,1]^2 \to [0,1]$ *is a uninorm if it is associative, commutative, monotone, and if it possesses a neutral element* $e \in [0,1]$.

A uninorm U is *proper* if its neutral element $e \in {]0,1[}$.

Every uninorm has an annihilator. A uninorm with the annihilator 0 is conjunctive, and a uninorm with annihilator 1 is disjunctive.

Lemma 2 ([20]). *Let* $U : [0,1]^2 \to [0,1]$ *be a uninorm whose neutral element is* e. *Then its dual operation*

$$U^d(x,y) = 1 - U(1 - x, 1 - y)$$

is a uninorm whose neutral element is $1 - e$. *Moreover,* U *is conjunctive if and only if* U^d *is disjunctive.*

Results in paper [20] imply the following assertion.

Lemma 3. *Let* $U : [0,1]^2 \to [0,1]$ *be a uninorm whose neutral element is* e. *Then there exists a t-norm* $T_U : [0,1]^2 \to [0,1]$ *and a t-conorm* $S_U : [0,1]^2 \to [0,1]$ *such that*

$$(\forall x, y \in [0,e]^2)(U(x,y) = e \cdot T_U(\tfrac{x}{e}, \tfrac{y}{e})),$$
$$(\forall x, y \in [e,1]^2)(U(x,y) = e + (1 - e) \cdot S_U(\tfrac{x-e}{1-e}, \tfrac{y-e}{1-e})).$$

Lemma 4 ([20]). *Assume* U *is a uninorm with neutral element* e. *Then:*

1. *for any* x *and all* $y > e$ *we get* $U(x,y) \geq x$,
2. *for any* x *and all* $y < e$ *we get* $U(x,y) \leq x$.

For more information on associative (and monotone) operations on $[0,1]$ refer to the monographs [5, 14, 18].

3   Uninorms with Given Underlying t-norm and t-conorm

Based on an arbitrary t-norm and a t-conorm defined on L, $[\mathbf{0}, \mathbf{e}]$ and $[\mathbf{e}, \mathbf{1}]$, respectively, a conjunctive and a disjunctive uninorm can be constructed [3]. However, in [3], the formulae were published with an error that resulted into the fact that the operations were without neutral element. Now, we provide the corrected formulae.

Proposition 1. *Let $T : L \times L \to L$ and $S : L \times L \to L$ be a t-norm and a t-conorm, respectively. Further, let $\mathbf{e} \in L \setminus \{0, 1\}$ be arbitrarily chosen. Then*

$$
U_c(x, y) = \begin{cases}
S(x, y) & \text{if } (x, y) \in \,]\mathbf{e}, \mathbf{1}]^2, \\
x & \text{if } y > \mathbf{e}, x \not\geq \mathbf{e}, \text{ or if } y = \mathbf{e}, \\
y & \text{if } x > \mathbf{e}, y \not\geq \mathbf{e}, \text{ or if } x = \mathbf{e}, \\
T(x, y) & \text{if } (x, y) \in [\mathbf{0}, \mathbf{e}[^2, \\
T(x \wedge \mathbf{e}, y \wedge \mathbf{e}) & \text{otherwise,}
\end{cases} \tag{1}
$$

$$
U_d(x, y) = \begin{cases}
T(x, y) & \text{if } (x, y) \in [\mathbf{0}, \mathbf{e}[^2, \\
x & \text{if } y < \mathbf{e}, x \not\geq \mathbf{e}, \text{ or if } y = \mathbf{e}, \\
y & \text{if } x < \mathbf{e}, y \not\geq \mathbf{e}, \text{ or if } x = \mathbf{e}, \\
S(x, y) & \text{if } (x, y) \in \,]\mathbf{e}, \mathbf{1}]^2, \\
S(x \vee \mathbf{e}, y \vee \mathbf{e}) & \text{otherwise.}
\end{cases} \tag{2}
$$

Then U_c is a conjunctive uninorms and U_d is a disjunctive uninorm, each of them with its neutral element $\mathbf{e}$.

Proof. The commutativity of both, U_c and U_d, is straightforward by the fact that T and S are commutative. The second and third items of (1) and (2) imply that $\mathbf{e}$ is the neutral element of both, U_C and U_d. Increasingness of the operations is due to the fact that $T(x, y) \leq x \wedge y$ and $S(x, y) \geq x \vee y$ are increasing operations.

   Let us prove the associativity of U_c. When looking at formula (1), we can see that, for $(x, y) \in (\|_\mathbf{e} \cup [\mathbf{0}, \mathbf{e}])^2$, we have

$$
U_c(x, y) = \begin{cases}
x & \text{if } y = \mathbf{e}, \\
y & \text{if } x = \mathbf{e}, \\
T(x \wedge \mathbf{e}, y \wedge \mathbf{e}) & \text{otherwise,}
\end{cases}
$$

i.e., $U_c \restriction (\|_\mathbf{e} \cup [\mathbf{0}, \mathbf{e}])^2$ behaves like a t-norm. $U_c \restriction [\mathbf{e}, \mathbf{1}]^2$ is a t-conorm, and all elements from $[\mathbf{e}, \mathbf{1}]$ serve like partial neutral elements for $\|_\mathbf{e}$, as well as for elements less than $\mathbf{e}$.

   Similarly we could prove the associativity of U_d.   □

   Now, we are going to investigate under which conditions, choosing incomparable $\mathbf{e}$ and $\mathbf{a}$, and a given t-norm $T_\mathbf{e} : [\mathbf{0}, \mathbf{e}]^2 \to [\mathbf{0}, \mathbf{e}]$ and a t-conorm

$S_{\mathbf{e}} : [\mathbf{e}, \mathbf{1}]^2 \to [\mathbf{e}, \mathbf{1}]$ there exists a uninorm U with the neutral element $\mathbf{e}$ and annihilator $\mathbf{a}$.

Following notation and assumption from paper [4], we adopt that Assumption (A) and (slightly modified) Notation (P).

Assumption (A) Let $(L, \wedge, \vee, \mathbf{0}, \mathbf{1})$ be a bounded lattice with two distinguished elements, $\mathbf{a}$, and $\mathbf{e}$, such that $\mathbf{a} \parallel \mathbf{e}$.

Notation (P) We denote $\varepsilon = \mathbf{e} \wedge \mathbf{a}$ and $\delta = \mathbf{e} \vee \mathbf{a}$.

Elements of $L \setminus \{\mathbf{a}, \mathbf{e}\}$ are partitioned into 9 different subsets as follows:
1. $P_1 = \{x \in L; \, x \le \varepsilon\} = [\mathbf{0}, \varepsilon]$,
2. $P_2 = \{x \in L \setminus P_1; \, x < \mathbf{e}\}$,
3. $P_3 = \{x \in L \setminus P_1; \, x < \mathbf{a}\}$,
4. $P_4 = \{x \in L; \, x \ge \delta\} = [\delta, \mathbf{1}]$,
5. $P_5 = \{x \in L \setminus P_4; \, x > \mathbf{e}\}$,
6. $P_6 = \{x \in L \setminus P_4; \, x > \mathbf{a}\}$,
7. $P_7 = \{x \in \|_{\mathbf{e}} \cap \|_{\mathbf{a}}; \varepsilon < x < \delta\}$,
8. $P_8 = \{x \in \|_{\varepsilon} \cap \|_{\delta}; ((\exists z \in P_2 \cup P_5)(z \lessgtr x) \text{ or} (\forall z \in P_3 \cup P_6)(z \parallel x))\}$,
9. $P_9 = L \setminus \{\mathbf{a}, \mathbf{e}\} \setminus (P_1 \cup P_2 \cup P_3 \cup P_4 \cup P_5 \cup P_6 \cup P_7 \cup P_8)$.

Assume L is a bounded lattice fulfilling Assumption (A). We are going to construct uninorms on L which are neither conjunctive nor disjunctive.

Definition 4. *Let L be a bounded lattice and $U : L \times L \to L$ be a uninorm whose absorbing element is $\mathbf{a} \in L \setminus \{\mathbf{0}, \mathbf{1}\}$. Then U is called the third type uninorm.*

The next proposition from [4] presents some necessary conditions to be fulfilled by a bounded lattice L, if there exists a uninorm of the third type on L.

Proposition 2 ([4]). *Let L be a lattice fulfilling Assumption (A), and let $U : L \times L \to L$ be a uninorm of the third type with the neutral element $\mathbf{e}$ and absorbing element $\mathbf{a}$. Then:*

(i) for all $x, y \in P_2$ we have $U(x, y) \in P_2$,
(ii) for all $x, y \in P_4$ we have $U(x, y) \in P_5$
(iii) if $P_3 = \emptyset$, then for all $x, y \parallel \mathbf{a}$ we have $U(x, y) \not\le \mathbf{a}$,
(iv) if $P_6 = \emptyset$, then for all $x, y \in \|_{\mathbf{a}}$ we have $U(x, y) \not\ge \mathbf{a}$.

Now, we are going to construct uninorms with given underlying t-norm and t-conorm. First, we handle the easiest case.

Proposition 3. *Let L be a lattice fulfilling Assumption (A). Assume all areas P_1–P_9 are non-empty and there exist elements $\zeta_2 \in P_2$ and $\zeta_5 \in P_5$ which are the least element of P_2 and the greatest element of P_5, respectively. Let $\zeta_8 \in P_8$ be arbitrarily chosen. Then $U_1 : L \times L \to L$, defined by*

$$U_1(x,y) = \begin{cases} x & \text{if } y = \mathbf{e}, \text{ or } x \in P_1 \cup P_3 \text{ and } y \in P_5 \cup P_7, \\ & \quad \text{or } y \in P_2 \text{ and } x \in P_3, \text{ or } y \in P_8 \text{ and } x \in P_3 \cup P_6, \\ & \quad \text{or } x \in L \setminus (P_1 \cup P_2 \cup P_8 \cup P_9 \cup \{\mathbf{e}\}) \text{ and } y \in P_2, \\ y & \text{if } x = \mathbf{e}, \text{ or } y \in P_1 \cup P_3 \text{ and } x \in P_5 \cup P_7, \\ & \quad \text{or } x \in P_2 \text{ and } y \in P_3, \text{ or } x \in P_8 \text{ and } y \in P_3 \cup P_6, \\ & \quad \text{or } y \in L \setminus (P_1 \cup P_2 \cup P_8 \cup P_9 \cup \{\mathbf{e}\}) \text{ and } x \in P_2, \\ \mathbf{a} & \text{if } x = \mathbf{a}, \text{ or } y = \mathbf{a}, \text{ or } x \leq \mathbf{a} \text{ and } y \geq \mathbf{a}, \\ & \quad \text{or } y \leq \mathbf{a} \text{ and } x \geq \mathbf{a}, \text{ or } x \in P_8 \text{ and } y \in P_9, \\ & \quad \text{or if } y \in P_8 \text{ and } x \in P_9, \text{ or } (x,y) \in P_9^2, \\ x \wedge \mathbf{a} & \text{if } y \in P_1 \cup P_2 \cup P_3 \text{ and } x \in P_9, \\ y \wedge \mathbf{a} & \text{if } x \in P_1 \cup P_2 \cup P_3 \text{ and } y \in P_9, \\ x \vee \mathbf{a} & \text{if } y \in P_4 \cup P_5 \cup P_6 \cup P_7 \text{ and } x \in P_9, \\ y \vee \mathbf{a} & \text{if } x \in P_4 \cup P_5 \cup P_6 \cup P_7 \text{ and } y \in P_9, \\ \zeta_8 & \text{if } (x,y) \in P_8^2, \\ \zeta_5 & \text{if } (x,y) \in (P_5 \cup P_7)^2, \text{ or } x \in P_5 \cup P_7 \text{ and } y \in P_8, \\ & \quad \text{or } y \in P_5 \cup P_7 \text{ and } x \in P_8, \\ \zeta_2 & \text{if } (x,y) \in P_2^2, \text{ or } x \in P_2 \text{ and } y \in P_8, \\ & \quad \text{or } y \in P_2 \text{ and } x \in P_8, \\ \mathbf{0} & \text{if } (x,y) \in (P_1 \cup P_3)^2, \text{ or } x \in P_1 \text{ and } y \in P_2 \cup P_8, \\ & \quad \text{or } y \in P_1 \text{ and } x \in P_2 \cup P_8, \\ \mathbf{1} & \text{if } (x,y) \in (P_4 \cup P_6)^2, \\ & \quad \text{or } x \in P_4 \cup P_6 \text{ and } y \in P_5 \cup P_7 \cup P_8, \\ & \quad \text{or } y \in P_4 \cup P_6 \text{ and } x \in P_5 \cup P_7 \cup P_8, \end{cases}$$

is a uninorm, whose neutral element is $\mathbf{e}$ and annihilator is $\mathbf{a}$.

Remark 1. The uninorm U_1 from Proposition 3 could be designed in several modifications. E.g., we could define $\tilde{U}_1(x,y) = \zeta_2$ for $(x,y) \in P_7$ and $\tilde{U}_1(x,y) = \min(x,y)$ for $(x,y) \in P_2 \times P_5 \cup P_5 \times P_2$.

Further, for $(x,y) \in P_2 \times P_3 \cup P_3 \times P_2$ we could define $\tilde{U}_1(x,y) = \mathbf{0}$, etc.

Now, we provide an idempotent uninorm.

Proposition 4. *Let L be a lattice fulfilling Assumption (A) and constraints (i) and (ii) form Proposition 2. Assume $P_8 = \emptyset$. Further assumptions are the following:*

(i) $x \wedge y \in P_9$ *for all* $(x,y) \in P_9^2$,
(ii) $x \wedge y \in P_3$ *for all* $(x,y) \in P_3^2$,
(iii) *there exists a sub-area of P_3 denoted by $\bar{P}_3$ such that*

$$(x = z \wedge \mathbf{a} \in \bar{P}_3) \, \& \, ([x, \mathbf{a}[\subset \bar{P}_3)$$

holds for all $z \in P_9$,

(iv) *there exists a sub-area of P_6 denoted by $\bar{P}_6$ such that*

$$(x = z \vee \mathbf{a} \in \bar{P}_6) \;\&\; (]\mathbf{a}, x] \subset \bar{P}_6)$$

holds for all $z \in P_9$,
(v) $x \vee y \in P_7$ *for all* $(x, y) \in P_7^2$,
(vi) $(x \wedge y) \vee \mathbf{a} = (x \vee \mathbf{a}) \wedge (y \vee \mathbf{a})$ *for all* $(x, y) \in P_9^2$,
(vii) $x \wedge \mathbf{a} \in P_3$ *for all* $x \in P_5 \cup P_7$.

We denote $\tilde{P}_3 = P_3 \setminus \bar{P}_3$ and $\tilde{P}_6 = P_6 \setminus \bar{P}_6$. Then $U_2 : L \times L \to L$, defined by

$$
U_2(x, y) =
\begin{cases}
x & \begin{aligned}&\text{if } y = \mathbf{e}, \text{ or } y \in P_1 \cup P_2 \cup P_5 \cup P_7 \text{ and } x \in \bar{P}_3,\\ &\text{or } y \in P_2 \text{ and } x \in L \setminus (P_1 \cup P_2 \cup \tilde{P}_3 \cup \mathbf{e}),\\ &\text{or } x \in \bar{P}_3 \text{ and } y \in \tilde{P}_3,\\ &\text{or } x \in \tilde{P}_3 \text{ and } y \in P_1 \cup P_2,\\ &\text{or } x \in \bar{P}_6 \text{ and } y \in P_4 \cup P_5 \cup \tilde{P}_6 \cup P_7,\end{aligned}\\[2pt]
y & \begin{aligned}&\text{if } x = \mathbf{e}, \text{ or } x \in P_1 \cup P_2 \cup P_5 \cup P_7 \text{ and } y \in \bar{P}_3,\\ &\text{or } x \in P_2 \text{ and } y \in L \setminus (P_1 \cup P_2 \cup \tilde{P}_3 \cup \mathbf{e}),\\ &\text{or } y \in \bar{P}_3 \text{ and } x \in \tilde{P}_3,\\ &\text{or } y \in \tilde{P}_3 \text{ and } x \in P_1 \cup P_2,\\ &\text{or } y \in \bar{P}_6 \text{ and } x \in P_4 \cup P_5 \cup \tilde{P}_6 \cup P_7,\end{aligned}\\[2pt]
\mathbf{a} & \begin{aligned}&\text{if } x \leq \mathbf{a} \text{ and } y \geq \mathbf{a}, \text{ or } x \geq \mathbf{a} \text{ and } y \leq \mathbf{a},\\ &\text{or } x = \mathbf{a}, \text{ or } y = \mathbf{a},\end{aligned}\\[2pt]
x \wedge y & \begin{aligned}&\text{if } (x, y) \in (P_1 \cup P_2)^2, \text{ or } (x, y) \in \bar{P}_6{}^2,\\ &\text{or } (x, y) \in (\bar{P}_3 \cup P_9)^2, \text{ or } (x, y) \in \tilde{P}_3{}^2,\\ &\text{or } x \in \tilde{P}_3 \text{ and } y \in P_5 \cup P_7,\\ &\text{or } y \in \tilde{P}_3 \text{ and } x \in P_5 \cup P_7,\end{aligned}\\[2pt]
x \wedge \mathbf{a} & \begin{aligned}&\text{if } x \in P_9 \text{ and } y \in P_1 \cup \tilde{P}_3,\\ &\text{or } x \in P_5 \cup P_7 \text{ and } y \in P_1,\end{aligned}\\[2pt]
y \wedge \mathbf{a} & \begin{aligned}&\text{if } y \in P_9 \text{ and } x \in P_1 \cup \tilde{P}_3,\\ &\text{or } y \in P_5 \cup P_7 \text{ and } x \in P_1 \cup P_3,\end{aligned}\\[2pt]
x \vee \mathbf{a} & \text{if } x \in P_9 \text{ and } y \in P_4 \cup P_5 \cup \tilde{P}_6 \cup P_7,\\[2pt]
y \vee \mathbf{a} & \text{if } y \in P_9 \text{ and } x \in P_4 \cup P_5 \cup \tilde{P}_6 \cup P_7,\\[2pt]
(x \wedge y) \vee \mathbf{a} & \text{if } x \in P_9 \text{ and } y \in \bar{P}_6, \text{ or } y \in P_9 \text{ and } x \in \bar{P}_6,\\[2pt]
x \vee y & \text{otherwise.}
\end{cases}
$$

Then U_2 is an idempotent uninorm on L, whose neutral element is $\mathbf{e}$ and annihilator is $\mathbf{a}$.

In the following proposition we construct again an idempotent uninorm on L. In this case we assume $P_8 \neq \emptyset$.

Proposition 5. *Let L be a lattice fulfilling Assumption (A) and constraints (i)–(vi) from Proposition 4 be valid. Assume $P_8 \neq \emptyset$. Denote $\bar{P}_5 = \{x \in P_5; x \wedge \mathbf{a} = \varepsilon\}$ and $\tilde{P}_5 = P_5 \setminus \bar{P}_5$. Instead of the constraint (vii) we will assume the following: (vii') $x \vee y \in \bar{P}_5$ for all $(x,y) \in \bar{P}_5$ and if for $x \in \bar{P}_5$ and $y \in P_5$ we have $x \geq y$, then $y \in \bar{P}_5$.*

Further assume there exists an associative commutative and monotone operation $* : (P_2 \cup \bar{P}_5 \cup P_8)^2 \to P_2 \cup \bar{P}_5 \cup P_8$ with the following properties:

(α) $x \wedge y \leq x * y \leq x \vee y$,
(β) $x * y = x \wedge y$ for $(x,y) \in P_2^2$,
(γ) $x * y = x \vee y$ for $(x,y) \in \bar{P}_5$,
(δ) $(x * y) \wedge \mathbf{e} = x \wedge y \wedge \mathbf{e}$,
(ϵ) $(x * y) \vee \mathbf{e} = x \vee y \vee \mathbf{e}$.

Then, $U_3 : L \times L \to L$, defined by

$$
U_3(x,y) = \begin{cases}
x * y & \text{for } (x,y) \in (P_2 \cup \bar{P}_5 \cup P_8)^2, \\
x & \text{if } y = \mathbf{e}, \text{ or } y \in P_1 \cup P_2 \cup P_5 \cup P_7 \cup P_8 \text{ and } x \in \bar{P}_3, \\
& \quad \text{or } y \in P_2 \text{ and } x \in L \setminus (P_1 \cup P_2 \cup \tilde{P}_3 \cup \bar{P}_5 \cup P_8 \cup \mathbf{e}), \\
& \quad \text{or } x \in \bar{P}_3 \text{ and } y \in \tilde{P}_3, \\
& \quad \text{or } x \in \tilde{P}_3 \text{ and } y \in P_1 \cup P_2 \cup \bar{P}_5 \cup P_8, \\
& \quad \text{or } x \in \bar{P}_6 \text{ and } y \in P_4 \cup P_5 \cup \tilde{P}_6 \cup P_7 \cup P_8, \\
y & \text{if } x = \mathbf{e}, \text{ or } x \in P_1 \cup P_2 \cup P_5 \cup P_7 \cup P_8 \text{ and } y \in \bar{P}_3, \\
& \quad \text{or } x \in P_2 \text{ and } y \in L \setminus (P_1 \cup P_2 \cup \tilde{P}_3 \cup \bar{P}_5 \cup P_8 \cup \mathbf{e}), \\
& \quad \text{or } y \in \bar{P}_3 \text{ and } x \in \tilde{P}_3, \\
& \quad \text{or } y \in \tilde{P}_3 \text{ and } x \in P_1 \cup P_2 \cup \bar{P}_5 \cup P_8, \\
& \quad \text{or } y \in \bar{P}_6 \text{ and } x \in P_4 \cup P_5 \cup \tilde{P}_6 \cup P_7 \cup P_8, \\
\mathbf{a} & \text{if } x \leq \mathbf{a} \text{ and } y \geq \mathbf{a}, \text{ or } x \geq \mathbf{a} \text{ and } y \leq \mathbf{a}, \\
& \quad \text{or } x = \mathbf{a}, \text{ or } y = \mathbf{a}, \\
x \wedge y & \text{if } x \wedge y \in P_1 \text{ and } x \vee y \in P_1 \cup P_2 \cup \bar{P}_5 \cup P_8, \\
& \quad \text{or } (x,y) \in \bar{P}_6^{\,2}, \text{ or } (x,y) \in (\bar{P}_3 \cup P_9)^2, \text{ or } (x,y) \in \tilde{P}_3^{\,2}, \\
& \quad \text{or } x \in \tilde{P}_3 \text{ and } y \in P_5 \cup P_7, \\
& \quad \text{or } y \in \tilde{P}_3 \text{ and } x \in P_5 \cup P_7, \\
x \wedge \mathbf{a} & \text{if } x \in P_9 \text{ and } y \in P_1 \cup \tilde{P}_3, \\
& \quad \text{or } x \in \tilde{P}_5 \cup P_7 \text{ and } y \in P_1, \\
y \wedge \mathbf{a} & \text{if } y \in P_9 \text{ and } x \in P_1 \cup \tilde{P}_3, \\
& \quad \text{or } y \in \tilde{P}_5 \cup P_7 \text{ and } x \in P_1 \cup P_3, \\
x \vee \mathbf{a} & \text{if } x \in P_9 \text{ and } y \in P_4 \cup P_5 \cup \tilde{P}_6 \cup P_7 \cup P_8, \\
y \vee \mathbf{a} & \text{if } y \in P_9 \text{ and } x \in P_4 \cup P_5 \cup \tilde{P}_6 \cup P_7 \cup P_8, \\
(x \wedge y) \vee \mathbf{a} & \text{if } x \in P_9 \text{ and } y \in \bar{P}_6, \text{ or } y \in P_9 \text{ and } x \in \bar{P}_6, \\
x \vee y & \text{otherwise,}
\end{cases}
$$

is an idempotent uninorm, whose neutral element is $\mathbf{e}$ and annihilator is $\mathbf{a}$.

Of course, there exist modifications of the uninorms U_2 and U_3 from Propositions 4 and 5, respectively. Also the constraints in those two propositions are just sufficient. So, there is still space for further generalizations of our results.

Example 1. Let $L = [0,1] \times [0,1]$ with the usual ordering. Set $\mathbf{e} = (0.2, 0.8)$ and $\mathbf{a} = (0.8, 0.2)$. Then

$P_1 = [0, 0.2]^2$, $P_2 = [0, 0, 2] \times]0.2, 0.8] \setminus \{(0.2, 0.8)\}$,

$P_3 =]0.2, 0.8] \times [0, 0.2] \setminus \{(0.8, 0.2)\}$, $\bar{P}_3 = \{0.8\} \times [0, 0.2[$, $P_4 = [0.8, 1]^2$,

$P_5 = [0.2, 0.8[\times [0.8, 1] \setminus \{(0.2, 0.8)\}$, $\bar{P}_5 = \{0.2\} \times]0.8, 1]$,

$P_6 = [0.8, 1] \times [0.2, 0.8[\setminus \{(0.8, 0.2)\}$, $\bar{P}_6 =]0.8, 1] \times \{0.2\}$, $P_7 =]0.2, 0.8[^2$,

$P_8 = [0, 0.2[\times]0.8, 1]$, $P_9 =]0.8, 1] \times [0, 0.2[$.

We can easily check that all the constraints from Proposition 5 are satisfied. Define the operation $* : (P_2 \cup \bar{P}_5 \cup P_8)^2 \to P_2 \cup \bar{P}_5 \cup P_8$ by the following

$$(x_1, x_2) * (y_1, y_2) = \begin{cases} (\min\{x_1, x_2\}, \min\{y_1, y_2\}) & \text{if } \max\{y_1, y_2\} \le 0.8, \\ (\min\{x_1, x_2\}, \max\{y_1, y_2\}) & \text{otherwise.} \end{cases}$$

Then U_3 from Proposition 5 is an idempotent uninorm on L.

In the next proposition we modify the uninorm from Proposition 5 in such a way that the resulting operation on L will yield a uninorm with arbitrarily chosen underlying t-norm and t-conorm.

Proposition 6. *Let L be a lattice fulfilling Assumption (Λ). Assume $P_0 \ne \emptyset$ and there exist $T : (P_1 \cup P_2 \cup \mathbf{e})^2 \to P_1 \cup P_2 \cup \mathbf{e}$ and $S : (P_4 \cup P_5 \cup \mathbf{e})^2 \to P_4 \cup P_5 \cup \mathbf{e}$, a t-norm and a t-conorm, respectively, with the following properties*

$$T(x, y) \in P_2 \quad \text{for all } (x, y) \in P_2^2, \qquad S(x, y) \in P_5 \quad \text{for all } (x, y) \in P_5^2.$$

The area P_5 is split into $\bar{P}_5$ and $\tilde{P}_5 = P_5 \setminus \bar{P}_5$. The lattice L is assumed to fulfil constraints (i)–(vi) from Proposition 4 and

(vii") $S(x, y) \in \bar{P}_5$ for all $(x, y) \in \bar{P}_5^2$ and for all $x \in \bar{P}_5$ and $y \in P_5$ if $y \le x$ then $y \in \bar{P}_5$.

Further assume there exists an associative commutative and monotone operation $* : (P_2 \cup \bar{P}_5 \cup P_8)^2 \to P_2 \cup \bar{P}_5 \cup P_8$ with the following properties:

(α) $T(x \wedge \mathbf{e}, y \wedge \mathbf{e}) \le a * y \le S(x \vee \mathbf{e}, y \vee \mathbf{e})$,

(β) $x * y = T(x, y)$ for $(x, y) \in P_2^2$,

(γ) $x * y = S(x, y)$ for $(x, y) \in \bar{P}_5^2$,

(δ) $(x * y) \wedge \mathbf{e} = T(x \wedge \mathbf{e}, y \wedge \mathbf{e})$,

(ϵ) $(x * y) \vee \mathbf{e} = S(x \vee \mathbf{e}, y \vee \mathbf{e})$.

Then, $U_4 : L \times L \to L$, defined by

$$U_4(x,y) = \begin{cases} x * y & \text{for } (x,y) \in (P_2 \cup \bar{P}_5 \cup P_8)^2, \\ x & \text{if } y = \mathbf{e}, \text{ or } y \in P_1 \cup P_2 \cup P_5 \cup P_7 \cup P_8 \text{ and } x \in \bar{P}_3, \\ & \text{or } y \in P_2 \text{ and } x \in L \setminus (P_1 \cup P_2 \cup \tilde{P}_3 \cup \bar{P}_5 \cup P_8 \cup \mathbf{e}), \\ & \text{or } x \in \tilde{P}_3 \text{ and } y \in \tilde{P}_3, \\ & \text{or } x \in \tilde{P}_3 \text{ and } y \in P_1 \cup P_2 \cup \bar{P}_5 \cup P_8, \\ & \text{or } x \in \bar{P}_6 \text{ and } y \in P_4 \cup P_5 \cup \tilde{P}_6 \cup P_7 \cup P_8, \\ y & \text{if } x = \mathbf{e}, \text{ or } x \in P_1 \cup P_2 \cup P_5 \cup P_7 \cup P_8 \text{ and } y \in \bar{P}_3, \\ & \text{or } x \in P_2 \text{ and } y \in L \setminus (P_1 \cup P_2 \cup \tilde{P}_3 \cup \bar{P}_5 \cup P_8 \cup \mathbf{e}), \\ & \text{or } y \in \tilde{P}_3 \text{ and } x \in \tilde{P}_3, \\ & \text{or } y \in \tilde{P}_3 \text{ and } x \in P_1 \cup P_2 \cup \bar{P}_5 \cup P_8, \\ & \text{or } y \in \bar{P}_6 \text{ and } x \in P_4 \cup P_5 \cup \tilde{P}_6 \cup P_7 \cup P_8, \\ \mathbf{a} & \text{if } x \leq \mathbf{a} \text{ and } y \geq \mathbf{a}, \text{ or } x \geq \mathbf{a} \text{ and } y \leq \mathbf{a}, \\ & \text{or } x = \mathbf{a}, \text{ or } y = \mathbf{a}, \\ T(x \wedge \mathbf{e}, y \wedge \mathbf{e}) & \text{if } x \wedge y \in P_1 \text{ and } x \vee y \in P_1 \cup P_2 \cup \bar{P}_5 \cup P_8, \\ x \wedge y & \text{if } (x,y) \in \bar{P}_6{}^2, \text{ or } (x,y) \in (\bar{P}_3 \cup P_9)^2, \\ & \text{or } (x,y) \in \tilde{P}_3{}^2, \text{ or } x \in \tilde{P}_3 \text{ and } y \in P_5 \cup P_7, \\ & \text{or } y \in \tilde{P}_3 \text{ and } x \in P_5 \cup P_7, \\ x \wedge \mathbf{a} & \text{if } x \in P_9 \text{ and } y \in P_1 \cup \tilde{P}_3, \\ & \text{or } x \in \bar{P}_5 \cup P_7 \text{ and } y \in P_1, \\ y \wedge \mathbf{a} & \text{if } y \in P_9 \text{ and } x \in P_1 \cup \tilde{P}_3, \\ & \text{or } y \in \tilde{P}_5 \cup P_7 \text{ and } x \in P_1 \cup P_3, \\ x \vee \mathbf{a} & \text{if } x \in P_9 \text{ and } y \in P_4 \cup P_5 \cup \tilde{P}_6 \cup P_7 \cup P_8, \\ y \vee \mathbf{a} & \text{if } y \in P_9 \text{ and } x \in P_4 \cup P_5 \cup \tilde{P}_6 \cup P_7 \cup P_8, \\ (x \wedge y) \vee \mathbf{a} & \text{if } x \in P_9 \text{ and } y \in \bar{P}_6, \text{ or } y \in P_9 \text{ and } x \in \bar{P}_6, \\ S(x \vee \mathbf{e}, y \vee \mathbf{e}) & \text{otherwise}, \end{cases}$$

is a uninorm, whose neutral element is **e** *and annihilator is* **a**.

4   Conclusions

In Bodjanova, Kalina [4] we have studied properties of bounded lattices where it is possible to construct a uninorm that is under which neither conjunctive nor disjunctive (i.e., are of the third type). We have shown there, that there exist lattices possessing incomparable elements on which it is not possible to construct any uninorm of the third type.

In the present paper we have provided 4 different uninorms of the third type on bounded lattices. In the first case the constraints are quite loose. This is just to demonstrate, under which conditions we are able to define a uninorm of the third type. Next we have provided two different idempotent uninorms (changing conditions the bounded lattice L has to fulfil). Finally, a uninorm with given underlying t-norm and t-conorm has been constructed. However, the underlying t-norm and t-conorm has to fulfil some conditions as it is shown, e.g., by Proposition 2.

There are, of course, also other possibilities how to design uninorms on bounded lattices that would yield some properties. From this point of view the paper is not exhaustive.

Acknowledgements. The work of Martin Kalina has been supported from the VEGA grant agency, grant No. 2/0069/16 and 1/0006/19.

References

1. De Baets, B.: Idempotent uninorms. Eur. J. Oper. Res. **118**, 631–642 (1999)
2. Birkhoff, G.: Lattice Theory. American Mathematical Society Colloquium Publishers, Providence (1967)
3. Bodjanova, S., Kalina, M.: Construction of uninorms on bounded lattices. In: IEEE 12th International Symposium on Intelligent Systems and Informatics, SISY 2014, Subotica, Serbia, pp. 61–66 (2014)
4. Bodjanova, S., Kalina, M.: Uninorms on bounded lattices - recent development. In: Kacprzyk, J., Szmidt, E., Zadrożny, S., Atanassov, K., Krawczak, M. (eds.) Advances in Fuzzy Logic and Technology 2017, EUSFLAT 2017, IWIFSGN : Advances in Intelligent Systems and Computing, vol. 641. Springer, Cham (2017)
5. Calvo, T., Kolesárová, A., Komorníková, M., Mesiar, R.: Aggregation operators: properties, classes and construction methods. In: Aggregation Operators, pp. 3–104. Physica-Verlag GMBH, Heidelberg (2002)
6. Çayli, G.D., Drygaś, P.: Some properties of idempotent uninorms on a special class of bounded lattices. Inf. Sci. **422**, 352–363 (2018)
7. Çayli, G.D., Karaçal, F.: Construction of uninorms on bounded lattices. Kybernetika **53**, 394–417 (2017)
8. Çayli, G.D., Karaçal, F., Mesiar, R.: On a new class of uninorms on bounded lattices. Inf. Sci. **367–368**, 221–231 (2016)
9. Czogała, E., Drewniak, J.: Associative monotonic operations in fuzzy set theory. Fuzzy Sets Syst. **12**, 249–269 (1984)
10. Deschrijver, G.: A representation of t-norms in interval valued L-fuzzy set theory. Fuzzy Sets Syst. **159**, 1597–1618 (2008)
11. Deschrijver, G.: Uninorms which are neither conjunctive nor disjunctive in interval-valued fuzzy set theory. Inf. Sci. **244**, 48–59 (2013)
12. Dombi, J.: Basic concepts for a theory of evaluation: the aggregative operator. Eur. J. Oper. Res. **10**, 282–293 (1982)
13. Dombi, J.: A general class of fuzzy operators, the DeMorgan class of fuzzy operators and fuzziness measures induced by fuzzy operators. Fuzzy Sets Syst. **8**, 149–163 (1982)
14. Grabisch, M., Marichal, J.-L., Mesiar, R., Pap, E.: Aggregation functions. In: Encyclopedia of Mathematics and Its Applications, vol. 127. Cambridge University Press, Cambridge (2009)
15. Kalina, M.: On uninorms and nullnorms on direct product of bounded lattices. Open Phys. **14**(1), 321–327 (2016)
16. Kalina, M., Král', P.: Uninorms on interval-valued fuzzy sets. In: Carvalho, J., Lesot, M.J., Kaymak, U., Vieira, S., Bouchon-Meunier, B., Yager, R. (eds.) Information Processing and Management of Uncertainty in Knowledge-Based Systems, IPMU 2016, Communications in Computer and Information Science, vol. 611, pp. 522–531. Springer, Cham (2016)

17. Karaçal, F., Mesiar, R.: Uninorms on bounded lattices. Fuzzy Sets Syst. **261**, 33–43 (2015)
18. Klement, E.P., Mesiar, R., Pap, E.: Triangular Norms. Springer, Berlin (2000)
19. Schweizer, B., Sklar, A.: Probabilistic Metric Spaces, North Holland, New York (1983)
20. Yager, R.R., Rybalov, A.: Uninorm aggregation operators. Fuzzy Sets Syst. **80**, 111–120 (1996)

Description and Properties
of Curve-Based Monotone Functions

Mikel Sesma-Sara[1,2]([⊠]), Laura De Miguel[1,2],
Antonio Francisco Roldán López de Hierro[3], Jana Špirková[4],
Radko Mesiar[5,6], and Humberto Bustince[1,2]

[1] Public University of Navarra, Pamplona, Spain
{mikel.sesma,laura.demiguel,bustince}@unavarra.es
[2] Institute of Smart Cities (UPNA), Pamplona, Spain
[3] University of Granada, Granada, Spain
aroldan@ugr.es
[4] Matej Bel University, Banská Bystrica, Slovakia
jana.spirkova@umb.sk
[5] Slovak University of Technology in Bratislava, Bratislava, Slovakia
mesiar@math.sk
[6] University of Ostrava, Ostrava, Czech Republic

Abstract. Curve-based monotonicity is one of the lately introduced relaxations of monotonicity. As directional monotonicity regards monotonicity along fixed rays, which are given by real vectors, curve based monotonicity studies the increase of functions with respect to a general curve α. In this work we study some theoretical properties of this type of monotonicity and we relate this concept with previous relaxations of monotonicity.

Keywords: Curve-based monotonicity · Weak monotonicity ·
Directional monotonicity · Aggregation function

1 Introduction

Aggregation operators are functions that aim at finding a single value to represent a collection of n numbers, and, since this is a desired feature in many processes, they have been largely studied and applied [6,8,14].

Aggregation functions need to be increasing with respect to every argument. However, according to various works in the literature [5,19], that condition may be too restrictive as it restricts some functions, that are fit to fuse information, to enter the framework of aggregation functions. That is the reason why, there is a trend towards relaxing the monotonicity constraint in the definition of aggregation function [7,13].

In that attempt, some generalizations of monotonicity have been proposed. For example, weak monotonicity [19] is a relaxed form of monotonicity, as it only asks for the value of a function to increase in the case where all the arguments

© Springer Nature Switzerland AG 2019
R. Halaš et al. (Eds.): AGOP 2019, AISC 981, pp. 195–204, 2019.
https://doi.org/10.1007/978-3-030-19494-9_18

have increased by the same amount. This notion can be seen as monotonicity defined by the vector $(1, \ldots, 1)$ and, considering any vector $\overrightarrow{r} \in \mathbb{R}^n$, instead, directional monotonicity was defined [5]. Directional monotonicity is the requirement that is demanded to functions in order to be considered pre-aggregation functions [11]. Lately, more different relaxations of monotonicity have appeared [1,4,16,18], some of which have been used to construct edge detectors [4,17] and fuzzy ruled-based classification systems [10,12].

In this work, we discuss a generalization of directional monotonicity: curve-based monotonicity [9]. Rather than directions given by vectors, curve-based monotonicity studies the monotonicity of functions along general curves $\alpha :$ $[0,1] \to \mathbb{R}^n$. Clearly, lines are particular instances of curves and, hence, curve-based monotonicity generalizes directional monotonicity.

We study some theoretical properties of curve-based monotonicity, including the situation of functions that are monotone with respect to two curves α and β, which are also monotone with respect to the combination of such curves.

This paper is organized in the following manner: first, we present some preliminaries, including the definitions of various relaxations of monotonicity. In Sect. 3, we recall the concept of curve-based monotonicity. In Sect. 4 we show a collection of properties of curve-based monotone functions and, in Sect. 5, we discuss the case of composition of two curves. Finally, we present some concluding remarks and future perspectives.

2   Preliminaries

Let $n \in \mathbb{N}$ such that $n > 1$. On the one hand, we use $\mathbf{x} = (x_1, \ldots, x_n)$ to denote points in $[0,1]^n$ and we set $\mathbf{0} = (0, \ldots, 0)$ and $\mathbf{1} = (1, \ldots, 1) \in [0,1]^n$. On the other hand, we use $\overrightarrow{r} \in \mathbb{R}^n$ to refer to vectors that denote directions in the real space.

In this work, we consider curves on $\mathbb{R}^n$ as functions defined on a closed real interval, i.e., $\alpha : [0,1] \to \mathbb{R}^n$. Note that the choice of the domain, $[0,1]$, could have been any other closed real interval $[a,b] \subset \mathbb{R}$, as any curve α defined on $[0,1]$ can be re-parametrized to be defined on $[a,b]$. Additionally, we only consider curves α such that $\alpha(0) = \mathbf{0}$.

Given a curve $\alpha : [0,1] \to \mathbb{R}^n$, all the components $\alpha_1, \ldots, \alpha_n$ of α can be seen as curves on $\mathbb{R}$: $\alpha_i : [0,1] \to \mathbb{R}$ for all $1 \leq i \leq n$.

The concept of monotonicity is intimately related to the notion of order. We use the standard partial order in $\mathbb{R}^n$ (and, hence, in $[0,1]^n$), i.e., given $\mathbf{x}, \mathbf{y} \in [0,1]^n$, we say that $\mathbf{x} \leq \mathbf{y}$ if $x_i \leq y_i$ for all $1 \leq i \leq n$. Thus, we can define the concept of standard monotonicity for a function $f : [0,1]^n \to [0,1]$.

Definition 1. *A function $f : [0,1]^n \to [0,1]$ is said to be increasing (resp. decreasing) if for all $\mathbf{x}, \mathbf{y} \in [0,1]^n$ such that $\mathbf{x} \leq \mathbf{y}$ it holds that $f(\mathbf{x}) \leq f(\mathbf{y})$ (resp. $f(\mathbf{x}) \geq f(\mathbf{y})$).*

Note that with the terms *increasing* and *decreasing* we do not refer to strict monotonicity.

As stated in the Introduction, we aim at relaxing the monotonicity condition of aggregation functions. An aggregation function is a function $A : [0,1]^n \rightarrow [0,1]$ such that $A(\mathbf{0}) = 0$, $A(\mathbf{1}) = 1$ and A is increasing in the sense of Definition 1. Let us now present some of relaxations of monotonicity.

Definition 2 ([19]). *A function $f : [0,1]^n \rightarrow [0,1]$ is said to be weakly increasing (resp. weakly decreasing) if for all $\mathbf{x} \in [0,1]^n$ and $c > 0$ such that $\mathbf{x} + c\mathbf{1} \in [0,1]^n$ it holds that $f(\mathbf{x}) \leq f(\mathbf{x} + c\mathbf{1})$ (resp. $f(\mathbf{x}) \geq f(\mathbf{x} + c\mathbf{1})$).*

Note that, although robust estimators of location [15], which are used in statistics, are generally not monotone, they are shift-invariant and shift-invariance implies weak monotonicity.

Nevertheless, although monotonicity with respect to all arguments may be too restrictive for certain applications, our expectation with respect to the behaviour of means requires that some monotonicity-like condition is satisfied, e.g., in the case of robust estimators of location shift-invariance is required.

Remark 1. If a function $f : [0,1]^n \rightarrow [0,1]$ is monotone, then it is also weakly monotone, i.e., standard increasingness (resp. decreasingness) implies weak increasingness (resp. decreasingness).

However, the converse does not hold. For example, the mode function, with the convention of taking the minimum if all the inputs are different, is not generally increasing. Indeed, $(0, 0.2, 0.3, 0.3, 0.3) \leq (0.2, 0.2, 0.3, 0.4, 0.5)$ but

$$modc(0, 0.2, 0.3, 0.3, 0.3) = 0.3 > 0.2 = mode(0.2, 0.2, 0.3, 0.4, 0.5).$$

But, the mode function satisfies a certain kind of monotonicity, as its value increases whenever all the inputs increase by the same amount, i.e., the mode is a weakly increasing function.

Weak monotonicity can be seen as monotonicity along the ray $(1, \ldots, 1)$. When, we consider a general vector $\overrightarrow{0} \neq \overrightarrow{r} \in \mathbb{R}^n$ instead, we obtain the notion of directional monotonicity.

Definition 3 ([5]). *Let $\overrightarrow{0} \neq \overrightarrow{r} \in \mathbb{R}^n$ and $f : [0,1]^n \rightarrow [0,1]$. We say that f is $\overrightarrow{r}$-increasing (resp. $\overrightarrow{r}$-decreasing) if for all $\mathbf{x} \in [0,1]^n$ and $c > 0$ such that $\mathbf{x} + c\overrightarrow{r} \in [0,1]^n$, it holds that $f(\mathbf{x}) \leq f(\mathbf{x} + c\overrightarrow{r})$ (resp. $f(\mathbf{x}) \geq f(\mathbf{x} + c\overrightarrow{r})$).*

A function f that is both $\overrightarrow{r}$-increasing and $\overrightarrow{r}$-decreasing for a certain $\overrightarrow{0} \neq \overrightarrow{r} \in \mathbb{R}^n$ is said to be $\overrightarrow{r}$-constant.

A function that satisfies the boundary conditions of aggregation functions and is directionally increasing with respect to some direction $\overrightarrow{0} \neq \overrightarrow{r} \in [0,1]^n$ is said to be a pre-aggregation function [10–12].

It is also interesting to study the directions for which a function is directionally increasing. If the set of such directions forms a cone, we say that a function is cone increasing. This concept was originally defined for positive cones $C \subset (\mathbb{R}^+)^n$, but the generalization to any cone $C \subset \mathbb{R}^n$ is straight. Recall that a subset $C \subset \mathbb{R}^n$ is said to be a cone if for each $\mathbf{x} \in C$ it holds that $a\mathbf{x} \in C$ for all $a \geq 0$.

Definition 4 ([1]). *Let $\emptyset \neq C \subset \mathbb{R}^n$ be a cone. A function $f : [0,1]^n \to [0,1]$ is said to be cone increasing with respect to C (resp. cone decreasing) if f is $\overrightarrow{r}$-increasing (resp. $\overrightarrow{r}$-decreasing) for all vectors $\overrightarrow{r} \in C$.*

Clearly, increasing functions in the standard sense are cone increasing with respect to the cone $(\mathbb{R}^+)^n$.

The interested reader can find numerous examples of functions that satisfy each of the monotonicity conditions in [1–3,5,19].

3   Curve-Based Monotonicity

In the same manner that directional monotonicity is a generalization of weak monotonicity considering general directions $\overrightarrow{r}$ instead of $\overrightarrow{1}$, we can think of an even more general concept by considering curves in the space.

Definition 5 ([9]). *Let $\alpha : [0,1] \to \mathbb{R}^n$ be a curve such that $\alpha(0) = \mathbf{0}$. A function $f : [0,1]^n \to [0,1]$ is said to be α-increasing (resp. α-decreasing) if $f(\mathbf{x}) \leq f(\mathbf{x}+\alpha(t))$ (resp. $f(\mathbf{x}) \geq f(\mathbf{x}+\alpha(t))$) for all $\mathbf{x} \in [0,1]^n$ and all $0 < t \leq 1$ such that $\mathbf{x} + \alpha(s) \in [0,1]^n$ for all $0 < s \leq t$.*

From this point forward, we assume that all curves $\alpha : [0,1] \to \mathbb{R}^n$ satisfy the condition $\alpha(0) = \mathbf{0}$, unless otherwise stated.

If a function f is both α-increasing and α-decreasing for a given curve $\alpha : [0,1] \to \mathbb{R}^n$, then f is said to be α-constant. For curves defined on an open interval, see [9].

Note that, by Definition 5, for a function f to be α-increasing, once the curve leaves the unit hypercube $[0,1]^n$, it has no influence in the property of α-monotonicity of f. Indeed, the condition that must hold is $f(\mathbf{x}) \leq f(\mathbf{x}+\alpha(t))$ provided that all the points $\mathbf{x} + \alpha(s) \in [0,1]^n$ for all $0 < s \leq t$. Therefore, the points $\mathbf{x} + \alpha(t) \notin [0,1]^n$, even in the case that the curve eventually returns to take values within $[0,1]^n$, do not influence the condition of α-monotonicity. This is shown in Fig. 1.

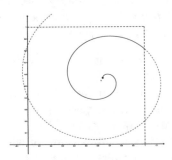

Fig. 1. Example of points $\mathbf{x} + \alpha(t)$ that have influence on the α-monotonicity of functions (in solid blue) and points that do not (dashed blue).

Remark 2. Straight lines (or segments of straight lines) are a particular instance of curve. Hence, curve-based monotonicity is a generalization of directional monotonicity. Indeed, let $\overrightarrow{0} \neq \overrightarrow{r} \in \mathbb{R}^n$ and $f : [0,1]^n \to [0,1]$ be an $\overrightarrow{r}$-increasing function. Then, if we set $\alpha : [0,1] \to \mathbb{R}^n$ to be the curve given by $\alpha(t) = t\overrightarrow{r}$, we obtain that f is α-increasing. In particular, weak monotonicity is also a particular case of curve-based monotonicity.

Further, we derive the following result from the definitions of curve-based monotonicity and directional monotonicity.

Proposition 1. *Let $\alpha : [0,1] \to \mathbb{R}^n$ be a curve such that, given $t_0 \in [0,1]$, $\alpha(t) = (r_1 t, \ldots, r_n t)$ for all $t \in [0, t_0]$, where $r_1, \ldots, r_n \in \mathbb{R}$, which defines a direction $\overrightarrow{r}$. If a function $f : [0,1]^n \to [0,1]$ is α-increasing, then f is $\overrightarrow{r}$-increasing.*

Let us stress that the concept of curve-based montonicity, presented in Definition 5, differs from monotonicity along a curve α. Namely, the fact that the value of a function f increases along the graph of a certain curve α implies that the function f is α-increasing, but the converse does not hold in general as the next example shows.

Example 1. The arithmetic mean, defined in $[0,1]^n$ by

$$A_n(x_1, \ldots, x_n) = \frac{1}{n} \sum_{i=1}^{n} x_i,$$

is an example of α-monotone function for a curve α that satisfies certain properties.

Specifically, the arithmetic mean is α-increasing for all curves $\alpha : [0,1] \to \mathbb{R}^n$ such that

$$\sum_{i=1}^{n} \alpha_i(t) \geq 0, \tag{1}$$

for all $0 \leq t \leq 1$. Furthermore, the arithmetic mean is α-decreasing for all curves α such that $\sum_{i=1}^{n} \alpha_i(t) \leq 0$, for all $0 \leq t \leq 1$. Consequently, A is α-constant for every curve α such that $\sum_{i=1}^{n} \alpha_i(t) = 0$, for all $0 \leq t \leq 1$.

The arithmetic mean serves also to show that α-monotonicity does not coincide with monotonicity along the graph of the curve α. Let $\alpha : [0,1] \to \mathbb{R}^2$ be a curve given by

$$\alpha(t) = \begin{cases} (t, 0), & \text{if } 0 \leq t \leq 0.5, \\ (0.5, 0.5 - t), & \text{if } 0.5 < t \leq 1. \end{cases} \tag{2}$$

Clearly, since α verifies condition (1), the arithmetic mean A_2 is α-increasing. However, the value of A_2 does not increase along the graph of α. Indeed, consider $(0,0) \in [0,1]^2$ and let $g : [0,1] \to [0,1]$ be the function that represents the values of A_2 along the graph of α. Thus, it is given by

$$g(t) = A_2(0 + \alpha_1(t), 0 + \alpha_2(t))$$

$$= \begin{cases} \frac{t}{2}, & \text{if } 0 \leq t \leq 0.5, \\ \frac{1-t}{2}, & \text{if } 0.5 < t \leq 1, \end{cases}$$

and, clearly, g is decreasing for $0.5 < t \leq 1$.

4 Properties of Curve-Based Monotone Functions

In this section we study some relevant properties of curve-based monotone functions and we discuss how curve-based monotonicity relates to other relaxations of monotonicity.

Proposition 2. *Let* $f : [0,1]^n \to [0,1]$ *be a function,* $\alpha : [0,1] \to \mathbb{R}^n$ *be a curve and* $\varphi : [0,1] \to [0,1]$ *be a strictly increasing one-to-one mapping. Then,* f *is* α-*increasing (resp.* α-*decreasing) if and only if* f *is* $(\alpha \circ \varphi)$-*increasing (resp.* $(\alpha \circ \varphi)$-*decreasing).*

The next result characterizes standard monotonicity in terms of curve-based monotonicity.

Proposition 3. *Let* $f : [0,1]^n \to [0,1]$. *Then,* f *is increasing (resp. decreasing) if and only if* f *is* α-*increasing (resp.* α-*decreasing) for all curves* $\alpha : [0,1] \to \mathbb{R}^n$ *such that* $\alpha(t) \geq \mathbf{0}$ *for all* $0 \leq t \leq 1$.

Proof. Let $f : [0,1]^n \to [0,1]$ be an increasing (resp. decreasing) function, $\alpha : [0,1] \to \mathbb{R}^n$ be a curve such that $\alpha(t) \geq \mathbf{0}$ for all $1 \leq t \leq n$ and let $\mathbf{x} \in [0,1]^n$ and $t \in [0,1]$ such that $\mathbf{x} + \alpha(s) \in [0,1]^n$ for all $0 < s \leq t$. Since $\alpha(t) \geq \mathbf{0}$, then $\mathbf{x} \leq \mathbf{x} + \alpha(t)$. Therefore, since f is increasing (resp. decreasing), it holds that $f(\mathbf{x}) \leq f(\mathbf{x} + \alpha(t))$ (resp. $f(\mathbf{x}) \geq f(\mathbf{x} + \alpha(t))$).

For the converse, assume that f is α-increasing (resp. α-decreasing) for every curve $\alpha : [0,1] \to \mathbb{R}^n$ such that $\alpha(t) \geq \mathbf{0}$ for all $1 \leq t \leq n$. Now, let $\mathbf{x}, \mathbf{y} \in [0,1]^n$ such that $\mathbf{x} \leq \mathbf{y}$. We can set a curve $\alpha : [0,1] \to \mathbb{R}^n$ given by $\alpha(t) = t\mathbf{y} - t\mathbf{x}$ for all $0 \leq t \leq 1$. Thus,

$$f(\mathbf{x}) \leq f(\mathbf{x} + \alpha(t)) = f((1-t)\mathbf{x} + t\mathbf{y})$$

$$(\text{resp. } f(\mathbf{x}) \geq f(\mathbf{x} + \alpha(t)) = f((1-t)\mathbf{x} + t\mathbf{y})).$$

In particular, for $t = 1$, we conclude that $f(\mathbf{x}) \leq f(\mathbf{y})$ (resp. $f(\mathbf{x}) \geq f(\mathbf{y})$) and therefore f is increasing (resp. decreasing).

As a consequence, in the case where all the components of the curve α are identical, we recover the notion of weak monotonicity.

Corollary 1. *Let* $f : [0,1]^n \to [0,1]$. *Then,* f *is weakly increasing (resp. weakly decreasing) if and only if* f *is* α-*increasing (resp.* α-*decreasing) for all curves* $\alpha : [0,1] \to \mathbb{R}^n$ *such that* $\alpha_i(t) = \alpha_j(t) \geq \mathbf{0}$ *for all* $i, j \in \{1, \ldots, n\}$ *and all* $0 \leq t \leq 1$.

The following is an example of a function that is α-increasing for a curve $\alpha : [0,1] \to \mathbb{R}^2$ but fails to be directionally monotone with respect to any direction $\overrightarrow{r} \in \mathbb{R}^2$.

Example 2. Let $f : [0,1]^2 \to [0,1]$ given by

$$f(x,y) = \begin{cases} 1, \text{ if } x = y = 0.25, \\ 0, \text{ otherwise;} \end{cases}$$

and let $\alpha : [0,1] \to \mathbb{R}^2$ be the curve given by

$$\alpha(t) = \begin{cases} (0.3,0), \text{ if } 0 < t \le 0.3, \\ (t,0), \quad \text{otherwise.} \end{cases}$$

The function f is not $\overrightarrow{r}$-increasing for any direction $\overrightarrow{r} \in \mathbb{R}^2$ since it has a strict global maximum at the point $(0.25, 0.25)$ and the value of f goes from 1 to 0 from the point $(0.25, 0.25)$ in any direction.

However, f is α-increasing as, clearly, there does not exist any point $(0.25, 0.25) \ne (x,y) \in \mathbb{R}^2$ such that $(x,y) + \alpha(t) = (0.25, 0.25)$ for all $0 \le t \le 1$.

Let us now present two examples of α-increasing functions for certain curves $\alpha : [0,1] \to \mathbb{R}^n$.

Example 3.(1) Let $f : [0,1]^2 \to [0,1]$ be the function given by

$$f(x,y) = \frac{1 + 3x - y^2}{4},$$

for all $x, y \in [0,1]$.
It is not hard to check that this function is α-increasing for any curve α . $[0,1] \to \mathbb{R}^2$ such that $\alpha_1 : [0,1] \to \mathbb{R}$ is increasing and $\alpha_2 : [0,1] \to \mathbb{R}$ is decreasing; or α_2 is increasing and $\alpha_1(t) \ge \dfrac{2\alpha_2(t) + \alpha_2^2(t)}{3}$ for all $0 \le t \le 1$; or the graph of α is located the fourth quadrant of the plane.
(2) Let $f : [0,1]^2 \to [0,1]$ be given by

$$f(x,y) = \frac{x}{y+1},$$

for all $x, y \in [0,1]$. This function is α-increasing for any curve $\alpha : [0,1] \to \mathbb{R}^2$ whose graph is located on the fourth quadrant of the plane.

The following result characterizes the notion of α-monotonicity for a function $f : [0,1]^n \to [0,1]$ in terms of the values that it takes in the proximity of each point of the domain. Specifically, it shows an upper and a lower bound for each point.

Theorem 1. *Let $f : [0,1]^n \to [0,1]$ and $\alpha : [0,1] \to \mathbb{R}^n$ be a curve that is continuous at $t = 0$. Then, f is α-increasing if and only if for all $\mathbf{x} \in [0,1]^n$ and all $r, s \in [0,1]$ such that*

(i) $\mathbf{x} - \alpha(r) + \alpha(t) \in [0,1]^n$, for all $0 \le t \le r$, and
(ii) $\mathbf{x} + \alpha(t) \in [0,1]^n$, for all $0 \le t \le s$;

it holds that

$$f(\mathbf{x} - \alpha(r)) \leq f(\mathbf{x}) \leq f(\mathbf{x} + \alpha(s)). \tag{3}$$

Proof. Given a curve $\alpha : [0,1] \to \mathbb{R}^n$, it is clear that if a function $f : [0,1]^n \to [0,1]$ verifies (3) for all $\mathbf{x} \in [0,1]^n$ and all $r, s \in [0,1]$ that satisfy (i) and (ii), then f is also α-increasing.

For the converse, let us suppose that f is α-increasing and let $r, s \in [0,1]$ such that (i) and (ii) hold. From (ii) and the fact that f is α-increasing, it is clear that $f(\mathbf{x}) \leq f(\mathbf{x} + \alpha(s))$. Similarly, from (i), since $\mathbf{x} - \alpha(r) + \alpha(t) \in [0,1]^n$, for all $0 \leq t \leq r$, in particular, for $t = 0$, it holds that $\mathbf{x} - \alpha(r) \in [0,1]^n$. Thus, since f is α-increasing, we obtain that

$$f(\mathbf{x} - \alpha(r) + \alpha(r)) \geq f(\mathbf{x} - \alpha(r)).$$

Hence, $f(\mathbf{x} - \alpha(r)) \leq f(\mathbf{x})$ and this completes the proof for the two inequalities in (3).

5   Curve-Based Monotonicity with Respect to the Composition of Curves

In this section, we study the conditions of curve-based monotonicity of functions with respect to the composition of two, or more, curves. By composition of two curves, we refer to the curve whose graph goes through the first curve and, then, through the second (see Fig. 2), i.e., given two curves $\alpha, \beta : [0,1] \to \mathbb{R}^n$, we define their composition $\alpha\beta : [0,1] \to \mathbb{R}^n$ by

$$\alpha\beta(t) = \begin{cases} \alpha(2t), & \text{if } 0 \leq t \leq 0.5, \\ \alpha(1) + \beta(2t - 1), & \text{if } 0.5 < t \leq 1. \end{cases} \tag{4}$$

The next result shows that if a function is curve-based monotone with respect to two different curves, then it is also curve-based monotone with respect to the combination of the two curves.

Theorem 2. *Let $\alpha : [0,1] \to \mathbb{R}^n$ and $\beta : [0,1] \to \mathbb{R}^n$ be two curves and let $f : [0,1]^n \to [0,1]$ be a function. If f is α-increasing (resp. α-decreasing) and β-increasing (resp. β-decreasing), then f is $\alpha\beta$-increasing (resp. $\alpha\beta$-decreasing).*

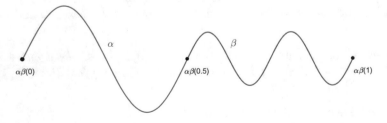

Fig. 2. Graph of a composition curve $\alpha\beta$, constructed from two curves α and β as in (4).

Proof. Let α, $\beta : [0,1] \to \mathbb{R}^2$ be two curves and let $f : [0,1]^n \to [0,1]$ be α- and β-increasing. Let $\mathbf{x} \in [0,1]^n$ and $t \in [0,1]$ such that $\mathbf{x} + \alpha\beta(s) \in [0,1]^n$ for all $0 \le s \le t$. If $t \le 0.5$, by α-increasingness of f it holds that

$$f(\mathbf{x} + \alpha\beta(t)) = f(\mathbf{x} + \alpha(2t)) \ge f(\mathbf{x}).$$

On the other hand, if $t > 0.5$, it holds that

$$f(\mathbf{x} + \alpha\beta(t)) = f(\mathbf{x} + \alpha(1) + \beta(2t - 1)), \tag{5}$$

and, since $\mathbf{x} + \alpha(1) \in [0,1]^n$ and $\mathbf{x} + \alpha(1) + \beta(2t - 1) \in [0,1]^n$, by the β-increasingness of f we derive that

$$f(\mathbf{x} + \alpha(1) + \beta(2t - 1)) \ge f(\mathbf{x} + \alpha(1)). \tag{6}$$

Therefore, from (5), (6) and the fact that f is α-increasing we derive that $f(\mathbf{x} + \alpha\beta(t)) \ge f(\mathbf{x})$.

6   Conclusion

We have discussed the concept of curve-based monotonicity, which is an extension of a recently introduced relaxation of monotonicity: directional monotonicity. We have presented some remarks and clarifications regarding the notion of α-monotonicity, for a certain curve α, as well as some theoretic properties. We have also shown some examples of functions that satisfy curve-based monotonicity for specific curves.

With respect to perspectives for future research, our aim is to investigate how this concept could benefit some applied problems. In particular, we intend to use curve-based monotone functions in the are of computer vision.

Acknowledgements. This work is supported by the research group FQM268 of *Junta de Andalucía*, by the project TIN2016-77356-P (AEI/FEDER, UE), by the Slovak Scientific Grant Agency VEGA no. 1/0093/17 Identification of risk factors and their impact on products of the insurance and savings schemes, by Slovak grant APVV-14-0013, and by Czech Project LQ1602 "IT4Innovations excellence in science".

References

1. Beliakov, G., Calvo, T., Wilkin, T.: Three types of monotonicity of averaging functions. Knowl.-Based Syst. **72**, 114–122 (2014). https://doi.org/10.1016/j.knosys.2014.08.028
2. Beliakov, G., Špirková, J.: Weak monotonicity of Lehmer and Gini means. Fuzzy Sets Syst. **299**, 26–40 (2016). https://doi.org/10.1016/j.fss.2015.11.006
3. Beliakov, G., Calvo, T., Wilkin, T.: On the weak monotonicity of Gini means and other mixture functions. Inf. Sci. **300**, 70–84 (2015). https://doi.org/10.1016/j.ins.2014.12.030

4. Bustince, H., Barrenechea, E., Sesma-Sara, M., Lafuente, J., Dimuro, G.P., Mesiar, R., Kolesárová, A.: Ordered directionally monotone functions. Justification and application. IEEE Trans. Fuzzy Syst. **26**(4), 2237–2250 (2018). https://doi.org/ 10.1109/TFUZZ.2017.2769486

5. Bustince, H., Fernandez, J., Kolesárová, A., Mesiar, R.: Directional monotonicity of fusion functions. Eur. J. Oper. Res. **244**(1), 300–308 (2015). https://doi.org/10. 1016/j.ejor.2015.01.018

6. Elkano, M., Sanz, J.A., Galar, M., Pekala, B., Bentkowska, U., Bustince, H.: Composition of interval-valued fuzzy relations using aggregation functions. Inf. Sci. **369**, 690–703 (2016). https://doi.org/10.1016/j.ins.2016.07.048

7. Gagolewski, M.: Data fusion: theory, methods, and applications. Institute of Computer Science Polish Academy of Sciences (2015)

8. García-Lapresta, J., Martínez-Panero, M.: Positional voting rules generated by aggregation functions and the role of duplication. Int. J. Intell. Syst. **32**(9), 926–946 (2017). https://doi.org/10.1002/int.21877

9. Roldán López de Hierro, A.F., Sesma-Sara, M., Špirková, J., Lafuente, J., Pradera, A., Mesiar, R., Bustince, H.: Curve-based monotonicity: a generalization of directional monotonicity. Int. J. General Syst. (2019). https://doi.org/10.1080/ 03081079.2019.1586684

10. Lucca, G., Sanz, J., Dimuro, G., Bedregal, B., Asiain, M.J., Elkano, M., Bustince, H.: CC-integrals: Choquet-like copula-based aggregation functions and its application in fuzzy rule-based classification systems. Knowl.-Based Syst. **119**, 32–43 (2017). https://doi.org/10.1016/j.knosys.2016.12.004

11. Lucca, G., Sanz, J.A., Dimuro, G.P., Bedregal, B., Mesiar, R., Kolesárová, A., Bustince, H.: Preaggregation functions: construction and an application. IEEE Trans. Fuzzy Syst. **24**(2), 260–272 (2016). https://doi.org/10.1109/TFUZZ.2015. 2453020

12. Lucca, G., Sanz, J.A., Dimuro, G.P., Bedregal, B., Bustince, H., Mesiar, R.: CF-integrals: a new family of pre-aggregation functions with application to fuzzy rule-based classification systems. Inf. Sci. **435**, 94–110 (2018). https://doi.org/10.1016/ j.ins.2017.12.029

13. Mesiar, R., Kolesárová, A., Stupňanová, A.: Quo vadis aggregation? Int. J. General Syst. **47**(2), 97–117 (2018). https://doi.org/10.1080/03081079.2017.1402893

14. Paternain, D., Fernandez, J., Bustince, H., Mesiar, R., Beliakov, G.: Construction of image reduction operators using averaging aggregation functions. Fuzzy Sets Syst. **261**, 87–111 (2015). https://doi.org/10.1016/j.fss.2014.03.008

15. Rousseeuw, P.J., Leroy, A.M.: Robust Regression and Outlier Detection, vol. 589. Wiley, Hoboken (2005)

16. Sesma-Sara, M., Lafuente, J., Roldán, A., Mesiar, R., Bustince, H.: Strengthened ordered directionally monotone functions. Links between the different notions of monotonicity. Fuzzy Sets Syst. **357**, 151–172 (2019). https://doi.org/10.1016/j.fss. 2018.07.007

17. Sesma-Sara, M., Bustince, H., Barrenechea, E., Lafuente, J., Kolesárová, A., Mesiar, R.: Edge detection based on ordered directionally monotone functions. In: Advances in Fuzzy Logic and Technology 2017, pp. 301–307. Springer (2017). https://doi.org/10.1007/978-3-319-66827-7_27

18. Sesma-Sara, M., Mesiar, R., Bustince, H.: Weak and directional monotonicity of functions on Riesz spaces to fuse uncertain data. Fuzzy Sets Syst. (in press). https://doi.org/10.1016/j.fss.2019.01.019

19. Wilkin, T., Beliakov, G.: Weakly monotonic averaging functions. Int. J. Int. Syst. **30**(2), 144–169 (2015). https://doi.org/10.1002/int.21692

A Construction Method for t-norms on Bounded Lattices

Funda Karaçal[1]([✉]) [iD], Ümit Ertuğrul[1] [iD], and M. Nesibe Kesicioğlu[2] [iD]

[1] Department of Mathematics, Karadeniz Technical University,
61080 Trabzon, Turkey
fkaracal@yahoo.com, uertugrul@ktu.edu.tr
[2] Department of Mathematics, Recep Tayyip Erdogan University,
53100 Rize, Turkey
m.nesibe@gmail.com

Abstract. In this paper, a construction method on a bounded lattice from a given t-norm on a subinterval of the bounded lattice is presented. The supremum distributivity of the constructed t-norm by the mentioned method has been proven under some special conditions. Giving an example, the constructed t-norm need not be supremum-distributive on any bounded lattice is shown. Moreover, some relationships between the mentioned construction method and the other construction methods in the literature are presented.

Keywords: t-norm · Bounded lattice · Construction method · Subinterval

1 Introduction

Triangular norms were first described by Menger as a generalization of the triangle inequality in [12]. Then, triangular norms were defined as we know today on $[0, 1]$ [15,16] and have been studied from many different perspectives and thus have been a study topic in itself [1,3,5,9,11]. In addition to its applicability to computer sciences and engineering, $[0, 1]$ is highly preferred by researchers considering the advantages of working on some important mathematical properties (topological structure, continuity on it, etc.)

Due to the presence of the incomparable elements and the lack of some important features provided on $[0, 1]$, working on bounded lattices is much more complex than working on $[0, 1]$. But it is still more attractive since bounded lattices are more general algebraic structures than the unit real interval $[0, 1]$. Considering these reasons, to define triangular norms on bounded lattices has been a current study topic for researchers and is still up to date.

In [11], there are some construction methods for t-norms via a subset A satisfying $(0, 1) \subseteq A \subseteq [0, 1]$ or h-additive generators. Some construction methods are also given on bounded lattices in [3,5,14]. It was investigated under which conditions T is a t-norm in [14]. In [5], when $[a, 1]$ is a subinterval of a bounded

© Springer Nature Switzerland AG 2019
R. Halaš et al. (Eds.): AGOP 2019, AISC 981, pp. 205–211, 2019.
https://doi.org/10.1007/978-3-030-19494-9_19

lattice L, a construction method for a t-norm on $[a, 1]$ to be a t-norm on a bounded lattice L is proposed.

The paper is organized as follows: In Sect. 1, we give some necessary definitions and previous results. The existence of a t-norm T on a subinterval $[a, b]$ of a bounded lattice is well known truth in the literature. Based on this, in Sect. 2, we extend a given t-norm on $[a, b]$ to a t-norm on L with no additional conditions. We present the t-norm obtained by our method coincides with the t-norm proposed by Saminger in [14], when a and b are comparable to all elements of L. Further, when $b = 1$, we observe that the t-norm obtained by mentioned method is coincidence with the method given in [5]. In our best knowledge, there is no construction method extending a t-norm on a subinterval $[0, b]$ of a bounded lattice L to be a t-norm on L. Our method fill this gap in the literature.

1.1   Preliminaries

Definition 1. *[2] Let $(L, \leq, 0, 1)$ be a bounded lattice. The elements x and y are called comparable if $x \leq y$ or $y \leq x$. Otherwise, x and y are called incomparable. In this situation, the notation $x \| y$ is used.*

We will use the notation I_a for the set of all incomparable elements with an element $a \in L$.

Definition 2. *[2] Let $(L, \leq, 0, 1)$ be a bounded lattice and $a, b \in L$ with $a \leq b$. The sublattice $[a, b]$ is defined as*

$$[a, b] = \{x \in L \mid \quad a \leq x \leq b\}.$$

Similarly, $(a, b] = \{x \in L \mid \quad a < x \leq b\}$, $[a, b) = \{x \in L \mid \quad a \leq x < b\}$ and $(a, b) = \{x \in L \mid \quad a < x < b\}$ can be defined.

Definition 3. *[11] An operation T (S) on a bounded lattice L is called a triangular norm (triangular conorm) if it is commutative, associative, increasing with respect to the both variables and has a neutral element 1 (0).*

Let T_1 and T_2 be two t-norms on L. T_1 is called smaller than T_2 if for any elements $x, y \in L$, $T_1(x, y) \leq T_2(x, y)$. Similar relation between two t-conorms can be given. In this sense, the smallest and the greatest t-norms on a bounded lattice L are given respectively as follows:

$$T_D(x, y) = \begin{cases} y & x = 1, \\ x & y = 1, \\ 0 & \text{otherwise} \end{cases} \quad \text{and } T_\wedge(x, y) = x \wedge y.$$

Similarly, the smallest t-conorm $S_\vee$ and the greatest t-conorm S_D on a bounded lattice L,

$$S_D(x, y) = \begin{cases} y & x = 0, \\ x & y = 0, \\ 1 & \text{otherwise} \end{cases} \quad \text{and } S_\vee(x, y) = x \vee y$$

are given.

Definition 4. *[8] A t-norm T (or t-conorm S) on a bounded lattice L is called $\vee$-distributive if for every $a, b_1, b_2 \in L$*

$$T(a, b_1 \vee b_2) = T(a, b_1) \vee T(a, b_2) \ (or \ S(a, b_1 \vee b_2) = S(a, b_1) \vee S(a, b_2))$$

holds.

Definition 5. *[14] Let $((a_i, b_i))_{i \in A}$ be a family of pairwise disjoint open subinterval of $[0, 1]$ and let $(T_i)_{i \in I}$ be a family of t-norms. Then, the ordinal sum $T = (\langle a_i, b_i, T_i \rangle)_{i \in I} : [0, 1]^2 \to [0, 1]$ is given by*

$$T(x, y) = \begin{cases} a_i + (b_i - a_i) T_i(\frac{x - a_i}{b_i - a_i}, \frac{y - a_i}{b_i - a_i}) & x, y \in [a_i, b_i], \\ \min(x, y) & otherwise. \end{cases}$$

2 Construction of Triangular Norm

In this section, a construction method is presented to obtain a t-norm on a bounded lattice L when W is a t-norm on $[a, b]$, where L is an arbitrary bounded lattice and $[a, b]$ is an arbitrary subinterval of L. Additionally, whether the triangular norm obtained by proposed method is a supremum-distributive t-norm is investigated and the relationship with the current methods are investigated.

Theorem 1. *Let $(L, \leq, 0, 1)$ be bounded lattice and W be a t-norm on a subinterval $[a, b]$ of L. Then, the function $T : L^2 \to L$ defined as follows*

$$T(x, y) = \begin{cases} x \wedge y \wedge a & x < a \ or \ y < a, \\ W(x, y) & (x, y) \in [a, b)^2, \\ x \wedge y \wedge a & x \in [a, b), \ (y \in I_a \ and \ y < b), \\ W(x, y \wedge b) & x \in [a, b), \ (a < y \ and \ y \in I_b), \\ x & x \in [a, b), \ 1 > y \geq b, \\ x \wedge y \wedge a & (x \in I_a \ and \ x < b), \ y \in [a, b), \\ x \wedge y \wedge a & (x \in I_a \ and \ x < b), \ (y \in I_a \ and \ y < b), \\ x \wedge y \wedge a & (x \in I_a \ and \ x < b), \ (a < y \ and \ y \in I_b), \\ x \wedge y \wedge a & (x \in I_a \ and \ x < b), 1 > y \geq b, \\ W(x \wedge b, y) & (a < x \ and \ x \in I_b), y \in [a, b), \\ x \wedge y \wedge a & (a < x \ and \ x \in I_b), (y \in I_a \ and \ y < b), \\ W(x \wedge b, y \wedge b) & (a < x \ and \ x \in I_b), (a < y \ and \ y \in I_b), \\ x \wedge y \wedge b & (a < x \ and \ x \in I_b), 1 > y \geq b, \\ y & 1 > x \geq b, \ y \in [a, b), \\ x \wedge y \wedge a & 1 > x \geq b, (y \in I_a \ and \ y < b), \\ x \wedge y \wedge b & 1 > x \geq b, (y > a \ and \ y \in I_b), \\ x \wedge y & 1 > x \geq b, 1 > y \geq b, \\ x \wedge y \wedge a & x \in I_a \cap I_b, \ y \in L \setminus \{1\} \ or \ y \in I_a \cap I_b, \ x \in L \setminus \{1\} \\ y & x = 1, \\ x & y = 1. \end{cases}$$

is a t-norm on L.

Remark 1. If $I_a = I_b = \emptyset$, the method presented in Theorem 1 coincides with the Saminger's construction method [14]. The method proposed by Saminger may not produce a t-norm on any bounded lattice. These two methods do not have to coincide on a bounded lattice that provide Saminger's constraints (see following example). Moreover, the method given in Theorem 1 coincides with the method given in [5] when $b = 1$.

Example 1. Consider the lattice L characterized by Hasse diagram, as shown in Fig. 1.

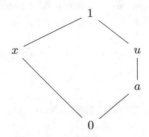

Fig. 1. Lattice diagram of L

Let take $T = T_\wedge$ on $[a, 1]$. We obtain the t-norm T_1 considering Saminger's method and the t-norm T_2 considering Theorem 1 on L, respectively (Tables 1 and 2).

Table 1. T-conorm T_1 on L

T_1	0	x	a	u	1
0	0	0	0	0	0
x	0	x	0	0	x
a	0	0	a	a	a
u	0	0	a	u	u
1	0	x	a	u	1

and

Table 2. T-conorm T_2 on L

T_2	0	x	a	u	1
0	0	0	0	0	0
x	0	0	0	0	x
a	0	0	a	a	a
u	0	0	a	u	u
1	0	x	a	u	1

It is clear that T_1 and T_2 do not coincide.

In the literature, there is no construction method which extends a t-norm on $[0,a]$ to be a t-norm on a bounded lattice L as our best knowledge. We also provided a method for constructing the t-norm on a bounded lattice L using given triangular norms on subintervals $[0,a]$ of L.

Corollary 1. *In Theorem 1, if a is taken as 0, the following t-norm is obtained:*

$$T(x,y) = \begin{cases} W(x,y) & (x,y) \in [0,b)^2, \\ W(x, y \wedge b) & x \in [0,b), \ (0 < y \ and \ y \in I_b), \\ x & x \in [0,b), \ 1 > y \geq b, \\ W(x \wedge b, y) & (0 < x \ and \ x \in I_b), y \in [0,b), \\ W(x \wedge b, y \wedge b) & (0 < x \ and \ x \in I_b), (0 < y \ and \ y \in I_b), \\ x \wedge y \wedge b & (0 < x \ and \ x \in I_b), 1 > y \geq b, \\ y & 1 > x \geq b, \ y \in [0,b), \\ x \wedge y \wedge b & 1 > x \geq b, (y > 0 \ and \ y \in I_b), \\ x \wedge y & (x,y) \in [b,1)^2, \\ y & x = 1, \\ x & y = 1. \end{cases}$$

Proposition 1. *Let $(L, \leq, 0, 1)$ be a distributive bounded lattice and W be a t-norm on a subinterval $[a,b]$ of L. If W is $\vee$-distributive and $I_a = I_b = \emptyset$, then the t-norm T given in Theorem 1 is $\vee$-distributive.*

Remark 2. Even if L is a distibutive lattice and W is a $\vee$-distributive t-norm on $[a,b] \subseteq L$, the t-norm T given in Theorem 1 need not be $\vee$-distributive. Let us look at the following example.

Example 2. Let L be a bounded lattice whose lattice diagram is depicted as follow (Fig. 2):

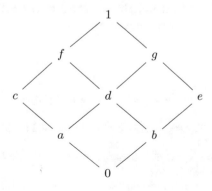

Fig. 2. $(L, \leq)$

Take the t-norm $W = T_\wedge$ on $[d, 1]$. Then, the t-norm T generated by the method given in Theorem 1 is as follow (Table 3).

Table 3. The function T

T	0	a	b	c	d	e	f	g	1
0	0	0	0	0	0	0	0	0	0
a	0	a	0	a	a	0	a	a	a
b	0	0	b	0	b	b	b	b	b
c	0	a	0	a	a	0	a	a	c
d	0	a	b	a	d	b	d	d	d
e	0	0	b	0	b	b	b	b	e
f	0	a	b	a	d	b	f	d	f
g	0	a	b	a	d	b	d	g	g
1	0	a	b	c	d	e	f	g	1

It is clear that, L is a distributive lattice and W is $\vee$-distributive on $[d, 1]$. But, t-norm T is not $\vee$-distributive since

$$T(c, f \vee g) = T(c, 1) = c \neq a = a \vee a = T(c, f) \vee T(c, g).$$

3   Concluding Remarks

Triangular norms are still a current subject matter. Therefore, the study of triangular norms on important algebraic structures continues to attract researchers' attention. In this study, we provide a way to obtain a triangular norm on L from a triangular norm defined on a subinterval of L. It should be noted that unlike existing methods in the literature, this method works without any restrictions and it is a more general construction method.

References

1. Aşıcı, E., Karaçal, F.: On the T-partial order and properties. Inf. Sci. **267**, 323–333 (2014)
2. Birkhoff, G.: Lattice Theory. American Mathematical Society Colloquium Publishers, Providence (1967)
3. Çaylı, G.D.: On a new class of t-norms and t-conorms on bounded lattices. Fuzzy Sets Syst. **332**, 129–143 (2018)
4. Durante, F., Sarkoci, P.: A note on the convex combinations of triangular norms. Fuzzy Sets Syst. **159**, 77–80 (2008)
5. Ertuğrul, Ü., Karaçal, F., Mesiar, R.: Modified ordinal sums of triangular norms and triangular conorms on bounded lattices. Int. J. Intell. Syst. **30**, 807–817 (2015)

6. Grabisch, M., Marichal, J.-L., Mesiar, R., Pap, E.: Aggregation Functions. Cambridge University Press, Cambridge (2009)
7. Karaçal, F.: On the direct decomposability of strong negations and S-implication operators on product lattices. Inf. Sci. **176**, 3011–3025 (2006)
8. Karaçal, F., Khadjiev, D.: ∨-Distributive and infinitely ∨-distributive t-norms on complete lattices. Fuzzy Sets Syst. **151**, 341–352 (2005)
9. Karaçal, F., Kesicioğlu, M.N.: A T-partial order obtained from t-norms. Kybernetika **47**, 300–314 (2011)
10. İnce, M.A., Karaçal, F.: t-closure operators. Int. J. General Syst. https://doi.org/10.1080/03081079.2018.1549041
11. Klement, E.P., Mesiar, R., Pap, E.: Triangular Norms. Kluwer Academic Publishers, Dordrecht (2000)
12. Menger, K.: Statistical metrics. Proc. Natl. Acad. Sci. U.S.A **8**, 535–537 (1942)
13. Mesiar, R., Mesiarová-Zemánková, A.: Convex combinations of continuous t-norms with the same diagonal function. Nonlinear Anal. **69**(9), 2851–2856 (2008)
14. Saminger, S.: On ordinal sums of triangular norms on bounded lattices. Fuzzy Sets Syst. **157**(10), 1403–1416 (2006)
15. Schweizer, B., Sklar, A: Espaces metriques aléatoires. C. R. Acad. Sci. Paris Sér. A **247** 2092–2094 (1958)
16. Schweizer, B., Sklar, A.: Statistical metric spaces. Pacific J. Math. **10**, 313–334 (1960)

Penalty-Based Aggregation of Strings

Raúl Pérez-Fernández[1,2(✉)] and Bernard De Baets[1]

[1] KERMIT, Department of Data Analysis and Mathematical Modelling,
Ghent University, Coupure links 653, 9000 Ghent, Belgium
{raul.perezfernandez,bernard.debaets}@ugent.be
[2] Department of Statistics and O.R. and Mathematics Didactics,
University of Oviedo, Oviedo, Spain
perezfernandez@uniovi.es

Abstract. Whereas the field of aggregation theory has historically studied aggregation on bounded posets (mainly the aggregation of real numbers), different aggregation processes have been analysed in different fields of application. In particular, the aggregation of strings has been a popular topic in many fields featuring computer science and bioinformatics. In this conference paper, we discuss different examples of aggregation of strings and position them within the framework of penalty-based data aggregation.

Keywords: Aggregation · Strings · Penalty functions

1 Introduction

Binary strings are ubiquitous in computer science [1], whereas DNA sequences are a prominent type of string arising naturally in the field of bioinformatics [2]. It is no surprise then that the aggregation of strings has been extensively studied. The computation of median strings and center strings, which respectively minimize the sum of the distances and the maximum distance to the strings to be aggregated, surely represents the core problem in the aggregation of strings [3]. Different distance metrics have been considered for defining median and center strings [4]. The two most prominent examples are the Hamming distance metric [5], popular in coding theory, and the Levenshtein distance metric [6], popular for code correction.

It is nonetheless surprising that there has been little interest in this topic from the field of aggregation theory, especially bearing in mind that the study of median strings and center strings certainly resembles some classical problems for aggregation theorists. This is probably due to the fact that there is no meaningful order when dealing with strings, letting aside the literature-oriented alphabetic/lexicographic order, and the field of aggregation theory has historically been linked to processes that aggregate elements of a bounded poset. In a recent work [7], the present authors introduced a new framework for penalty-based data aggregation that does not restrict to the aggregation on ordered

© Springer Nature Switzerland AG 2019
R. Halaš et al. (Eds.): AGOP 2019, AISC 981, pp. 212–222, 2019.
https://doi.org/10.1007/978-3-030-19494-9_20

structures and that embraces the aggregation on structures equipped with a betweenness relation. In this paper, we position the search for median and center strings with respect to different distance metrics within this framework. We discuss both cases in which the strings are or are not restricted to have a fixed length.

2   The Framework of Penalty-Based Data Aggregation

Penalty functions have been used for decades in the context of aggregation theory [8]. Mostly confined to the aggregation of real numbers [9], the current understanding of a penalty function is more or less as follows [10] (up to a positive additive constant).

Definition 1. *Consider $n \in \mathbb{N}$ and a closed interval $I \subseteq \mathbb{R}$. A function $P : I \times I^n \to \mathbb{R}^+$ is called a penalty function if:*

(i) $P(y; \mathbf{x}) \geq 0$, for any $y \in I$ and any $\mathbf{x} \in I^n$;
(ii) $P(y; \mathbf{x}) = 0$ if and only if $\mathbf{x} = (y, \ldots, y)$;
(iii) $P(\cdot; \mathbf{x})$ is quasi-convex[1] and lower semi-continuous[2] for any $\mathbf{x} \in I^n$.

In a recent paper [7], the present authors proposed a generalization of the definition of a penalty function based on the compatibility with a betweenness relation.

Definition 2. *A ternary relation B on a non-empty set X is called a betweenness relation if it satisfies the following three properties:*

(i) Symmetry in the end points: for any $x, y, z \in X$, it holds that

$$(x, y, z) \in B \Leftrightarrow (z, y, x) \in B.$$

(ii) Closure: for any $x, y, z \in X$, it holds that

$$\big((x, y, z) \in B \wedge (x, z, y) \in B\big) \Leftrightarrow y = z.$$

(iii) End-point transitivity: for any $o, x, y, z \in X$, it holds that

$$\big((o, x, y) \in B \wedge (o, y, z) \in B\big) \Rightarrow (o, x, z) \in B.$$

Two betweenness relations of importance to this paper are the betweenness relation induced by a given distance metric and the product betweenness relation.

---

[1] Consider a closed interval $I \subseteq \mathbb{R}$. A function $f : I \to \mathbb{R}$ is called quasi-convex if, for any $u, v \in I$ and any $\lambda \in {]}0, 1{[}$, it holds that $f(\lambda u + (1 - \lambda)v) \leq \max(f(u), f(v))$.
[2] Consider a closed interval $I \subseteq \mathbb{R}$. A function $f : I \to \mathbb{R}$ is called lower semi-continuous if, for any $u \in I$, it holds that $\liminf_{v \to u} f(v) = f(u)$.

Proposition 1. *Consider a distance metric d on a set X. The ternary relation B_d on X defined as*

$$B_d = \left\{ (x,y,z) \in X^3 \mid d(x,z) = d(x,y) + d(y,z) \right\},$$

is a betweenness relation on X, called the betweenness relation induced by d.

Proposition 2. *Consider $n \in \mathbb{N}$ and a betweenness relation B on a set X. The ternary relation $B^{(n)}$ on X^n defined as*

$$B^{(n)} = \left\{ (\mathbf{x}, \mathbf{y}, \mathbf{z}) \in (X^n)^3 \mid (\forall i \in \{1,\dots,n\})((x_i, y_i, z_i) \in B) \right\},$$

is a betweenness relation on X^n, called the product betweenness relation.

Given a betweenness relation, we have proposed a definition of a penalty function in which the original third property aiming at providing some desirable semantics to the penalty is substituted by the requirement of the set of minimizers to be non-empty and by the compatibility with a betweenness relation aiming again at providing the penalty with some desirable semantics.

Definition 3. *Consider $n \in \mathbb{N}$, a set X and a betweenness relation B on X^n. A function $P : X \times X^n \to \mathbb{R}^+$ is called a penalty function (compatible with B) if the following four properties hold:*

(P1) $P(y; \mathbf{x}) \geq 0$, for any $y \in X$ and any $\mathbf{x} \in X^n$;
(P2) $P(y; \mathbf{x}) = 0$ if and only if $\mathbf{x} = (y, \dots, y)$;
(P3) The set of minimizers of $P(\cdot; \mathbf{x})$ is non-empty, for any $\mathbf{x} \in X^n$.
(P4) $P(y; \mathbf{x}) \leq P(y; \mathbf{x}')$, for any $y \in X$ and any $\mathbf{x}, \mathbf{x}' \in X^n$ such that $((y, \dots, y), \mathbf{x}, \mathbf{x}') \in B$.

Remark 1. If a betweenness relation on X is given instead of a betweenness relation on X^n, it is assumed that the product betweenness relation is considered.

The process of aggregation is then understood as a process of minimizing a penalty function given the list of objects to be aggregated. Different existing procedures coming from different fields are discussed in [7] and are shown to fit within this framework.

Definition 4. *Consider $n \in \mathbb{N}$, a set X, a betweenness relation B on X^n and a penalty function $P : X \times X^n \to \mathbb{R}^+$ compatible with B. The function $f : X^n \to \mathcal{P}(X)$ defined by*

$$f(\mathbf{x}) = \arg\min_{y \in X} P(y; \mathbf{x}),$$

for any $\mathbf{x} \in X^n$, is called the penalty-based function associated with P.

It is important to note that any aggregation process characterized as above is idempotent, i.e., the result of aggregating a list of n times the same object needs to be this very object. Additionally, one should note that more than one minimizer could be obtained. This is often the case in the setting of this paper in which we deal with the aggregation of strings.

The two most common examples of penalty-based functions are defined by means of the sum of distances or the maximum distance to the objects to be aggregated [7].

Corollary 1. *Consider* $n \in \mathbb{N}$, *and a metric space* (X, d). *The function* $P :$ $X \times X^n \to \mathbb{R}^+$ *defined by*

$$P(y; \mathbf{x}) = \sum_{i=1}^{n} d(y, x_i),$$

for any $(y; \mathbf{x}) \in X^{n+1}$, *is a penalty function (compatible with* $B_d^{(n)}$).

Corollary 2. *Consider* $n \in \mathbb{N}$, *and a metric space* (X, d). *The function* $P :$ $X \times X^n \to \mathbb{R}^+$ *defined by*

$$P(y; \mathbf{x}) = \max_{i=1}^{n} d(y, x_i),$$

for any $(y; \mathbf{x}) \in X^{n+1}$, *is a penalty function (compatible with* $B_d^{(n)}$).

Functions of the former type are usually referred to as *medians* and functions of the latter type are usually referred to as *centers*[3]. For some types of object such as real numbers and real vectors, *centroids* (which minimize the sum of squared distances to the objects to be aggregated) have also been extensively studied. However, centroid strings are to the best of our knowledge way less popular than median and center strings. For this very reason, centroid strings will not be discussed in this paper, although they would perfectly fit within the framework of penalty-based aggregation.

In case the set X is finite, there is a large literature in the field of operations research on how to compute the minimizers of $P(\cdot; \mathbf{x})$ (for any given $\mathbf{x} \in X^n$). More specifically, we refer to the minimum facility location problem for the computation of medians and to the minmax facility location problem for the computation of centers [11].

3 Strings of the Same Length

Given an alphabet (set of characters) Σ, any list of m elements of Σ is called a string of length m. The set of all strings of length m (on an alphabet Σ) is denoted by Σ_m. For any $S \in \Sigma_m$ and any $j \in \{1, \dots, m\}$, we denote by $S(j)$ the j-th element of S. In this section, we fix the value of m and we discuss four natural examples of distance metrics on the set of strings of the same length (m).

3.1 The Discrete Distance Metric

The first and most trivial example of distance metric on the set of strings of length m is the discrete distance metric, defined as $\delta(S, S') = 0$, if $S = S'$, and $\delta(S, S') = 1$, otherwise. This distance metric induces the betweenness relation

---

[3] In the context of the aggregation of strings, both terms 'center string' and 'closest string' are found to carry the same meaning.

B_δ on Σ_m illustrated in Fig. 1, known as the minimal betweenness relation. Note that the minimal betweenness relation is contained in any possible betweenness relation [7]. For this reason, the semantics brought to any penalty function by the betweenness relation B_δ are negligible.

Fig. 1. Illustration of the strings (in grey) that are in between the strings *cat* and *dog* (in red) according to the betweenness relation B_δ for the English alphabet $\Sigma = \{a, b, c, \dots, z\}$. Please note that there is no string in grey.

Still, one could think of identifying the median string(s) and the center string(s) of a given list of strings **S** with respect to the discrete distance metric. A median string of **S** is characterized as a string appearing with the highest frequency in **S**, whereas any possible string in Σ_m is a center string of **S** unless there exists a string $S \in \Sigma_m$ such that $\mathbf{S} = (S, \dots, S)$, a case in which S is the unique center string (due to the idempotence of any penalty-based function).

Example 1. Consider the following list of strings of length 3 on the English alphabet $\Sigma = \{a, b, c, \dots, z\}$: $\mathbf{S} = \{cat, dog, cat, dot, cog\}$. The unique median string is *cat*, whereas any possible string in Σ_3 is a center string.

3.2   The Hamming Distance Metric

The Hamming distance is a popular distance metric for strings of the same length [5]. Intuitively, this distance metric assigns to each couple of strings of the same length m the number of positions at which both strings differ, i.e.,

$$H(S, S') = |\{j \in \{1, \dots, m\} \mid S(j) \neq S'(j)\}|.$$

This distance metric induces a betweenness relation B_H on Σ_m, illustrated in Fig. 2. As can be seen, the semantics induced by any penalty function compatible with B_H is richer than that induced by any penalty function compatible with B_δ.

The search for the median string(s) and the center string(s) of a given list of strings $\mathbf{S} = (S_1, \dots, S_n)$ with respect to the Hamming distance metric is a popular topic in computer science. Median strings with respect to the Hamming distance metric are easily characterized as the strings S such that $S(j)$ appears with the highest frequency in $(S_1(j), \dots, S_n(j))$ for all $j \in \{1, \dots, m\}$. Unfortunately, center strings with respect to the Hamming distance metric are not characterizable and they are known to be NP-hard to compute [1]. The latter problem of finding the center string (usually referred to as the closest string problem) is of interest to many scientific disciplines such as molecular biology and coding theory [12] and thus has led to many works aiming at finding an efficient algorithm/approximation for the closest string problem [13].

Example 2. Consider again the list of strings of length 3 on the English alphabet $\Sigma = \{a, b, c, \ldots, z\}$: **S** $= \{cat, dog, cat, dot, cog\}$. The unique median string is *cot* since *c* appears with the highest frequency (three) at the first position, *o* appears with the highest frequency (three) at the second position and *t* appears with the highest frequency (three) at the third position. Center strings are *dag* and all strings of the form *co∗* or *∗ot*, where ∗ represents any element in Σ.

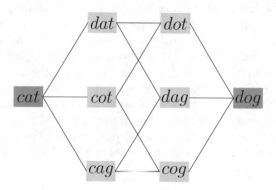

Fig. 2. Illustration of the strings (in grey) that are in between the strings *cat* and *dog* (in red) according to the betweenness relation B_H for the English alphabet $\Sigma = \{a, b, c, \ldots, z\}$.

3.3   The Lexicographic Distance Metric

The use of lexicographic orders arises naturally when dealing with strings [14]. In this setting, the most common lexicographic order is the widely-known alphabetical order. Because of its popularity, both terms are often used interchangeably when talking about strings. Obviously, the set of strings of length m is linearly ordered if we consider the lexicographic order $\leq$. Thus, as in every linearly ordered set, we can define the lexicographic distance metric:

$$L(S, S') = |\{S'' \in \Sigma_m \mid \min(S, S') \leq S'' \leq \max(S, S')\}| - 1.$$

This distance metric induces the betweenness relation B_L on Σ_m (which amounts to $B_{\leq}$ if we consider the betweenness relation induced by the alphabetical order $\leq$ – for more details see [7]), illustrated in Fig. 3. Note that this betweenness relation B_L carries a totally different semantics than B_H. For this very reason, a penalty-based function compatible with B_L will almost surely lead to quite different results than a penalty-based function compatible with B_H.

Fig. 3. Illustration of the strings (in grey) that are in between the strings *cat* and *dog* (in red) according to the betweenness relation B_L for the English alphabet $\Sigma = \{a, b, c, \ldots, z\}$.

The search for the median string(s) and the center string(s) of a given list of strings **S** with respect to the lexicographic distance metric is similar to the search for the median(s) and center(s) of a list of real numbers. Actually, it suffices to identify each string with its position in the alphabetical order and compute the median of these positions for identifying the median string(s) (as usual, only assured to be unique if the number of strings is odd). If, instead of the median of the positions, we compute the arithmetic mean of the smallest position and the greatest position, then we would obtain the center string (in case this arithmetic mean of the smallest position and the greatest position is not a natural number, there will be two center strings identified by the floor and the ceiling of said value).

Example 3. Consider again the list of strings of length 3 on the English alphabet $\Sigma = \{a, b, c, \ldots, z\}$: **S** $= \{cat, dog, cat, dot, cog\}$. We identify the string *cat* with the position $2 \cdot 26^2 + 0 \cdot 26 + 19 \cdot 1 + 1 = 1372$, the string *dog* with the position $3 \cdot 26^2 + 14 \cdot 26 + 6 \cdot 1 + 1 = 2399$, the string *dot* with the position $3 \cdot 26^2 + 14 \cdot 26 + 19 \cdot 1 + 1 = 2412$ and the string *cog* with the position $2 \cdot 26^2 + 14 \cdot 26 + 6 \cdot 1 + 1 = 1723$. The median string is the string identified with the median of $\{1372, 2399, 1372, 2412, 1723\}$, which is 1723 – thus being the median string *cog*. The center string is the string identified with the position $\frac{1372+2412}{2} = 1892$. Since $1892 = 2 \cdot 26^2 + 20 \cdot 26 + 19 \cdot 1 + 1$, the center string is *cut*.

In the particular case in which we are dealing with existent words and not with just any possible string, both the median string(s) and the center string(s) could be nonexistent words even though all strings to be aggregated are perfectly fine existent words (for the case of the median string this is only possible if an even number of strings is to be aggregated). A potential way of solving this issue would require to compute medoid [15] (or set median) strings: instead of considering the set of minimizers of some certain $P(\cdot; \mathbf{S})$ in the whole Σ_m, we would just restrict our attention to the minimizers among those strings in **S**.

Example 4. Consider the list of strings of length 3 on the English alphabet $\Sigma = \{a, b, c, \ldots, z\}$: **S** $= \{cat, car\}$. The center string is *cas*. Since this word does not appear in the English dictionary, one might think of computing a medoid-like center string (sometimes called set center string). We would thus obtain both *cat* and *car* as solutions.

3.4   The Baire Distance Metric

The Baire distance metric [16] (also referred to as Generalized Cantor distance metric) is an ultrametric[4] on the set of strings of length m that is defined as $B(S, S') = 0$, if $S = S'$, and $B(S, S') = \alpha^r$, otherwise, where $\alpha \in]0, 1[$ is a parameter to be fixed and r is the first position at which S and S' differ.

Like the discrete distance metric, the betweenness relation induced by the Baire distance metric is the minimal betweenness relation. This is actually a common result for all ultrametrics. Suppose that three pairwisely different elements x, y, z of the ultrametric space are such that $d(x, y) + d(y, z) = d(x, z)$. A contradiction then arises from the fact that $d(x, z) \leq \max(d(x, y), d(y, z))$ and being both $d(x, y)$ and $d(y, z)$ greater than zero.

Nevertheless, the fact that the induced betweenness relation is the minimal one (and thus does not bring any interesting semantics to the penalty) does not mean that penalty-based functions compatible with the betweenness relation induced by an ultrametric are not interesting. In this case, both the search for the median string(s) and the search for the center string(s) are of interest. The former is easily characterizable when a small enough value of α is considered. In particular, if $\alpha < \frac{1}{n-1}$ (where n is the number of strings to be aggregated), the median string(s) is(are) obtained in an iterative manner by first computing the most frequent first element and eliminating all the strings with a different first element, subsequently, among the remaining strings computing the most frequent second element and eliminating all the strings with a different second element, and so on. The computation of the center string(s) is easier. Independently of the value of α, we compute the longest common prefix among all strings to be aggregated. Any string starting with this prefix is a center string.

Example 5. Consider again the list of strings of length 3 on the English alphabet $\Sigma = \{a, b, c, \ldots, z\}$: $\mathbf{S} = \{cat, dog, cat, dot, cog\}$. The most common first element is c, thus we consider the strings cat, cat and cog. Among those three strings, the most common second element is a, thus we consider the strings cat and cat. Among those two strings, the most common third element is t. Thus, the median string is cat. Since there is no common prefix among all strings in $\mathbf{S}$, all strings in Σ_3 are center strings.

4   Strings of Different Length

Given an alphabet Σ, we denote by Σ_* the set of all possible strings of any length $m \in \mathbb{N}$, i.e., $\Sigma_* = \cup_{i=0}^{\infty} \Sigma_i$ ($\Sigma_0 = \{\emptyset\}$).

An edit operation is a basic change that allows to transform one string into another. Typically, edit operations are related to spelling errors. According to Damerau [17], the most common spelling errors (and thus examples of edit operations) are addition of a single character, removal of a single character, substitution of a single character and transposition of two consecutive characters, which

---

[4] We recall that an ultrametric on X is a distance metric on X satisfying that $d(x, z) \leq \max(d(x, y), d(y, z))$, for any $x, y, z \in X$.

amount to more than 95% of the spelling errors found in different texts. Other possible edit operations could include inversion of the whole string or, when dealing with proper words, substitution by a homophonous word.

Edit distance metrics [4] are defined relative to a set of edit operations $\mathbb{E}$ and a cost function $c : \mathbb{E} \to [0, \infty[$:

$$E_{\mathbb{E},c}(S, S') = \min_{\mathbf{E} \in \mathbb{E}(S,S')} \sum_{e \in \mathbf{E}} c(e),$$

where $\mathbb{E}(S, S')$ represents the set of all lists of edit operations in $\mathbb{E}$ that turn S into S'. Undoubtedly, the most common edit distance – which is almost considered a standard – is the Levenshtein distance metric [6] (just denoted by E, without subscripts). For the Levenshtein distance metric, the set of edit operations is formed by addition of a single character, removal of a single character and substitution of a character into another one, and the cost function is any constant function (e.g., $c(e) = 1$, for any $e \in \mathbb{E}$). Note that the Hamming distance metric is a special case of edit distance metric in which the unique edit operation allowed is substitution.

The betweenness relation B_E on Σ_* (illustrated in Fig. 4) is quite interesting for error detection. Note that we use two different strings compared to the previous cases, otherwise we would obtain the same diagram as displayed in Fig. 2.

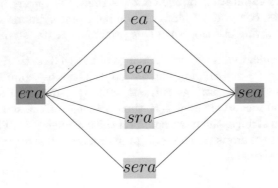

Fig. 4. Illustration of the strings (in grey) that are in between the strings *era* and *sea* (in red) according to the betweenness relation B_E for the English alphabet $\Sigma = \{a, b, c, \ldots, z\}$.

Obtaining the median string(s) and the center string(s) with respect to the Levenshtein distance metric is known to be an NP-complete problem [3]. Some approximation techniques have been proposed. For instance, Kohonen [18] proposed to compute the set median string (or medoid string), which is straightforward, and then proceeding in a hill-climbing-like style of making small changes until (hopefully) the median is found.

Example 6. Consider the list of strings on the English alphabet $\Sigma = \{a, b, c, \ldots, z\}$: $\mathbf{S} = \{era, sea\}$. Note that any string S such that $(era, S, sea) \in B_E$ is a median string. This means that all among era, ea, eea, sra, $sera$ and sea are median strings. Obviously, only era and sea are medoid strings. Finally, center strings are only ea, eea, sra and $sera$.

5 Conclusions

In this paper, we have discussed the problem of finding median and center strings for five popular distance metrics on the set of strings within the framework of penalty-based data aggregation. Some other popular distance metrics such as the Damerau-Levenshtein distance metric (in which all edit operations proposed by Damerau [17] instead of those proposed by Levenshtein [6] are considered), the Jaccard distance metric (based on the Jaccard index [19]), the Jaro distance metric [20] and the Jaro-Winkler distance metric [21] have also been used for measuring distances between strings – thus they could potentially be considered for searching for median and center strings. Future research concerns a more in-depth study of the aggregation of strings in which all these different distance metrics are analysed. The study of centroid strings and an analysis of the property of monotonicity from a betweenness-based (instead of order-based) perspective are also highlighted as interesting future research subjects.

Acknowledgments. Raúl Pérez-Fernández acknowledges the support of the Research Foundation of Flanders (FWO17/PDO/160) and the Spanish MINECO (TIN2017-87600-P).

References

1. Lanctot, J.K., Li, M., Ma, B., Wang, S., Zhang, L.: Distinguishing string selection problems. Inf. Comput. **185**, 41–55 (2003)
2. Gusfield, D.: Algorithms on Strings, Trees and Sequences: Computer Science and Computational Biology. Cambridge University Press, Cambridge (1997)
3. Nicolas, F., Rivals, E.: Complexities of the centre and median string problems. In: Proceedings of the 14th Annual Conference on Combinatorial Pattern Matching, pp. 315–327. Springer, Heidelberg (2003)
4. Gagolewski, M.: Data Fusion. Theory, Methods and Applications. Institute of Computer Science, Polish Academy of Sciences, Warsaw (2015)
5. Hamming, R.W.: Error detecting and error correcting codes. Bell Syst. Tech. J. **29**(2), 147–160 (1950)
6. Levenshtein, V.I.: Binary codes capable of correcting deletions, insertions, and reversals. Soviet Phys. Doklady **10**(8), 707–710 (1966)
7. Pérez-Fernández, R., De Baets, B.: On the role of monometrics in penalty-based data aggregation. IEEE Trans. Fuzzy Syst. (in press). https://doi.org/10.1109/TFUZZ.2018.2880716
8. Yager, R.R.: Toward a general theory of information aggregation. Inf. Sci. **68**, 191–206 (1993)

9. Calvo, T., Beliakov, G.: Aggregation functions based on penalties. Fuzzy Sets Syst. **161**, 1420–1436 (2010)
10. Bustince, H., Beliakov, G., Dimuro, G.P., Bedregal, B., Mesiar, R.: On the definition of penalty functions in data aggregation. Fuzzy Sets Syst. **323**, 1–18 (2017)
11. Owen, S.H., Daskin, M.S.: Strategic facility location: a review. Eur. J. Oper. Res. **111**, 423–447 (1998)
12. Li, M., Ma, B., Wang, L.: On the closest string and substring problems. J. ACM **49**(2), 157–171 (2002)
13. Ma, B., Sun, X.: More efficient algorithms for closest string and substring problems. In: Vingron, M., Wong, L. (eds.) Research in Computational Molecular Biology, pp. 396–409. Springer, Berlin (2008)
14. Fishburn, P.C.: Lexicographic orders, utilities and decision rules: a survey. Manag. Sci. **20**(11), 1442–1471 (1974)
15. Kaufman, L., Rousseeuw, P.J.: Finding Groups in Data: An Introduction to Cluster Analysis. Wiley, New York (2009)
16. Deza, M.M., Deza, E.: Encyclopedia of Distances. Springer, Heidelberg (2009)
17. Damerau, F.J.: A technique for computer detection and correction of spelling errors. Commun. ACM **7**(3), 171–176 (1964)
18. Kohonen, T.: Median strings. Pattern Recogn. Lett. **3**, 309–313 (1985)
19. Jaccard, P.: The distribution of the flora in the Alpine zone. New Phytol. **11**(2), 37–50 (1912)
20. Jaro, M.A.: Advances in record-linkage methodology as applied to matching the 1985 census of Tampa, Florida. J. Am. Stat. Assoc. **84**, 414–420 (1989)
21. Winkler, W.E.: String comparator metrics and enhanced decision rules in the Fellegi-Sunter model of record linkage. In: Proceedings of the Section on Survey Research Methods of the American Statistical Association, pp. 354–359 (1990)

On Some Properties of Generalized Convex Combination of Triangular Norms

Funda Karaçal$^{1(\boxtimes)}$ ⓘ, M. Nesibe Kesicioğlu^2 ⓘ, and Ümit Ertuğrul^1 ⓘ

^{1} Department of Mathematics, Karadeniz Technical University,
61080 Trabzon, Turkey
fkaracal@yahoo.com, uertugrul@ktu.edu.tr
^{2} Department of Mathematics, Recep Tayyip Erdoğan University,
53100 Rize, Turkey
nesibe.kesicioglu@erdogan.edu.tr

Abstract. In this paper, we define linear and generalized convex combination of triangular norms on bounded lattice. We investigate its some algebraic properties like unit element and zero-divisor element by putting some conditions.

Keywords: Triangular norm · Triangular conorm ·
Linear combination · Convex combination · Bounded lattice

1 Introduction

A binary operation with a unit element 1 called as triangular norm if it satisfies the boundary condition, commutativity and associativity properties on bounded lattices. After Menger, Schweizer and Sklar's works [11,14,15], many researchers have studied on triangular norms and lots of properties of triangular norms have been investigated.

At first, triangular norms were studied on the unit real interval $[0,1]$ and then triangular norms were studied on more general structures, i.e., bounded lattice [5,9]. In literature, fuzzy logical operations on the unit real interval were studied in the sense of its convex combinations of $[1,4,7,12]$. In this paper, we deal with the linear and generalized convex combination of triangular norms on bounded lattice.

The present paper consists of two main parts. After some literature information is given in Sect. 1, we give some necessary definitions, which they are crucial for our study in Sect. 2. In Sect. 3, linear and generalized convex combination of two t-norms on bounded lattice L is defined and we research its unit element and zero divisor element properties under different conditions. Finally, some concluding remarks are added.

A bounded lattice $(L, \leq)$ is a lattice which has the top and bottom elements, which are written as 1 and 0, respectively, i.e., there exist two elements $0, 1 \in L$ such that $0 \leq x \leq 1$, for all $x \in L$.

© Springer Nature Switzerland AG 2019
R. Halaš et al. (Eds.): AGOP 2019, AISC 981, pp. 223–231, 2019.
https://doi.org/10.1007/978-3-030-19494-9_21

2   Basic Notions

In this section, some necessary definitions are recalled.

Definition 1 *([3]). Let $(L, \leq, 0, 1)$ be a bounded lattice and $a, b \in L$. If neither $a \leq b$ nor $b \leq a$ is hold, a and b are called incomparable elements, in this case we use the notation $a \| b$.*

Definition 2 *([3]). Given a bounded lattice $(L, \leq, 0, 1)$, and $a, b \in L$, $a \leq b$, a subinterval $[a, b]$ of L is a sublattice of L defined as*

$$[a, b] = \{x \in L \mid a \leq x \leq b\}.$$

Similarly, $(a, b] = \{x \in L \mid a < x \leq b\}$, $[a, b) = \{x \in L \mid a \leq x < b\}$ and $(a, b) = \{x \in L \mid a < x < b\}$.

Definition 3 *([10]). An operation T (S) on a bounded lattice L is called a triangular norm (triangular conorm) if it is commutative, associative, increasing with respect to the both variables and has a unit element 1 (0).*

Example 1 ([10]). Let $(L, \leq, 0, 1)$ be a bounded lattice. The following are the two basic t-norms $T_D, T_\wedge$ on bounded lattice L respectively given by

$$T_D(x, y) = \begin{cases} y & \text{if } x = 1 \\ x & \text{if } y = 1 \\ 0 & \text{otherwise,} \end{cases}$$

$$T_\wedge(x, y) = x \wedge y.$$

The following are two basic t-conorms $S_D, S_\vee$ on bounded lattice L are respectively given by

$$S_D(x, y) = \begin{cases} y & \text{if } x = 0 \\ x & \text{if } y = 0 \\ 1 & \text{otherwise,} \end{cases}$$

$$S_\vee(x, y) = x \vee y.$$

When $L = [0, 1]$, Lukasiewicz t-norm and t-conorm and product t-norm are respectively given by

$$T_L(x, y) = \max(x + y - 1, 0),$$

$$S_L(x, y) = \min(x + y, 1),$$

$$T_P(x, y) = x \cdot y.$$

Definition 4 *([6]). Let T be a t-norm on a bounded lattice L. An element $x \in L$ is called an idempotent element of T if $T(x, x) = x$. If $T(x, x) = x$ is satisfied for all $x \in L$, T is called idempotent t-norm.*

Definition 5 *([6]). Let T be a t-norm on a bounded lattice L. An element $x \in L \setminus \{0\}$ is called an zero divisor of T if there exist $y \in L \setminus \{0\}$, $x \wedge y \neq 0$ such that $T(x, y) = 0$.*

Definition 6 *([6]). Let $(L, \leq, 0, 1)$ be a bounded lattice. A decreasing function $n : L \to L$ is called a negation if $n(0) = 1$ and $n(1) = 0$. A negation n on L is called strong if it is an involution, i.e., $n(n(x)) = x$, for all $x \in L$.*

3   Generalized Convex Combination of Triangular Norms

In this section, the notions of linear combination and generalized convex combination of triangular norms have been introduced and its unit element and zero divisor element properties are researched.

In the following definition, the linear combination notion for uninorms is presented.

Definition 7. *Let* $(L, \leq, 0, 1)$ *be a bounded lattice,* T, T_1, T_2 *be t-norms,* S *be a t-conorm,* n *be a negation on* L *and* $a \in L$. *Linear combination of t-norms* T_1 *and* T_2 *is defined as follows*

$$K_{a,n}^{T,S}(x,y) = S(T(a, T_1(x,y)), T(n(a), T_2(x,y)))$$

for all $(x,y) \in L^2$.

Definition 8. *Let* $(L, \leq, 0, 1)$ *be a bounded lattice,* T, T_1, T_2 *be t-norms,* S *be a t-conorm,* n *be a negation on* L, $a \in L$ *and* $S(a, n(a)) = 1$. *Generalized convex combination (shortly g-convex combination) of t-norms* T_1 *and* T_2 *is defined as follows*

$$K_{a,n}^{T,S}(x,y) = S(T(a, T_1(x,y)), T(n(a), T_2(x,y)))$$

for all $(x,y) \in L^2$.

One can easily observe that $K_{a,n}^{T,S}$ is a commutative and monotone operator.

Remark 9. *The equation* $S(x, n(x)) = 1$ *for all* $x \in L$ *mentioned in the foregoing Definition 8 is known in the literature as the law of excluded middle property feature (shortly LEM) (see page 52 Definition 2.3.8 in [2]). Here, we have used a much weaker condition than LEM property in view of* $S(a, n(a)) = 1$ *is hold for only* $a \in L$.

Remark 10. *Let us take* $L = [0, 1]$, $T = T_p$, $S = S_L$ *and* $n(x) = n_C(x) = 1 - x$ *in Definition 7. Then, it is seen that the Definition 7 on the bounded lattices coincides with the definition of convex combination of t-norms on* $[0, 1]$ *given in [13]. Indeed, since* $a \cdot T_1(x,y) + (1-a) \cdot T_2(x,y) \leq a \cdot 1 + (1-a) \cdot 1 = 1$, *we have that*

$$K_{a,n_C}^{T_P,S_L}(x,y) = min(a \cdot T_1(x,y) + (1-a) \cdot T_2(x,y), 1)$$
$$= a \cdot T_1(x,y) + (1-a) \cdot T_2(x,y).$$

Proposition 11. *Let* $(L, \leq, 0, 1)$ *be a bounded lattice,* $K_{a,n}^{T,S}$ *be the linear combination of* T_1, T_2 *such that* T, T_1, T_2 *are t-norms,* S *is a t-conorm,* n *is a strong negation on* L *and* $a \in L$.

(i) *If* $a \in L \setminus \{0, 1\}$ *and* $K_{a,n}^{T,S}$ *has the unit element 1, then* $T \neq T_D$.

(ii) *If* $a \in L \setminus \{0, 1\}$, $K_{a,n}^{T,S}$ *has the unit element 1,* T *is a t-norm without zero divisor and* $a \wedge n(a) \neq 0$, *then* $S \neq S_D$.

Proposition 12. *Let* $(L, \leq, 0, 1)$ *be a bounded lattice,* $K_{a,n}^{T,S}$ *be the linear combination of* T_1, T_2 *such that* T, T_1, T_2 *are t-norms, S is a t-conorm, n is a strong negation on L and* $a \in L$. *If* $S = S_D$, *neither a nor* $n(a)$ *are zero divisor elements of T and* T_1, T_2 *are t-norms without zero divisors, then*

$$K_{a,n}^{T,S_D}(x,y) = \begin{cases} 0, & \text{if } x \wedge y = 0 \\ 1, & \text{otherwise.} \end{cases}$$

Corollary 13. *Let* $(L, \leq, 0, 1)$ *be a bounded lattice,* $K_{a,n}^{T,S}$ *be the linear combination of* T_1, T_2 *such that* T, T_1, T_2 *are t-norms, S is a t-conorm, n is a strong negation on L and* $a \in L \setminus \{0, 1\}$. *If* $S = S_D$, *neither a nor* $n(a)$ *are zero divisor elements of T and* T_1, T_2 *are t-norms without zero divisors. Then,*

(i) $K_{a,n}^{T,S}$ *can not be a t-norm.*

(ii) $K_{a,n}^{T,S}$ *is not an associative operator iff there exist* $x, y, z \in L \setminus \{0\}$ *such that* $y \wedge z \neq 0$ *when* $x \wedge y = 0$.

Proposition 14. *Let* $(L, \leq, 0, 1)$ *be a bounded lattice,* $K_{a,n}^{T,S}$ *be the linear combination of* T_1, T_2 *such that* T, T_1, T_2 *are t-norms, S is a t-conorm, n is a strong negation on L and* $a \in L \setminus \{0, 1\}$. *If a is a zero divisor of the t-norm T and* $T_2 = T_D$, *then* $K_{a,n}^{T,S}$ *has zero divisor elements.*

Proposition 15. *Let* $(L, \leq, 0, 1)$ *be a bounded lattice,* $K_{a,n}^{T,S}$ *be the linear combination of* T_1, T_2 *such that* T, T_1, T_2 *are t-norms, S is a t-conorm, n is a strong negation on L and* $a \in L \setminus \{0, 1\}$. *If* $n(a)$ *is a zero divisor of the t-norm T and* $T_1 = T_D$, *then* $K_{a,n}^{T,S}$ *has zero divisor elements.*

Proposition 16. *Let* $(L, \leq, 0, 1)$ *be a bounded lattice,* $K_{a,n}^{T,S}$ *be the linear combination of* T_1, T_2 *such that* T, T_1, T_2 *are t-norms, S is a t-conorm, n is a negation on L and* $a \in L$. *If* $a \in L \setminus \{0, 1\}$ *and* $K_{a,n}^{T,S}$ *has the unit element* 1, $S(a, n(a)) = 1$, *namely,* $K_{a,n}^{T,S}$ *is a g-convex combination of* T_1, T_2.

Proof. Since $K_{a,n}^{T,S}$ has the unit element 1, $K_{a,n}^{T,S}(x, 1) = x$ for all $x \in L$. Then,

$$x = K_{a,n}^{T,S}(x, 1) = S(T(a, T_1(x, 1)), T(n(a), T_2(x, 1)))$$

$$= S(T(a, x), T(n(a), x))$$

$$\leq S(T(a, 1), T(n(a), 1)) = S(a, n(a)).$$

It follows $S(a, n(a)) = 1$ from that $x \leq S(a, n(a))$ is satisfied for all $x \in L$.

Remark 17. *The converse of Proposition 16 may not be valid. See following example.*

Example 2. Let $L = [0, 1]$, $a = \frac{1}{2}$, $S = S_L$, $T = min$ and $n(x) = 1 - x$ for all $x \in L$.

$$K_{\frac{1}{2},n}^{min,S_L}\left(\frac{1}{2}, 1\right) = S_L(T(\frac{1}{2}, T_1(\frac{1}{2}, 1)), T(n(\frac{1}{2}), T_2(\frac{1}{2}, 1)))$$

$$= S_L(T(\frac{1}{2}, \frac{1}{2}), T(\frac{1}{2}, \frac{1}{2}))$$

$$= S_L(\frac{1}{2}, \frac{1}{2}) = 1 \neq \frac{1}{2}.$$

Although $S_L(a, n(a)) = 1$ for $a = \frac{1}{2}$ since $S_L(\frac{1}{2}, n(\frac{1}{2})) = S_L(\frac{1}{2}, \frac{1}{2}) = 1$, $K_{a,n}^{T,S_L}$ has not the unit element 1.

Proposition 18. *Let $(L, \leq, 0, 1)$ be a bounded lattice, $K_{a,n}^{T,S}$ be the linear combination of T_1, T_2 such that T, T_1, T_2 are t-norms, S is a t-conorm, n is a negation on L and $a \in L$. If T is a $\vee$-distributive t-norm on L, $S = \vee$ and $a \vee n(a) = 1$ (namely, when $K_{a,n}^{T,S}$ is a g-convex combination of T_1, T_2), then $K_{a,n}^{T,S}$ has the unit element 1.*

Corollary 19. *Let $(L, \leq, 0, 1)$ be a bounded distributive lattice, $K_{a,n}^{T,S}$ be the linear combination of T_1, T_2 such that T, T_1, T_2 are t-norms, S is a t-conorm, n is a negation on L and $a \in L$. If $T = \wedge$, $S = \vee$ and $a \vee n(a) = 1$ (namely, when $K_{a,n}^{T,S}$ is a g-convex combination of T_1, T_2), $K_{a,n}^{T,S}$ has the unit element 1.*

Proof. Let L be a bounded distributive lattice, $T = \wedge$, $S = \vee$ and $a \vee n(a) = 1$. For all $x \in L$,

$$
\begin{aligned}
K_{a,n}^{\wedge,\vee}(x, 1) &= (a \wedge T_1(x, 1)) \vee (n(a) \wedge T_2(x, 1)) \\
&= (a \wedge x) \vee (n(a) \wedge x) \\
&= (a \vee n(a)) \wedge x \\
&= 1 \wedge x = x.
\end{aligned}
$$

Corollary 20. *$(L, \leq, 0, 1)$ be a Boolean algebra, $K_{a,n}^{T,S}$ be the linear combination of T_1, T_2 such that T, T_1, T_2 are t-norms, S is a t-conorm, n is a strong negation on L, i.e., $n(x) = x'$ for all $x \in L$, and $a \in L$. If $S = \vee$ and $T = \wedge$ (namely, when $K_{a,n}^{T,S}$ is a g-convex combination of T_1, T_2), $K_{a,n}^{T,S}$ has the unit element 1.*

Proposition 21. *$(L, \leq, 0, 1)$ be a bounded lattice with involution n, $K_{a,n}^{T,S}$ be the g-convex combination of T_1, T_2 such that T, T_1, T_2 are t-norms, S is a t-conorm, n is a negation on L and $a \in L$. If T is a $\vee$-distributive t-norm, S is a $\wedge$-distributive t-conorm, $a \vee n(a) = 1$ (namely, when $K_{a,n}^{T,S}$ is a q-convex combination of T_1, T_2) and $a \wedge n(a) = 0$, then $K_{a,n}^{T,S}$ has the unit element 1.*

Proposition 22. *$(L, \leq, 0, 1)$ be a bounded lattice, $K_{a,n}^{T,S}$ be the linear combination of T_1, T_2 such that T, T_1, T_2 are t-norms, S is a t-conorm, n is a negation on L and $a \in L$. If T is a $\vee$-distributive t-norms, $S = \vee$, then $K_{a,n}^{T,\vee}$ has the unit element 1 iff $a \vee n(a) = 1$.*

Corollary 23. *If one of the equivalent condition of Proposition 22 is provided, the $K_{a,n}^{T,S}$ linear combination of T_1, T_2 is a g-convex combination T_1, T_2.*

Proposition 24. *$(L, \leq, 0, 1)$ be a Boolean algebra, $K_{a,n}^{T,S}$ be the linear combination of T_1, T_2 such that T, T_1, T_2 are t-norms, S is a t-conorm, $n(x) = x'$ for all $x \in L$ is a strong negation on L and $a \in L$. If L is Boolean algebra and $T = \wedge$, then $K_{a,n}^{\wedge,S}(x, 1) \geq x$ for all $x \in L$.*

Corollary 25. *If the negation n is taken as complement in Boolean algebras, the $K_{a,n}^{T,S}$ linear combination of T_1, T_2 is a g-convex combination T_1, T_2.*

Proposition 26. $(L, \leq, 0, 1)$ be a bounded lattice, $K_{a,n}^{T,S}$ be linear combination of T_1, T_2 such that T, T_1, T_2 are t-norms such that $T_1 = T_2$, S is a t-conorm, n is a strong negation on L and $a \in L$. Then, $K_{a,n}^{T,S} = K_{n(a),n}^{T,S}$.

Corollary 27. $(L, \leq, 0, 1)$ be a bounded lattice, $K_{a,n}^{T,S}$ be the linear combination of T_1, T_2 such that T, T_1, T_2 are t-norms, S is a t-conorm, n is a strong negation on L and $a \in L$. Then, $K_{n(a),n}^{T,S}$ is linear combination of T_2, T_1.

Proposition 28. $(L, \leq, 0, 1)$ be a bounded lattice, $K_{a,n}^{T,S}$ be the linear combination of T_1, T_2 such that T, T_1, T_2 are t-norms, S is a t-conorm, n is a strong negation on L and $a \in L$. Then, $K_{a,n}^{T,S}(x, 1) = K_{n(a),n}^{T,S}(x, 1)$.

Corollary 29. $(L, \leq, 0, 1)$ be a bounded lattice, $K_{a,n}^{T,S}$ be the linear combination of T_1, T_2 such that T, T_1, T_2 are t-norms, S is a t-conorm, n is a strong negation on L and $a \in L$. Then, $K_{a,n}^{T,S}$ has the unit element 1 iff $K_{n(a),n}^{T,S}$ has the unit element 1.

Proposition 30. Let $(L, \leq, 0, 1)$ be a bounded lattice, $a \in L \setminus \{0\}$ such that $a \wedge x \neq 0$ for all $x \in L \setminus \{0\}$ and $K_{a,n}^{T,S}$ linear combination of T_1, T_2 such that T, T_1, T_2 t-norms, S t-conorm and n strong negation on L. If T and T_1 (or T_2) t-norms without zero divisor on L, then $K_{a,n}^{T,S}$ is an operator on L without zero divisor.

Proof. Let T and T_1 be t-norms without zero divisor on L. Suppose that $x \wedge y \neq 0$ and $K_{a,n}^{T,S}(x, y) = 0$. Then, it follows $T(a, T_1(x, y)) = T(n(a), T_2(x, y)) = 0$ from that $S(T(a, T_1(x, y)), T(n(a), T_2(x, y))) = 0$. Since T is t-norm without zero divisor and $a \neq 0$, it is obtained that $T_1(x, y) = 0$. It contradicts to T_1 be t-norm without zero divisor since $x \wedge y \neq 0$. Therefore, $K_{a,n}^{T,S}(x, y) \neq 0$ if $x \wedge y \neq 0$, namely, $K_{a,n}^{T,S}$ is an operator on L without zero divisor.

Proposition 31. Let $(L, \leq, 0, 1)$ be a bounded lattice such that $x \wedge y \neq 0$ for all $x, y \in L \setminus \{0\}$, $K_{a,n}^{T,S}$ be the linear combination of T_1, T_2 such that T, T_1, T_2 are t-norms, S is a t-conorm, n is a strong negation on L and $a \in L$. If a and $n(a)$ are zero divisor elements of T, then $K_{a,n}^{T,S}$ have zero divisor elements.

Corollary 32. Let $(L, \leq, 0, 1)$ be a chain, $K_{a,n}^{T,S}$ be the linear combination of T_1, T_2 such that T, T_1, T_2 are t-norms, S is a t-conorm, n is a strong negation on L and $a \in L$. If a and $n(a)$ are zero divisor elements of T, then $K_{a,n}^{T,S}$ have zero divisor elements.

Proposition 33. Let $(L, \leq, 0, 1)$ be a complemented bounded lattice, $K_{a,n}^{T,S}$ linear combination of T_1, T_2 such that T, T_1, T_2 are t-norms, S is a t-conorm such that $S(a, n(a)) = 1$, n is a strong negation on L, i.e., $n(x) = x'$ for all $x \in L$ and $a \in L$. Let

$$L_{K_{a,n}^{T,S}}^* = \{x \in L : K_{a,n}^{T,S}(x, 1) = x\}.$$

i. $L^*_{K^{T,S}_{a,n}}$ is a bounded partially ordered set.

ii. $a \in L^*_{K^{T,S}_{a,n}}$ iff a is an idempotent element of T.

iii. $n(a) \in L^*_{K^{T,S}_{a,n}}$ iff $n(a)$ is an idempotent element of T.

iv. $L^*_{K^{T,S}_{a,n}} = L^*_{K^{T,S}_{n(a),n}}$.

Theorem 34. *Let* $(L, \leq, 0, 1)$ *be a complete lattice,* $K^{T,S}_{a,n}$ *be the linear combination of* T_1, T_2 *such that* T, T_1, T_2 *are t-norms,* S *is a t-conorm,* n *is a strong negation on* L *and* $a \in L$. *If* T *is a* $\vee$-*distributive t-norm and* $a \vee n(a) = 1$, *then* $L^*_{K^{T,S}_{a,n}}$ *is complete lattice.*

Example 3. Consider the lattice L characterized by Hasse diagram, as shown in Fig. 1.

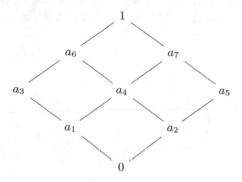

Fig. 1. Lattice diagram of L

One can easily check that L is a distributive lattice. Take $T = T_1 = T_2 = \wedge$, let the t-conorm S and the strong negation n on L be respectively, as follows (Table 1);

Table 1. T-conorm S on L

S	0	a_1	a_2	a_3	a_4	a_5	a_6	a_7	1
0	0	a_1	a_2	a_3	a_4	a_5	a_5	a_7	1
a_1	a_1	a_1	a_4	1	a_4	1	1	1	1
a_2	a_2	a_4	a_2	1	a_4	1	1	1	1
a_3	a_3	1	1	1	1	1	1	1	1
a_4	a_4	a_4	a_4	1	a_4	1	1	1	1
a_5	a_5	1	1	1	1	1	1	1	1
a_6	a_6	1	1	1	1	1	1	1	1
a_7	a_7	1	1	1	1	1	1	1	1
1	1	1	1	1	1	1	1	1	1

and

$$
n\left(x\right) = \begin{cases}
1 & , & \text{if } x = 0 \\
a_6 & , & \text{if } x = a_1 \\
a_7 & , & \text{if } x = a_2 \\
a_5 & , & \text{if } x = a_3 \\
a_4 & , & \text{if } x = a_4 \\
a_3 & , & \text{if } x = a_5 \\
a_1 & , & \text{if } x = a_6 \\
a_2 & , & \text{if } x = a_7 \\
0 & , & \text{if } x = 1.
\end{cases}
\tag{1}
$$

Since L is a distributive lattice, $\wedge$ is $\vee$-distributive t-norm on L. Moreover, $a_3 \vee n(a_3) = 1$. Therefore, hypothesis of Theorem 34 is satisfied. We obtain $L^*_{K^{T,S}_{a,n}} = \{0, a_1, a_2, a_3, a_5, 1\}$ and its lattice diagram as follows (Fig. 2).

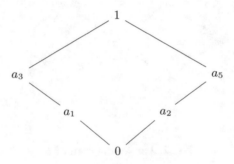

Fig. 2. Lattice diagram of $L^*_{K^{T,S}_{a,n}}$

Observe that the supremum of a_1 and a_2 in L is different from the supremum of a_1 and a_2 in $L^*_{K^{T,S}_{a,n}}$.

Lemma 1. *Let $(L, \leq, 0, 1)$ be a bounded lattice, $K^{T,S}_{a,n}$ be the linear combination of T_1, T_2 such that T, T_1, T_2 are t-norms, S is a t-conorm, n is a negation on L and $a \in L$. If $S = \vee$, then $K^{T,S}_{a,n}(x, y) \leq x \wedge y$.*

Proposition 35. *Let $(L, \leq, 0, 1)$ be a bounded lattice, $K^{T,S}_{a,n}$ be the linear combination of T_1, T_2 such that T, T_1, T_2 are t-norms, S is a t-conorm, n is a strong negation on L and $a \in L$. If $S = \vee$, T is a $\vee$-distributive t-norm and $a \vee n(a) = 1$, then $L^*_{K^{T,\vee}_{a,n}} = L$.*

4   Concluding Remarks

In literature, some paper can be seen on convex combination of triangular norms on $[0, 1]$. But, in our best knowledge, there is no work on convex combination of triangular norms on bounded lattice (we called it g-convex combination). Aim of this paper is to fill this gap in literature. We deal with some properties of the linear and g-convex combination of triangular norms like unit element and zero divisor element under different conditions. Moreover, our definition of g-convex combination of triangular norms on bounded lattice coincides with the definition of convex combination of triangular norms on unit real interval ($[0,1]$) when $L = [0,1]$, $T = T_p$, $S = S_L$ and $n(x) = n_C(x) = 1 - x$.

References

1. Alsina, C., Frank, M.J., Schweizer, B.: Problems on associative functions. Aequationes Math. **66**(1–2), 128–140 (2003)
2. Baczyński, M., Jayaram, B.: Fuzzy Implications. Studies in Fuzziness and Soft Computing, vol. 231. Springer, Heidelberg (2008)
3. Birkhoff, G.: Lattice Theory. American Mathematical Society Colloquium Publishers, Providence (1967)
4. Durante, F., Sarkoci, P.: A note on the convex combinations of triangular norms. Fuzzy Sets Syst. **159**, 77–80 (2008)
5. Ertuğrul, Ü., Karaçal, F., Mesiar, R.: Modified ordinal sums of triangular norms and triangular conorms on bounded lattices. Int. J. Intell. Syst. **30**, 807–817 (2015)
6. Grabisch, M., Marichal, J.-L., Mesiar, R., Pap, E.: Aggregation Functions. Cambridge University Press, Cambridge (2009)
7. Jenei, S.: On the convex combination of left-continuous t-norms. Aequationes Math. **72**(1–2), 47–59 (2006)
8. Karaçal, F.: On the direct decomposability of strong negations and S-implication operators on product lattices. Inf. Sci. **176**, 3011–3025 (2006)
9. Karaçal, F., Kesicioğlu, M.N.: A T-partial order obtained from t-norms. Kybernetika **47**, 300–314 (2011)
10. Klement, E.P., Mesiar, R., Pap, E.: Triangular Norms. Kluwer Academic Publishers, Dordrecht (2000)
11. Menger, K.: Statistical metrics. Proc. Natl. Acad. Sci. USA **8**, 535–537 (1942)
12. Mesiar, R., Mesiarová-Zemánková, A.: Convex combinations of continuous t-norms with the same diagonal function. Nonlinear Anal. **69**(9), 2851–2856 (2008)
13. Ouyang, Y., Fang, J.: Some observations about the convex combinations of continuous triangular norms. Nonlinear Anal. **69**(11), 3382–3387 (2008)
14. Schweizer, B., Sklar, A.: Espaces metriques aléatoires. C.R. Acad. Sci. Paris Sér. A. **247** 2092–2094 (1958)
15. Schweizer, B., Sklar, A.: Statistical metric spaces. Pac. J. Math. **10**, 313–334 (1960)

Analysis of Relationship Among V4 Countries and Germany by Their Gross Domestic Products and Copula Models

Tomáš Bacigál[1]([⊠])[iD], Magdaléna Komorníková[1][iD], and Jozef Komorník[2][iD]

[1] Department of Mathematics and Constructive Geometry,
Faculty of Civil Engineering, Slovak University of Technology in Bratislava,
Radlinského 11, 81005 Bratislava, Slovakia
{tomas.bacigal,magdalena.komornikova}@stuba.sk
[2] Faculty of Management, Comenius University,
Odbojárov 10, P.O. BOX 95, 820 05 Bratislava, Slovakia
Jozef.Komornik@fm.uniba.sk

Abstract. We analyzed quarterly seasonally-filtered time series (OECD link) of GDP in EURO per capita for the V4 countries (Visegrad treaty - Czech republic, Hungary, Poland, Slovakia) and Germany in the period 1996/Q1 – 2018/Q1. First, ARIMA models were used to clear the temporal dependence, then marginal distribution functions were utilized to standardize the data such that it is U(0,1) distributed. Dependence among the 5 random variables was analyzed in terms of correlation strength and joint distribution modeled by elliptical, vine and factor copulas, both in the whole 22 years period and 6-year rolling windows with 3-year overlaps. The choice of such different model classes allows us study the nature of underlying dependence structure from different standpoints.

Keywords: Elliptic copulas · Vine copulas · Factor copulas · Gross domestic product

1 Introduction

Gross domestic product is the final result of the production activity resident production units established in the observation territory for a given period. GDP is used in economics to compare the performance of the economies of the States. Factors affecting the development of GDP are mainly households, companies and the state but also natural disasters, structural changes in the economy and also political instability in the country. The development of GDP is therefore individual and influenced by different factors. But we can also find common features in development such as sharp decline in the year 2008 due to the great financial crisis. This crisis has struck the whole of Europe like all other countries with an advanced economy. We follow a sharp drop in GDP also in the development of

© Springer Nature Switzerland AG 2019
R. Halaš et al. (Eds.): AGOP 2019, AISC 981, pp. 232–243, 2019.
https://doi.org/10.1007/978-3-030-19494-9_22

Germany's GDP, which is the largest Central European economy affecting the surrounding landscape (see e.g. [8,12,14]).

We analyzed quarterly seasonally-filtered time series (OECD link) of GDP in EURO per capita for the V4 countries (Visegrad treaty - Czech republic, Hungary, Poland, Slovakia) and Germany in the period 1996/Q1 – 2018/Q1, altogether 5×89 records. The time series has strong trend component and residual auto-correlation, therefore we used ARIMA models to clear the temporal dependence and get the data of time-invariant random shocks. Since we are interested mainly in investigation of their mutual dependence, observations of the five random variables of GDP shocks were transformed by their individual (univariate) distribution functions, in particular we used non-parametric estimate, i.e., empirical distribution function, to obtain so-called pseudo-observations uniformly distributed on $[0,1]$. If plotted, the pairwise scatter plots displays pure dependence, deprived of individual behavior. This dependence can be modeled by copulas described in the next section, in particular we used the parametric class of elliptical copulas, vine-copulas and factor copulas, all of which are suitable for modeling dependence in multi-dimensional random vector. We analyze both the whole period and 6-year spans with 3-year overlaps (1996/Q1–2002/Q4, 1999/Q1–2005/Q4, ...), first by calculating rank correlation coefficient, then by fitting several parametric model candidates and comparing their goodness of fit. In Sect. 3 we report and discuss the results and in Sect. 4 we conclude.

2   Theory

In this section we briefly introduce copulas, the particular classes used in the present paper, their estimation and performance evaluation.

2.1   Copula

A copula C is a restriction, to the multi-cube $[0,1]^d$, of the distribution function of the random vector $U = (U_1, \ldots, U_n)$, such that all margins U_i are uniformly distributed on $[0,1]$. The strongest copula $M(u_1, \ldots, u_n) = \min(u_1, \ldots, u_n)$ is known as Frechet-Hoefding upper bound and represents comonotonicity in the random vector $\mathbf{U}$, the product copula $\Pi(u_1, \ldots, u_n) = \prod_{i=1}^{n} u_i$ the copula of random vector with independent components, and finally the lower Frechet-Hoefding bound $W(u_1, \ldots, u_n) = max\left(0, \sum_{i=1}^{n} u_i - (n-1)\right)$ represents counter-monotonicity for $n = 2$ but in higher dimensions lacks the n-increasingness, thus it is not a copula. For further details on copula theory see, e.g, the monograph [19].

From the most prominent classes – Elliptical copulas (Student, normal), Extreme-Value copulas and Archimedean copulas – only the first one is flexible and simple enough to model dependence structure of real data in more than two dimensions. They are the copulas of elliptically contoured distribution, for example the normal (Gaussian) copulas $C_\Phi(u_1, ..., u_n) = \Phi\left[\Phi_1^{-1}(u_1), ..., \Phi_n^{-1}(u_n)\right]$ and (Student) t-copulas $C_t(u_1, ..., u_n) = t\left[t_1^{-1}(u_1), ..., t_n^{-1}(u_n)\right]$, where Φ and t are joint distribution functions of multivariate normal and Student t-distributions, respectively (similarly Φ_i^{-1} and t_i^{-1}, $i = 1, ..., n$, are quantile functions of univariate distributions related to U_i).

However, there are construction methods to build up multivariate copulas ($n \geq 3$) from lower dimensional ones. One such construction leads to the class of so called hierarchical (or nested) Archimedean copulas, which nicely balance parsimony and flexibility, but its building block copulas are limited to the class of Archimedean copulas with constraints imposed upon their parameters, see, e.g., [10,11,17,20] for theoretical background and our recent contribution [3] for application to modeling relationship among international financial market indexes, where their performance is compared with the performance of yet another construction, vine-copulas. While here we do not consider the former class, we do so with the later one. Vine-copulas construction [1,6,22] uses arbitrary bivariate copulas and a vine tree structure [5] to form a multivariate dependence model, which can be estimated, visualized and interpreted. To picture out the idea, consider the simplest case $n = 3$, when there are only three possible ways of constructing vine graphical model of dependence and one of them – in terms of copula density c – is given by the following conditional chain

$$c_{123}(u_1, u_2, u_3) = 1 \times c_{2|1}(u_1, u_2) \times c_{3|12}(u_1, u_2, u_3)$$
$$= 1 \times c_{12}(u_1, u_2) \cdot 1 \times c_{31|2}\left(C_{3|2}(u_3, u_2), C_{1|2}(u_1, u_2)\right) c_{23}(u_2, u_3) \cdot 1 \quad (1)$$

where $C_{i|j}(u_i, u_j) = \frac{\partial C_{ij}(u_i, t)}{\partial t}|_{t=u_j} = \int_0^{u_i} c_{ij}(t, u_j) dt$ and adopting the simplifying assumption, that the conditional copula does not depend on conditioning variables (see [2] for a critical review), here

$$C_{ij|k}(u_i, u_j; u_k) = C_{ij|k}(u_i, u_j). \quad (2)$$

The binary operators $\times$ and $\cdot$ stand for ordinary multiplication used here just for separating the derived terms as well as the density of (uniform) marginal distribution. Graphically, it is represented by the vine trees shown in Fig. 1. There the first level tree (the upper one) contains variables of $\mathbf{U}$ as nodes and unconditional copulas between some of them as edges. The second level tree is a so-called line graph of the first level tree, i.e., edges become nodes and those are connected by an edge, that represents conditional copula between U_1 and U_3 given observation of U_2. The other two ways of construction just interchange the root node (in the middle) with one of the others at the first level. In the higher dimensions the graph structure gets complicated with $\binom{n}{2}(n-2)! 2^{\binom{n-2}{2}}$ possible combinations, generally denoted as regular vines. The most popular kinds of regular vines – C-vines (canonical) and D-vines (drawable) – count $\frac{n!}{2}$ members each, see [18], which is, e.g., 12 C-vines from 24 regular vines in $n = 4$ dimensions. This is a pretty difficult task to find the most appropriate one and, further, to estimate each of the $\frac{n^2-n}{2}$ bivariate copulas forming up every n-dimensional vine copula. In practice, usually the pairs with the strongest rank correlation are selected to be connected by edge, see [7]. When there is a reason to believe that the nodes (variables) in higher-level trees are conditionally

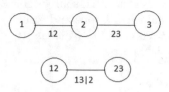

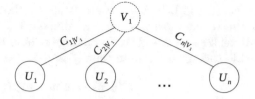

Fig. 1. Vine tree of the construction (1) having the 2nd variable as a root node.

Fig. 2. 1-factor copula graphical model of dependence corresponding to construction (3)

independent (preferably supported by a test for independence), then product copula is assigned to the corresponding edges and thus the estimation process becomes easier. Any vine copula containing exclusively the product copulas from $m+1$-th tree level up is called a m-truncated vine copula ([11, p. 117]) and such a subclass becomes very popular in applications with very-high-dimensional random vectors.

Yet another subclass within pair-copula construction approach is getting considerable attention: factor copulas. However, it is fair to remark that there is more than one concept attributed to this name. According to [13] factor copula models are conditional independence models where observed variables $(U_1, \ldots, U_n)$ are conditionally independent given one or more latent variables $(V_1, \ldots, V_p)$. These models extend the multivariate Gaussian model with factor correlation structure. They can be also viewed as p-truncated C-vine copulas rooted at the latent variables, one just needs to integrate out latent variables in the joint copula density to get density of observables while admitting the simplifying assumption similar to (2). The most popular are 1-factor copulas defined as

$$C(u_1, \ldots, u_n) = \int_0^1 \prod_{i=1}^n C_{i|V_1}(u_i, v_1) dv_1 \qquad (3)$$

with the density $c(u_1, \ldots, u_n) = \int_0^1 \prod_{i=1}^n c_{iV_1}(u_i, v_1) dv_1$ (because $\frac{\partial}{\partial u} C_{i|V_1}(u, v_1)$ $= \frac{\partial^2}{\partial u \partial v_1} C_{iV_1}(u, v_1) = c_{iV_1}(u, v_1)$) thus the dependence among U_is is induced by the so-called linking copulas C_{iV_1}, $i = 1, \ldots, n$, and there are no constraints among the bivariate copulas. The dependence structure of n-dimensional 1-facor copula is graphically illustrated in Fig. 2.

It is interesting to note, that the Archimedean copulas generated by universal generators (those based on Laplace transform, see e.g. [4] for some examples) are special case of 1-factor copulas, with exceptionally simple form. A main advantage of factor copula models comparing to Archimedean and Gaussian copulas is that it allows for asymmetric dependence structure (both reflection asymmetry and non-exchangeability) among observables. Later we will see that they are flexible enough to compete with more complex class of vine copulas while keeping relative parsimony and interpretability. The main drawback nowadays, however, is the lack of software implementation. Commercial programs are rather conservative in bringing new statistical methods and from the open source tools, only in R, the most popular environment for statistical calculations and visualizations [21], we found single package related to factor copula: *FDGcopulas* [15]. In this package a Durante class of bivariate copulas defined by

$$C(u, v) = \min(u, v) f\left(\max(u, v)\right),$$

are used as linking copulas, where the generator $f \colon [0, 1] \to [0, 1]$ is differentiable and increasing function such that f $(1) = 1$ and $t \to f(t)/t$ is decreasing. There may be chosen four different parametric families, such as Cuadras-Augé $f(t) = t^{1-\theta}$, $\theta \in [0, 1]$, Fréchet $f(t) = (1-\theta)t + \theta$, $\theta \in [0, 1]$, Durante-sinus $f(t) = \frac{\sin(\theta t)}{\sin(\theta)}$, $\theta \in (0, \pi/2]$, and Durante-exponential $f(t) = \exp\left(\frac{t^\theta - 1}{\theta}\right)$ with parameter $\theta > 0$, please refer to [16] for finer details. For our analysis we used only the first three families, due to numerical failure of the last one. The downside of this copula class is a singular component present in the model, which is not natural for most economic, hydrologic or other frequently analyzed phenomenons. However, as the authors argue, it is not of that much importance in higher dimensions, where just certain features of distribution is preferred (such as critical levels or tail behavior) instead of its overall shape. On the other hand, the class of 1-factor copulas with Durante generators reduce the computational burden of general 1-factor copulas while giving a good fit to observed data, as we will see in the results.

Within the class of vine copulas the following parametric families of well-known bi-variate copulas (available in R package *VineCopula*) were considered: Gumbel, Clayton, Frank, Joe, BB1, BB6, BB7 (from the Archimedean class; including their rotations) and Gaussian, Student t-copula (from the elliptical class). The reflections C_α are defined through formulas $C_{90}(u_1, u_2) = u_2 - C(1 - u_1, u_2)$, $C_{180}(u_1, u_2) = u_1 + u_2 - 1 + C(1 - u_1, 1 - u_2)$ which is survival copula, and $C_{270}(u_1, u_2) = u_1 - C(u_1, 1 - u_2)$. They allow one to easily extend variety of properties of the basic copulas set.

In the following subsection we briefly outline standard procedures of fitting copulas to observed data used in the paper and how we compared goodness of fit among several competing models.

2.2 Inference

The optimal models among the considered classes of 2–dimensional copulas as well as n-dimensional elliptical copulas, are selected using the well-known Maximum likelihood estimation (MLE) method, which in this context is defined as follows. For given m observations $\{X_{ji}\}_{i=1,\ldots,m}$ of j-th random variable X_j, $j = 1, \ldots, n$, (which need not be uniformly distributed) the parameters θ of all copulas under consideration were estimated by maximizing the likelihood function

$$L(\theta) = \sum_{i=1}^{m} \log\left[c(U_{1i}, \ldots, U_{ni}; \boldsymbol{\theta})\right], \tag{4}$$

where $c(\cdot; \theta)$ denotes density of a parametric copula family $C(\cdot; \theta)$, and the so-called pseudo-observations $U_{ji} = \hat{F}_j(X_{ji})$ are calculated through the corresponding empirical marginal distribution functions $\hat{F}_j(x) = \frac{1}{m+1} \sum_{k=1}^{m} \mathbf{1}(X_{jk} \le x)$.

The higher dimensional structures of vine and factor copulas were estimated as described in [7] and [16], respectively. Briefly, vine copulas – thanks to the simplifying assumption (2) – are estimated stage-wise by applying maximum likelihood method in each tree separately. First, the copulas in the lowest level tree are fitted, subsequently their parameters are fixed while fitting the second level copulas, and so on. The tree structure is determined either by expert's opinion or (as in our case) by choosing a criterion, such as a rank correlation with which the more correlated pairs are assigned an edge. In the case of factor copula, the problem is to tackle integration in copula density definition. Authors in [13] recommend numerical integration using Gauss-Legendre quadrature, such that for 1-factor the copula density term in (4) is approximated as follows,

$$c(U_{1,i}, \ldots, U_{n,i}; \boldsymbol{\theta}) \approx \sum_{k=1}^{n_q} w_k \prod_{j=1}^{n} c_{jV_1}(U_{ji}, x_k; \theta_j)$$

where x_k and w_k are the quadrature nodes and weights, respectively, with $20 < n_q \le 25$ giving a good approximation. The same nodes also helps with smooth numerical derivatives for numerical minimization of likelihood function (4), e.g., by Newton-Raphson algorithm. However, the package [15] implements a weighted least squares estimator based on dependence coefficients, such as that of Kendall's or Spearman's rank correlation, since it does not assume existence and continuity of partial derivatives on the unit hypercube (which is not fulfilled with Durante class). On the other hand, the estimator rely on one-to-one correspondence between dependence coefficient and a copula parameter. Formally, $\hat{\boldsymbol{\theta}} = \arg\min_{\boldsymbol{\theta}} \|\hat{\mathbf{r}} - \mathbf{r}(\boldsymbol{\theta})\|^2$ where $\hat{\mathbf{r}} = (\hat{r}_{1,2}, \ldots, \hat{r}_{n-1,n})$ are empirical estimates and $\mathbf{r}(\boldsymbol{\theta}) = (r_{1,2}(\theta_1, \theta_2), \ldots, r_{n-1,n}(\theta_1, \theta_2))$ values of dependence coefficient r_{ij}, $i = j - 1 = 1, \ldots, n - 1$, calculated from bivariate margin $C_{ij}(u_i, u_j) = \min(u_i, u_j) f_{ij}(\max(u_i, u_j))$ with the generator $f_{ij}(t) = f_i(t) f_j(t) + t \int_t^1 f_i'(x) f_j'(x) dx$. Then, after derivation from the general relation between copula and popular dependence coefficients, there may be used the corresponding formulas for Durante class copula, namely

$$\rho_{ij} = 12 \int_0^1 x^2 f_i(x) f_j(x) dx + 3 \int_0^1 x^4 f_i'(x) f_j'(x) dx - 3 \qquad \text{(Spearman's)}$$

$$\tau_{ij} = 4 \int_0^1 x \left(f_i(x) f_j(x) + x \int_0^1 f_i'(t) f_j'(t) dt \right)^2 dx - 1 \qquad \text{(Kendall's tau)}$$

Unfortunately, implementation of the package *FDGcopulas* restricts all $f_{i,j}$'s in the same factor copula to be from the same parametric family.

To compare goodness-of-fit among competing models, we used a test proposed by [9] and based on empirical copula process using Cramer-von Misses test statistic $S_{CM} = \sum_{i=1}^{m} [C(U_{1i}, \ldots, U_{ni}; \boldsymbol{\theta}) - C_m(U_{1i}, \ldots, U_{ni})]^2$ with empirical copula $C_m(\mathbf{x}) = \frac{1}{m} \sum_{i=1}^{m} \prod_{j=1}^{n} \mathbf{1}(X_{ji} \le x_j)$ and indicator function $\mathbf{1}(A) = 1$

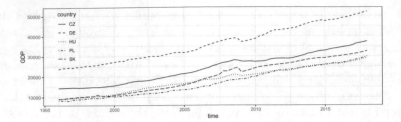

Fig. 3. Time series of original GDP data.

whenever A is true, otherwise $\mathbf{1}(A) = 0$. This is a standard test included in most of the packages used for our study, however, their implementation varies slightly, thus to unify the testing procedure across copula classes – while circumventing nonexistence of closed form of CDF in some cases and utilizing the easiness of random sampling – we approximated $C(\cdot; \theta)$ by empirical copula of the random samples generated from the corresponding copulas each counting 100 000 realizations.

3   Application

In this section we apply the above described models of dependence to quarterly time series of gross domestic product in order to analyze economic relation of five European countries during the last two decades. After introductory details about data and their pre-processing an overall dependence structure is explored in terms of Kendall's rank correlation coefficient, partial linear correlation and qualities of best fitting multidimensional copulas. Afterwards we estimate copulas of the same three classes in 6-years rolling window to detect structural changes of dependence.

3.1   Data Overview

As mentioned in the introduction, the data comes from seasonally adjusted quarterly records of GDP in EURO per capita[1] for Germany (DE) and the so-called Visegrad treaty countries, namely Czech Republic (CZ), Hungary (HU), Poland (PL) and Slovakia (SK) from 1996/Q1 to 2018/Q1, see Fig. 3.

All series follow an obvious trend, interestingly not all of them are disrupted by the 2009 crisis. The serial dependence is satisfactorily captured by ARIMA class model (with drift term allowing a deterministic trend) indicated with the residuals in Fig. 4 so that residual auto-correlation estimates are insignificant. As one may seen, there is quite a number of extremal values in the residuals of all but Polish GDP which can be further observed from density plot on the diagonal of Fig. 5.

---

[1] EUROSTAT main GDP aggregates per capita, http://appsso.eurostat.ec.europa.eu/nui/show.do?dataset=namq_10_pc.

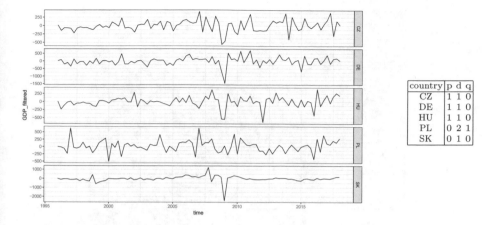

country	p	d	q
CZ	1	1	0
DE	1	1	0
HU	1	1	0
PL	0	2	1
SK	0	1	0

Fig. 4. Residuals of the ARIMA(p,d,q) models with the specified orders.

3.2 Overall Dependence

To inspect relationship among random shocks that passed the ARIMA filter one may start with pair-wise scatterplot such as in the lower triangle of Fig. 5, however the presence of individual qualities of the two samples crimps any attempt to see pure dependence. Therefore the upper triangle of the same figure offers scatter-plots of data "standardized" to have uniform marginal distribution supplemented by estimate of Kendall's correlation coefficient. Three stars (***) indicate rejection of zero correlation hypothesis with confidence 99%, further (**) 95% and (*) 90%. We may observe the strongest connection between Poland and Hungary, as well as between Czech and Slovak Republic.

Knowing the countries, this may be no surprise at all, however with the tools given we are better to dig deeper and see whether the links are straightforward or they are affected by a third economy.

Figure 6 reveals that the later couple appears to be in "love triangle" with the stronger Germany[2]. On the other hand Poland does not seem to get emotional over its western neighbor[3].

Partial correlation serves well as summary information about structure encoded in copula of elliptically contoured distributions, however since such copulas (Gaussian or Student t) capture just relatively homogeneous dependence, other alternatives are desirable.

For instance, Fig. 8 shows a best fitting vine copula represented by tree structure and contour plot from which we see unconditional dependence models in Tree 1, namely symmetric relation of Frank family type in couples CZ-SK and

---

[2] That means that the significant correlation between SK and CZ from Fig. 5 is explained by their common tie to DE in Fig. 6. Individuals SK and CZ would comove in much lesser extent if not accompanying their common friend DE.

[3] Though PL is slightly correlated with DE, this is fully explained by their membership to one group, i.e. PL (DE equivalently) chat mainly with the other three.

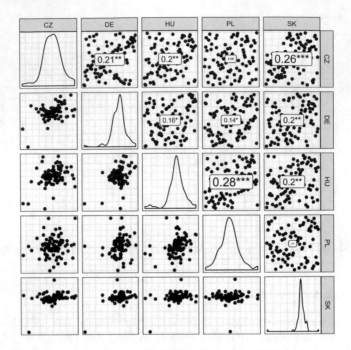

Fig. 5. Composition of pair plots (lower triangle), density plots (diagonal) and pair plots of re-scaled data accompanied by the corresponding Kendall's tau and its significance.

HU-PL and of t-copula type in CZ-DE, while asymmetric rotated Gumbel copula underlying relation of HU-SK. The later means that negative shocks of Slovak and Hungarian GDP are related more closely than on average. Higher level trees and the corresponding densities reveals conditional independence between other countries given their connections to other one, two or three countries. Such a vine tree is called 1-truncated and offers a great reduction in number of parameters needed to describe the system.

Another such a sparse model is provided by factor copulas and for the GDP data, their fit in terms of GOF measure test statistic S_{CM} (in parenthesis) is super, compare: from elliptical Gaussian (=0.030) and Student (0.027), Vine copula from Fig. 8 (0.054) and finally from 1-factor copulas Fréchet (0.029) and Cuadras-Auge (0.027). Notice the singular component present in simulated pseudo-observations in Fig. 7 with higher density in the upper tail in case of the winning Cuadras-Auge family.

3.3   Dependence Evolution

In this subsection we focus on evolution of dependence and watch the preference of the above mentioned classes of models across 6 time periods.

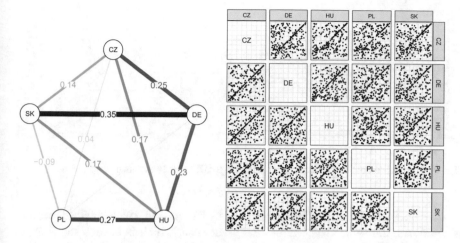

Fig. 6. Graph of partial correlations.

Fig. 7. Simulations from FDG copulas of Frechet (lower triangle) and Cuadras-Auge family (upper).

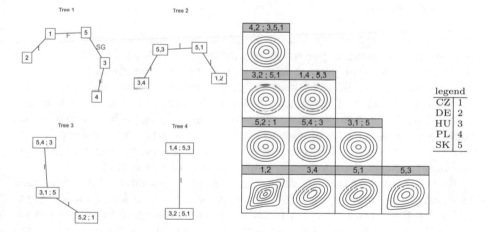

Fig. 8. Globally best fitting vine copula - trees and density contour plot.

The data were divided into 6 years-spanning intervals such that boundaries coincide with neighbor's center. First, Kendall's tau is checked for significance in each period, see Fig. 9. Clearly, the strength varies wildly and no couple retain its comovement for long.

Then all two factor and two elliptical copula families in question are fitted to those periods. In case of vines one candidate structure per each period is selected and then estimated on all the other periods so that there is max 6 unique vine structures. Finally, their goodness-of-fit measure is compared as portrayed in Fig. 10. Vine copula V2 serves as benchmark corresponding to full independence.

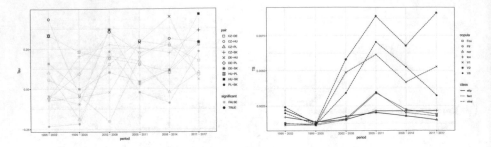

Fig. 9. Kendall's tau evolution.   **Fig. 10.** Goodness of fit evolution.

As a consequence of no significant relation in the period 1999–2005 all copula types perform the same (they model independence, which is always the same), afterwards elliptical and factor copulas outperform vines, especially about the 2009 crisis when also factor nature is weaker.

4   Conclusion

We analyzed relation of four Central European countries and a dominant player of that region in terms of one economy performance measure. Several tools were recommended for that purpose and few observations were given. For instance, graph of partial correlations discloses that correlation of Czech and Slovak GDP shocks are caused by the common factor - Germany - which is not the case with the pair Hungary–Poland. Or that the superb fitting performance of factor copulas suggests presence of some general economic component affecting all the countries in the given European region such that inclusion of certain relevant variables could improve predictive power of such stochastic model.

In the future research we would like to improve potential of factor copulas by developing a software package that would extend the choice of 1-factor parametric families, even at the price of losing computational performance compared to the elegant FDG class. Also properties of the fitted copulas should be studied more thoroughly to give better interpretation of the evolving dependence structure.

Acknowledgement. This work was supported by Slovak Research and Development Agency under contracts No. APVV–14–0013 and by VEGA 1/0006/19.

References

1. Aas, K., Czado, C., Frigessi, A., Bakken, H.: Pair-copula constructions of multiple dependence. Insur.: Math. Econ. **44**(2), 182–198 (2009)
2. Acar, E.F., Genest, C., Nešlehová, J.: Beyond simplified pair-copula constructions. J. Multivar. Anal. **110**, 74–90 (2012)

3. Bacigál, T., Komorníková, M., Komorník, J.: Multidimensional copula models of dependencies between selected international financial market indexes. In: 3rd Conference on Information Technology, Systems Research and Computational Physics, IRSRCP 2018. Springer, Heidelberg (2019, to appear)
4. Bacigál, T., Ždímalová, M.: Convergence of linear approximation of archimedean generator from williamson's transform in examples. Tatra Mt. Math. Publ. **69**, 1–18 (2017)
5. Bedford, T., Cooke, R.M.: Vines: a new graphical model for dependent random variables. Ann. Stat. **30**, 1031–1068 (2002)
6. Czado, C.: Pair-copula constructions of multivariate copulas. In: Copula Theory and Its Applications, pp. 93–109. Springer, Heidelberg (2010)
7. Dissmann, J., Brechmann, E.C., Czado, C., Kurowicka, D.: Selecting and estimating regular vine copulae and application to financial returns. Comput. Stat. Data Anal. **59**, 52–69 (2013)
8. Galor, O.: From stagnation to growth: unified growth theory. In: Aghion, P., Durlauf, S. (eds.) Handbook of Economic Growth, vol. 1. Elsevier, Amsterdam (2005)
9. Genest, C., Rémillard, B., Beaudoin, D.: Goodness-of-fit tests for copulas: a review and a power study. Insur.: Math. Econ. **44**(2), 199–213 (2009)
10. Hofert, M.: Construction and sampling of nested Archimedean copulas. In: Copula Theory and Its Applications, pp. 147–160. Springer, Heidelberg (2010)
11. Joe, H.: Dependence Modeling with Copulas. Monographs on Statistics and Applied Probability, vol. 134 (2015)
12. Johnson, N.D., Koyama, M.: States and economic growth: capacity and constraints. Explor. Econ. Hist. **64**(C), 1–20 (2017). https://EconPapers.repec.org/RePEc:eee:exehis:v:64:y:2017:i:c:p:1 20
13. Krupskii, P., Joe, H.: Factor copula models for multivariate data. J. Multivar. Anal. **120**, 85 101 (2013)
14. Marcellino, M., Stock, J., Watson, M.: Macroeconomic forecasting in the Euro area: country specific versus Euro wide information. Eur. Econ. Rev. **47**(1), 1–18 (2003). https://EconPapers.repec.org/RePEc:eee:eecrev:v:47:y:2003:i:1:p:1-18
15. Mazo, G., Girard, S.: FDGcopulas: multivariate dependence with FDG copulas (2014). R package version 1.0: https://CRAN.R-project.org/package=FDGcopulas
16. Mazo, G., Girard, S., Forbes, F.: A flexible and tractable class of one-factor copulas. Stat. Comput. **26**(5), 965–979 (2016)
17. McNeil, A.J.: Sampling nested archimedean copulas. J. Stat. Comput. Simul. **78**(6), 567–581 (2008)
18. Morales Napoles, O., Cooke, R.M., Kurowicka, D.: About the number of vines and regular vines on n nodes. Technical report, Delft University of Technology (2010). http://resolver.tudelft.nl/uuid:912abf55-8112-48d2-9cca-323f7f6aecc7
19. Nelsen, R.B.: An Introduction to Copulas. Springer, Heidelberg (2007)
20. Okhrin, O., Ristig, A.: Hierarchical Archimedean copulae: the HAC package. J. Stat. Softw. **58**(4), 1–20 (2014). https://doi.org/10.18637/jss.v058.i04. https://www.jstatsoft.org/v058/i04
21. R Core Team: R: A Language and Environment for Statistical Computing. R Foundation for Statistical Computing, Vienna (2018). https://www.R-project.org/
22. Schirmacher, D., Schirmacher, E.: Multivariate dependence modeling using pair-copulas. Technical report (2008)

Aggregation via Clone Theory Approach

Radomír Halaš[1]([⊠]) and Jozef Pócs[1,2]

[1] Department of Algebra and Geometry, Faculty of Science,
Palacký University Olomouc, 17. listopadu 12, 771 46 Olomouc, Czech Republic
radomir.halas@upol.cz
[2] Mathematical Institute, Slovak Academy of Sciences,
Grešákova 6, 040 01 Košice, Slovakia
pocs@saske.sk

Abstract. In our recent papers we have observed that the set of aggregation functions (on any bounded poset) contains all the projections and it is composition-closed. These classes of functions, called clones, are very intensively studied for decades in many different branches of mathematics.

The aim of our paper is to give an extended overview of results concerning certain important clones of aggregation functions on bounded lattices. In particular, we focus on the full clone of aggregation functions, the clone of idempotent aggregation functions, the clone of Sugeno integrals and the clone of polynomial functions.

Keywords: Clone · Aggregation function · Generating set

1 Introduction

Aggregation theory is a rapidly growing field of mathematics with roots in all domains where a merging of several inputs from a considered ordered scale into one output, reflecting some natural constraints, is considered. Mathematically, the process of aggregation is realized by functions, called aggregation functions. They appear in many different branches of science, among the most widely used one in all experimental sciences is the arithmetic mean but aggregation functions appear also in various parts of mathematics (functional equations, theory of means and averages, theory of measure and integration). The reader can consult an extensive literature devoted to the study of aggregation functions, e.g. [2,4,8].

Originally, aggregation functions were introduced on real interval scales, as well as discrete scales of type $\{0, 1, \ldots, n\}$ or $N = \{1, 2, \ldots\}$ were also considered, see e.g. [19]. It is important to note that all the above scales are based on a linear order. However, several applications deal with possibly incomparable inputs, such as intervals or products of chains. Only recently more abstract scales were considered, in particular lattice (poset) scales. For information sciences, and in particular for decision problems, typical scales deal with bounded (distributive) lattices. Not going into details, among different papers dealing with aggregation functions acting on lattices, we recall e.g. the seminal papers [6,18], our recent papers [13,16,17], or the papers on nullnorms and uninorms on bounded lattices.

© Springer Nature Switzerland AG 2019
R. Halaš et al. (Eds.): AGOP 2019, AISC 981, pp. 244–254, 2019.
https://doi.org/10.1007/978-3-030-19494-9_23

As the aggregation theory is roughly about aggregation functions, it can be also recognized as a part of algebra devoted to function algebras. At the core of this theory is a natural composition of functions and the sets of functions being closed under this composition. Closed classes of functions which contain a natural class of all the projections are called clones. Particular classes of aggregation functions can be seen as special clones, ranging from the least one (all projections) to the greatest one (all aggregation functions of a considered lattice L). Among the most important cases, a prominent role have idempotent aggregation functions on L, and its subclass of idempotent lattice polynomials (i.e., Sugeno integrals, see [6,18]). The possible complexity of mentioned clones can be reduced when we look at their generating sets, i.e., subsets of aggregation functions from the considered clone such that their compositions generates all members of that clone. This problem was discussed for idempotent lattice polynomials on a distributive bounded lattice L in [6,18], see also [11,12], and for all idempotent aggregation functions on a bounded lattice L in [3,10].

The main aim of this paper is to present an extended overview of our recent results on generating sets of certain aggregation clones on bounded lattices. A new result (Theorem 2) characterizing finite lattices for which idempotent aggregation functions form a subset of their lattice polynomials is presented.

2   Clones

In this section we briefly recall some important concepts from universal algebra and clone theory. As our main aim is to investigate aggregation functions on bounded lattices, we recall that a *lattice* is an algebra $(L; \vee, \wedge)$, where L is a nonempty set (called *support*) with two binary operations $\vee$ and $\wedge$ representing suprema and infima, cf. [9]. To simplify expressions, we usually do not distinguish between the lattice and its support.

The investigation of clones originates in the field of logic, namely in the study of composition of truth functions, and also in universal algebra from the observation that a lot of important properties of algebraic structures depend on their term operations rather than on the selection of their basic operations.

Particularly, a *clone* on a set A is a set of (finitary) operations on A which contains all of the projection operations on A and is closed under the composition (superposition), where projections and composition are formally defined as follows:

Let A be a set and $n \in \mathbb{N}$ be a positive integer. For any $i \leq n$, the *i-th n-ary projection* is for all $x_1, \ldots, x_n \in A$ defined by

$$p_i^n(x_1, \ldots, x_n) := x_i.$$

Composition forms from one k-ary operation $f : A^k \to A$ and k n-ary operations $g_1, \ldots, g_k : A^n \to A$, an n-ary operation $f(g_1, \ldots, g_k) : A^n \to A$ defined by

$$f(g_1, \ldots, g_k)(x_1, \ldots, x_n) := f(g_1(x_1, \ldots, x_n), \ldots, g_k(x_1, \ldots, x_n)),$$

for all $x_1, \ldots, x_n \in A$.

Clones as sets of functions can be viewed by another equivalent way, namely, as the sets of (finitary) relations on A that are preserved by all of the functions from the clone. More precisely, let $\rho \subseteq A^d$ be a d-ary relation on A, and $f : A^n \to A$ an n-ary operation on A. We say that f *preserves* ρ (or ρ is *invariant* with respect to f), if for any $d \times n$ matrix (b_{ij}) of elements of A all the columns of which belong to ρ, then so does the d-tuple when applying f to its rows, i.e.,

$$\begin{pmatrix} b_{11} \\ b_{21} \\ \vdots \\ b_{d1} \end{pmatrix}, \begin{pmatrix} b_{12} \\ b_{22} \\ \vdots \\ b_{d2} \end{pmatrix}, \ldots, \begin{pmatrix} b_{1n} \\ b_{2n} \\ \vdots \\ b_{dn} \end{pmatrix} \in \rho \implies \begin{pmatrix} f(b_{11}, b_{12}, \ldots, b_{1n}) \\ f(b_{21}, b_{22}, \ldots, b_{2n}) \\ \vdots \\ f(b_{d1}, b_{d2}, \ldots, b_{dn}) \end{pmatrix} \in \rho.$$

The fact that a function f preserves ρ will be denoted by $f \vartriangleleft \rho$. Clearly, $\vartriangleleft$ represents a binary relation between the set $\mathcal{O}_A$ of all (finitary) functions on A and $\mathcal{R}_A$, the set of all (finitary) relations on A. In algebra, one of the most important concepts is that of a Galois connection, which is associated to a binary relation. The Galois connection is given by the two operators Pol and Inv as follows: for $F \subseteq \mathcal{O}_A$ and $R \subseteq \mathcal{R}_A$,

$$\mathsf{Inv}\, F := \{\rho \in \mathcal{R}_A : f \vartriangleleft \rho \text{ for all } f \in F\},$$

and

$$\mathsf{Pol}\, R := \{f \in \mathcal{O}_A : f \vartriangleleft \rho \text{ for all } \rho \in R\}.$$

In other words, the set $\mathsf{Inv}\, F$ consists of all relations invariant with respect to all functions $f \in F$, and dually, $\mathsf{Pol}\, R$, called the set of *polymorphisms* of R, consists of all operations which are invariant to all of the relations belonging to R. Combining the maps Inv and Pol, we obtain a pair of closure operators defined on $\mathcal{O}_A$ and $\mathcal{R}_A$ respectively, defined by

$$F \mapsto \mathsf{Pol}\,\mathsf{Inv}\, F,$$

$$R \mapsto \mathsf{Inv}\,\mathsf{Pol}\, R.$$

It is well known that the clones (of functions) are just the closed sets with respect to the closure operator Pol Inv, i.e. $\mathcal{C}$ is a clone if and only if $\mathcal{C} = \mathsf{Pol}\,\mathsf{Inv}\,\mathcal{C}$. Hence clones can be viewed as sets of functions invariant with respect to appropriate sets of finitary relations. The clone $\mathsf{Pol}\,\mathsf{Inv}\, F$ is the least clone containing the set F of functions, and thus called the clone *generated* by F. We call a clone *finitely generated* if it has a finite generating set of functions.

Many important properties of clones on finite sets are closely related to the existence of so-called near-unanimity functions. Recall that a *near-unanimity function* on a set A is any function f (of arity at least 3) fulfilling the equalities

$$f(x, \ldots, x, y) = f(x, \ldots, y, \ldots, x) = f(y, x, \ldots, x) = x$$

for any elements $x, y \in A$. Considering a lattice $(L; \vee, \wedge)$, the median functions given by

$$\text{med}_\wedge(x, y, z) := (x \wedge y) \vee (y \wedge z) \vee (x \wedge z) \tag{1a}$$

or

$$\text{med}_\vee(x, y, z) := (x \vee y) \wedge (y \vee z) \wedge (x \vee z) \tag{1b}$$

are apparently (monotone) ternary near-unanimity functions on L, called *majority functions* on L.

We recall the following famous Baker-Pixley theorem, see [1]:

Proposition 1. *Let A be a finite set, $F \subseteq \mathcal{O}_A$ and assume that the clone $\mathcal{C} = \text{Pol Inv}\, F$ contains a $(d+1)$-ary near-unanimity function. Then for any $f \in \mathcal{O}_A$ it holds $f \in \mathcal{C}$ if and only if $f \vartriangleleft \rho$ for all d-ary relations $\rho \in \text{Inv}\, F$.*

Remark 1. (1) Considering a finite lattice L, Baker-Pixley theorem can be stated in the following form. Let $F \subseteq \mathcal{O}_L$ and $\mathcal{C} = \text{Pol Inv}\, F$ be a clone containing the lattice operations $\vee \wedge$. Then $\mathcal{C}$ possesses the median functions given by (1a) and (1b) respectively, and $f \in \mathcal{C}$ if and only if $f \vartriangleleft \rho$ for all binary relations $\rho \in \text{Inv}\, F$.

(2) In the sequel we will use the Baker-Pixley Theorem for proving negative results, i.e., in order to show that $f \notin \mathcal{C}$, $\mathcal{C} = \text{Pol Inv}\, F$ containing the lattice operations, it suffices to find a relation $\rho \subseteq L^2$, $\rho \in \text{Inv}\, F$ such that f does not preserve ρ.

(3) Let $\mathcal{C}$ be a clone on a finite set A containing a near-unanimity function. Then it is the well known fact that $\mathcal{C}$ is finitely generated. Hence any clone on a finite lattice, which contains the lattice operations, is finitely generated.

3   Aggregation Clones and Their Generators

First, we briefly recall the notion of aggregation function on a bounded lattice L. The bottom and the top element of a bounded lattice will be denoted by 0 and 1, respectively. A function $f : L^n \to L$ is called an *aggregation function* (on L) if it fulfills the following two conditions:

(i) is nondecreasing (in each variable), i.e. for any $\mathbf{x}, \mathbf{y} \in L^n$:

$$f(\mathbf{x}) \le f(\mathbf{y}) \quad \text{whenever} \quad \mathbf{x} \le \mathbf{y},$$

(ii) fulfills the boundary conditions

$$f(0, \dots, 0) = 0 \quad \text{and} \quad f(1, \dots, 1) = 1.$$

The integer n represents the arity of the aggregation function.

An important family of aggregation functions form (bounds-preserving) lattice polynomials. By an n-ary *polynomial* ($n \in \mathbb{N} \cup \{0\}$) on a lattice L is meant any function $p : L^n \to L$ obtained by composing lattice operations $\vee, \wedge$ and constants. More precisely, n-ary polynomials are defined inductively as follows:

– any variable x_i, $i \in \{1, \dots, n\}$, is a polynomial
– any constant $a \in L$ (considered as an n-ary function) is a polynomial
– if p_1 and p_2 are polynomials, then so does the functions $p_1 \vee p_2$ and $p_1 \wedge p_2$
– any polynomial is obtained by finitely many of the preceding steps.

The family of all polynomials on a bounded lattice L satisfying the boundary condition (ii) will be denoted through the paper by $\mathsf{Pol}_{0,1}(L)$.

In this paper we focus our attention on the following sets of functions on a bounded lattice L:

– $\mathsf{Agg}(L)$, the set of all *aggregation functions* on L,
– $\mathsf{Id}(L)$, the set of all *idempotent aggregation functions* on L, i.e. all functions with $f(x, \dots, x) = x$ for any $x \in L$,
– $\mathsf{Pol}_{0,1}(L)$, the set of all polynomials on L satisfying the boundary condition (ii).
– $\mathsf{Sug}(L)$, the set of all *Sugeno integrals* on a distributive lattice L.

It is not difficult to observe that all of the above mentioned sets of functions form a clone, and the following inclusions are valid:

$$\mathsf{Id}(L) \subseteq \mathsf{Agg}(L), \quad \mathsf{Pol}_{0,1}(L) \subseteq \mathsf{Agg}(L), \quad \mathsf{Sug}(L) = \mathsf{Pol}_{0,1}(L) \cap \mathsf{Id}(L).$$

Note that for the last equality, we refer the reader to [6,18].

In order to find generating sets of some clones, the following theorem is helpful:

Theorem 1. *Let L be a finite lattice, and $C \subseteq \mathsf{Agg}(L)$ be a subclone containing the lattice operations $\vee$ and $\wedge$. Then for any $f \in C^n$ and for any n-tuple $\mathbf{a} \in L^n$ the function*

$$h^C_{f(\mathbf{a})} = \bigwedge \{g \in C^n \mid g(\mathbf{a}) = f(\mathbf{a})\} \tag{2}$$

belongs to the clone C. Moreover, for any $\mathbf{x} \in L^n$ the following expression holds:

$$f(\mathbf{x}) = \bigvee_{\mathbf{a} \in L^n} h^C_{f(\mathbf{a})}(\mathbf{x}). \tag{3}$$

Proof. Obviously, the set $\{g \in C^n \mid g(\mathbf{a}) = f(\mathbf{a})\}$ is finite and non-void, since f belongs to this set. Consequently, we obtain that the function

$$h^C_{f(\mathbf{a})} = \bigwedge \{g \in C^n \mid g(\mathbf{a}) = f(\mathbf{a})\}$$

is a composition of functions from C, i.e. $h^C_{f(\mathbf{a})} \in C$.

Clearly $f \geq h^C_{f(\mathbf{a})}$ as $f \in \{g \in C^n \mid g(\mathbf{a}) = f(\mathbf{a})\}$. Thus we obtain the inequality $f(\mathbf{x}) \geq \bigvee_{\mathbf{a} \in L^n} h^C_{f(\mathbf{a})}(\mathbf{x})$ for all $\mathbf{x} \in L^n$.

Conversely, for any $\mathbf{x} \in L^n$ we have the inequality

$$\Big(\bigwedge_{\mathbf{a} \in L^n} h^C_{f(\mathbf{a})} \Big)(\mathbf{x}) \geq h^C_{f(\mathbf{x})}(\mathbf{x}) = f(\mathbf{x}),$$

completing the proof. $\qquad\square$

Note that the previous theorem guarantees the existence of the least function, namely $h^C_{f(\mathbf{a})}(\mathbf{a})$, belonging to the clone $C \subseteq \mathsf{Agg}(L)$, and satisfying $f(\mathbf{a}) = h^C_{f(\mathbf{a})}(\mathbf{a})$.

In the following two lemmas we describe such functions for the clones $\mathsf{Agg}(L)$ and $\mathsf{Id}(L)$, respectively.

Lemma 1. *Let* $f\colon L^n \to L$ *be an aggregation function. Then for each* $\mathbf{a} \in L^n$

$$h^{\mathsf{Agg}(L)}_{f(\mathbf{a})}(\mathbf{x}) = \begin{cases} 1, & \text{if } \mathbf{x} = (1, \ldots, 1); \\ f(\mathbf{a}), & \text{if } \mathbf{x} \geq \mathbf{a}, \mathbf{x} \neq (1, \ldots, 1); \\ 0, & \text{if } \mathbf{x} \not\geq \mathbf{a}. \end{cases} \tag{4}$$

Proof. It can be easily seen that the function $h^{\mathsf{Agg}(L)}_{f(\mathbf{a})}$ is an aggregation function. Also if $h \in \mathsf{Agg}(L)$ satisfies $h(\mathbf{a}) = f(\mathbf{a})$, then $h(\mathbf{x}) \geq f(\mathbf{a})$ for all $\mathbf{x} \geq \mathbf{a}$. Hence, it follows $h^{\mathsf{Agg}(L)}_{f(\mathbf{a})} \leq h$, i.e., $h^{\mathsf{Agg}(L)}_{f(\mathbf{a})}$ is the least aggregation function with $f(\mathbf{a}) = h^{\mathsf{Agg}(L)}_{f(\mathbf{a})}(\mathbf{a})$. $\square$

Lemma 2. *Let* $n \geq 2$ *be a nonnegative integer and* $f\colon L^n \to L$ *be an idempotent aggregation function. Then for all* $\mathbf{a} \in L^n$

$$h^{\mathsf{Id}(L)}_{f(\mathbf{a})}(\mathbf{x}) = \begin{cases} f(\mathbf{a}) \vee \bigwedge_{i=1}^{n} x_i, & \text{if } \mathbf{x} \geq \mathbf{a}, \\ \bigwedge_{i=1}^{n} x_i, & \text{otherwise.} \end{cases} \tag{5}$$

Proof. Obviously, $\bigwedge \mathbf{x} \leq h^{\mathsf{Id}(L)}_{f(\mathbf{a})} \leq \bigvee \mathbf{x}$, thus $h^{\mathsf{Id}(L)}_{f(\mathbf{a})}$ is an idempotent aggregation function. If $h \in \mathsf{Id}(L)$ fulfills $f(\mathbf{a}) = h(\mathbf{a})$ then $h(\mathbf{x}) \geq f(\mathbf{a})$ for all $\mathbf{x} \geq \mathbf{a}$ due to monotonicity and also $h(\mathbf{x}) \geq \bigwedge \mathbf{x}$ for all $\mathbf{x} \in L^n$ since h is idempotent. This implies $h \geq h^{\mathsf{Id}(L)}_{f(\mathbf{a})}$, i.e., $h^{\mathsf{Id}(L)}_{f(\mathbf{a})}$ is the least idempotent aggregation function with $f(\mathbf{a}) = h^{\mathsf{Id}(L)}_{f(\mathbf{a})}(\mathbf{a})$. $\square$

Hence, with respect to (3) of Theorem 1, in order to find generating sets of the clones $\mathsf{Agg}(L)$ and $\mathsf{Id}(L)$, respectively, it suffices to find sets of functions such that the functions $h^{\mathsf{Agg}(L)}_{f(\mathbf{a})}$ and $h^{\mathsf{Id}(L)}_{f(\mathbf{a})}$ can be expressed as their composition.

For the generation of the aggregation clone $\mathsf{Agg}(L)$, beside the lattice operations, we use the following unary aggregation functions:

For any element $a \in L$ we define $\chi_a \colon L \to L$ by

$$\chi_a(x) = \begin{cases} 1, & \text{if } x \geq a, x \neq 0; \\ 0, & \text{otherwise.} \end{cases} \tag{6}$$

Further, we use the so-called full system of functions, cf. [14]. Recall, that a system $\mathcal{G} = \{g_j \colon L^{n_j} \to L : j \in J\}$ of aggregation functions on a bounded lattice L is *full*, provided the set $M = \bigcup_{j \in J} \{g_j(\mathbf{1}_I) : I \subseteq [n_j]\}$ generates the lattice L, where $\mathbf{1}_I$ denotes the (L-valued) characteristic function of a subset $I \subseteq \{1, \ldots, n\} = [n]$.

The main result concerning the generation of the full aggregation clone is the following proposition, cf. [14].

Proposition 2. *Let L be a finite lattice and $\mathcal{G}$ be a full system of aggregation functions on L. Then the set $\mathcal{G}_\chi = \mathcal{G} \cup \{\chi_a : a \in J(L)\} \cup \{\wedge, \vee\}$ generates the aggregation clone $\mathsf{Agg}(L)$.*

The proof of this proposition is based on the fact, that given an arbitrary aggregation function $f \colon L^n \to L$, the definition of a full system yields, that for any $\mathbf{a} \in L^n$ there is a system of m functions $g_j \colon L^{n_j} \to L$, $g_j \in \mathcal{G}$, $j = 1, \ldots, m$, an m-ary lattice polynomial q with $q(a_1, \ldots, a_m) = f(\mathbf{a})$, where for all $j = 1, \ldots, m$, $a_j = g_j(\mathbf{1}_{I_j})$ for some $I_j \subseteq [n_j]$. Consequently, for $\mathbf{x} \in L^n$ we obtain

$$h_{f(\mathbf{a})}^{\mathsf{Agg}(L)}(\mathbf{x}) = H_{\mathbf{a}}(\mathbf{x}) \wedge q\big(H_{g_1}^{I_1}(\mathbf{x}), \ldots, H_{g_m}^{I_m}(\mathbf{x})\big),$$

where

$$H_{\mathbf{a}}(\mathbf{x}) = \bigwedge_{i \in J_{\mathbf{a}}} \chi_{a_i}(x_i), \quad J_{\mathbf{a}} = \{i \in \{1, \ldots, n\} : a_i \neq 0\}$$

and

$$H_g^I(x_1, \ldots, x_n) = g\big(y_1(\mathbf{x}), \ldots, y_k(\mathbf{x})\big), \quad y_i(x_1, \ldots, x_n) = \begin{cases} \bigvee\limits_{l=1}^{n} \chi_0(x_l), & \text{if } i \in I; \\ \bigwedge\limits_{l=1}^{n} \chi_1(x_l), & \text{if } i \notin I. \end{cases}$$

Hence every function $h_{f(\mathbf{a})}^{\mathsf{Agg}(L)}$ is a composition of some functions from the set $\mathcal{G}_\chi$, i.e., with respect to (3), the set $\mathcal{G}_\chi$ generates the aggregation clone $\mathsf{Agg}(L)$.

In the following example we present a full system consisting of binary aggregation functions.

Example 1. Let $a, b \in L$ be arbitrary elements. We put

$$g_{a,b}(x_1, x_2) = (x_1 \wedge a) \vee (x_2 \wedge b) \vee (x_1 \wedge x_2).$$

Using direct computation, it can be easily seen that $g_{a,b}(1, 0) = a$ and $g_{a,b}(0, 1) = b$. Hence, if $T \subseteq L$ is a generating subset of a finite lattice L, one can take any two subsets $A = \{a_1, \ldots, a_m\} \subseteq T$, $B = \{b_1, \ldots, b_m\} \subseteq T$ with $m = \left\lceil \frac{|T|}{2} \right\rceil$, $A \cup B = T$ and put $\mathcal{G} = \{g_{a_i, b_i} : 1 \leq i \leq m\}$. Evidently, $\mathcal{G}$ represents a full system.

With respect to the previous result, we obtain the following estimation for the cardinality of a generating set of the aggregation clone. Let L be an n-element lattice and let T be a generating set of L with the minimal cardinality. Then

$\mathsf{Agg}(L)$ can be generated by the set S consisting of at most binary aggregation functions with

$$|S| \leq |\mathsf{J}(L)| + \left\lceil \frac{|T|}{2} \right\rceil + 2 \leq \left\lceil \frac{3n}{2} \right\rceil + 1.$$

We have presented the generating set of the aggregation clone consisting of certain unary and binary aggregation functions. Concerning the arities of generating functions, a natural question can be raised: can $\mathsf{Agg}(L)$ be generated using the lattice operations and by unary aggregation functions only? In [17] this question was answered negatively, i.e., if L is a finite lattice with at least three elements, then the lattice operations together with any subset of unary aggregation functions do not generate the aggregation clone $\mathsf{Agg}(L)$.

In order to describe a generating set for the idempotent clone $\mathsf{Id}(L)$, we use the following binary functions:

For $a \in L$ define the binary functions $\mathsf{med}_a \colon L^2 \to L$ and $\chi_a \colon L^2 \to L$ by stipulation

$$\mathsf{med}_a(x,y) = (x \wedge y) \vee \big(a \wedge (x \vee y)\big), \tag{7}$$

$$\chi_a(x,y) = \begin{cases} x \vee y, & \text{if } x \geq a, \\ x \wedge y, & \text{otherwise.} \end{cases} \tag{8}$$

Proposition 3. *For any finite lattice L, the set of binary functions $\{\vee, \wedge\} \cup \{\chi_a(x,y) \mid a \in L \setminus \{0\}\} \cup \{\mathsf{med}_a(x,y) \mid a \in L \setminus \{0,1\}\}$ generates the idempotent clone $\mathsf{Id}(L)$.*

In this case, see [10], it can be shown that for an arbitrary idempotent f, the function $h_{f(\mathbf{a})}^{\mathsf{Id}(L)}$ can be expressed in the form:

$$h_{f(\mathbf{a})}^{\mathsf{Id}(L)}(x_1,\ldots,x_n) = \mu_{\mathbf{a}}(x_1,\ldots,x_n) \wedge \mathsf{med}_{f(\mathbf{a})}\Big(\bigwedge_{i=1}^{n} x_i, \bigvee_{i=1}^{n} x_i \Big),$$

where for $\mathbf{a} = (a_1,\ldots,a_n)$, $\mu_{\mathbf{a}}(x_1,\ldots,x_n) = \bigwedge_{i=1}^{n} \chi_{a_i}^n(x_i,\ldots,x_1,\ldots,x_n)$ and $\chi_a^n(x_1,\ldots,x_n) = \chi_a(\chi_a^{n-1}(x_1,\ldots,x_{n-1}),x_n)$ for $n \geq 3$ and an arbitrary element $a \in L$. Consequently, the idempotent clone $\mathsf{Id}(L)$ can be generated by using the lattice operations and functions given by (7) and (8), respectively. As it can be easily seen that $\mathsf{med}_0(x,y) = x \wedge y$ and $\mathsf{med}_1(x,y) = x \vee y$ we obtain the estimation $|S| \leq 2n - 1$ for the minimal cardinality of a generating set of the idempotent clone.

Further, we briefly discuss the so-called Sugeno clone $\mathsf{Sug}(L)$, which is defined on bounded distributive lattices. In what follows L denotes a bounded distributive lattice with 0 and 1 as its bottom and the top element respectively. Recall that discrete Sugeno integrals are defined with respect to an L-valued capacities, i.e., monotone L-valued set functions $\mu \colon \mathbf{P}([n]) \to L$ such that $\mu(\emptyset) = 0$ and $\mu([n]) = 1$. For a given L-valued capacity μ there are two equivalent expressions for the n-ary Sugeno integral

$$\mathsf{Su}_\mu^n(\mathbf{x}) = \bigvee_{I \subseteq [n]} \Big(\mu(I) \wedge \bigwedge_{i \in I} x_i \Big), \tag{9}$$

$$\mathsf{Su}_{\mu}^{n}(\mathbf{x}) = \bigwedge_{I \subseteq [n]} \left(\mu([n] \smallsetminus I) \vee \bigvee_{i \in I} x_i \right), \tag{10}$$

It can be shown (see [15]) that the family $\mathsf{Sug}(L)$ is closed under composition and contains all the projections, i.e., it forms a clone.

In particular, for $n = 2$ and $\mu \colon \mathbf{P}(\{1, 2\}) \to L$ with $\mu(\{1\}) = a$ and $\mu(\{2\}) = b$ we denote the binary Sugeno integral by $\mathsf{Su}_{a,b}^{2}$. Consequently, for any pair $(x_1, x_2) \in L^2$ we have

$$\mathsf{Su}_{a,b}^{2}(x_1, x_2) = (a \wedge x_1) \vee (b \wedge x_2) \vee (x_1 \wedge x_2).$$

Observe that the lattice operations are binary Sugeno integrals as well, particularly $\mathsf{Su}_{0,0}^{2}(x_1, x_2) = x_1 \wedge x_2$ and $\mathsf{Su}_{1,1}^{2}(x_1, x_2) = x_1 \vee x_2$.

Further, we use the following notation: for $\mu \colon \mathbf{P}([n]) \to L$ and for $j \in [n]$ we put $\mu_j \colon \mathbf{P}([n] \smallsetminus \{j\}) \to L$, where $\mu_j(A) = \mu(A)$ for $A \subsetneq [n] \smallsetminus \{j\}$ and $\mu_j([n] \smallsetminus \{j\}) = 1$. The set function μ_j is a capacity. Also denote by $\mu^{c(j)} = \mu([n] \smallsetminus \{j\})$ the value corresponding to the complement of the singleton $\{j\}$ with respect to μ, and for $\mathbf{x} \in L^n$ we put $\mathbf{x}^j = (x_1, \ldots, x_{j-1}, x_{j+1}, \ldots, x_n)$.

Using induction with respect to the arity n of a Sugeno integral, the following formula is valid, cf. [15] for the detailed proof.

Proposition 4. *For any $n \geq 3$ and $\mathbf{x} \in L^n$ the following holds:*

$$\mathsf{Su}_{\mu}^{n}(\mathbf{x}) = \bigvee_{j \in [n]} \mathsf{Su}_{\mu^{c(j)}, \mu(\{j\})}^{2} \left(\mathsf{Su}_{\mu_j}^{n-1}(\mathbf{x}^j), x_j \right).$$

Hence, using this formula it can be easily seen that any Sugeno integral is in fact a composition of binary ones. Hence the clone $\mathsf{Sug}(L)$ is generated by its binary part involving $|L|^2$ generators.

In the last part we will briefly discuss questions related to the problem of order polynomial completeness. As the lattice operations are monotone, it is obvious that polynomial functions are monotone as well. Therefore it is natural to ask, whether they are the only monotone functions. Lattices for which this is the case are referred to as order polynomially complete.

In [16] the similar problem concerning aggregation functions was studied. It was shown that each aggregation function is a lattice polynomial if and only if all functions χ_a, $a \in \mathsf{J}(L)$, defined by (6) are polynomials. Consequently, finite Agg-polynomially complete lattices, i.e., lattices for which any aggregation function can be expressed as a lattice polynomial fulfilling the boundary conditions, were characterized by means of tolerance relations.

Recall that a tolerance relation on a lattice L is a reflexive, symmetric binary relation, which is compatible with the lattice operations, i.e., for all $a, b, c, d \in L$ the conditions $(a, b) \in T$ and $(c, d) \in T$ imply $(a \vee c, b \vee d) \in T$ as well as $(a \wedge c, b \wedge d) \in T$. More on tolerance relations can be found in [5]. It can be easily verified that the identity relation id_L and $L \times L$ are tolerance relations on any lattice L. These are called trivial tolerances. The above mentioned characterization is the following: a finite lattice is Agg-polynomially complete if and only if it has no nontrivial tolerances.

In connection with idempotent aggregation functions we show that in a finite case, the characterization of the Id-polynomially complete lattices (lattices L such that each function in $\mathsf{Id}(L)$ is a polynomial) is the same as in the case of aggregation functions.

Theorem 2. *A finite lattice is Id-polynomially complete if and only if it has no nontrivial tolerances.*

Proof. Obviously, if a finite lattice L is tolerance free (it has only trivial tolerances), then every idempotent aggregation function is a polynomial.

Since $\mathsf{Id}(L)$ is a subset of the set of all aggregation functions, potentially there can be a finite lattice with nontrivial tolerance relations, which is Id-polynomially complete. However, we show that this is not the case.

Let L be a finite lattice and $T \subseteq L \times L$ be a tolerance such that $(b, 1) \in T$ for some $b \neq 1$ and $(0, 1) \notin T$. Note that the first condition is equivalent with $T \neq \mathrm{id}_L$, while the second one with $T \neq L \times L$. Consider the ternary aggregation function $\iota \colon L^3 \to L$ given by

$$
\iota(x_1, x_2, x_3) := \begin{cases} 0, & \text{if } x_1 = 0, \ x_2 \leq b; \\ \bigvee_{i=1}^{3} x_i, & \text{otherwise.} \end{cases}
$$

Obviously $\bigwedge_{i=1}^{3} x_i < \iota(x_1, x_2, x_3) \leq \bigvee_{i=1}^{3} x_i$, i.e., it is an idempotent aggregation function. We show that this function does not preserve the tolerance T, which yields that ι is not a polynomial, since polynomials preserve (are compatible with) any tolerance relation. Indeed, we have $(0, 0) \in T$, $(b, 1) \in T$ and $(1, 1) \in T$, but

$$
\iota(0, b, 1) = 0 \wedge (0 \vee b \vee 1) = 0
$$

and

$$
\iota(0, 1, 1) = 0 \vee 1 \vee 1 = 1.
$$

As $(0, 1) \notin T$, the function ι is not a polynomial, hence L is not Id-polynomially complete. $\qquad\square$

Let us mention that it was proved in [7] that within ZFC, there is no infinite order polynomially complete lattice. Currently we are not able to decide whether there exist infinite Agg-polynomially complete (Id-polynomially complete) lattices or not.

Acknowledgments. The authors were supported by the project of Grant Agency of the Czech Republic (GAČR) no. 18-06915S. The second author was also supported by the Slovak Research and Development Agency under the contract No. APVV-16-0073.

References

1. Baker, K.A., Pixley, A.F.: Polynomial interpolation and the Chinese remainder theorem for algebraic systems. Math. Zeitschrift **143**, 165–174 (1975)
2. Beliakov, G., Pradera, A., Calvo, T.: Aggregation Functions: A Guide for Practitioners. Studies in Fuzziness and Soft Computing, vol. 221. Springer, Heidelbeg (2007)
3. Botur, M., Halaš, R., Mesiar, R., Pócs, J.: On generating of idempotent aggregation functions on finite lattices. Inf. Sci. **430–431**, 39–45 (2018)
4. Calvo, T., Mayor, G., Mesiar, R. (eds.): Aggregation Operators. Physica Verlag, Heidelberg (2002)
5. Chajda, I.: Algebraic Theory of Tolerance Relations. Monograph Series. Palacký University Olomouc, Olomouc (1991)
6. Couceiro, M., Marichal, J.-L.: Characterizations of discrete Sugeno integrals as polynomial functions over distributive lattices. Fuzzy Sets Syst. **161**, 694–707 (2010)
7. Goldstern, M., Shelah, S.: There are no infinite order polynomially complete lattices, after all. Algebra Universalis **42**(1–2), 49–57 (1999)
8. Grabisch, M., Marichal, J.-L., Mesiar, R., Pap, E.: Aggregation Functions. Cambridge University Press, Cambridge (2009)
9. Grätzer, G.: Lattice Theory: Foundation. Birkhäuser, Basel (2011)
10. Halaš, R., Kurač, Z., Mesiar, R., Pócs, J.: Binary generating set of the clone of idempotent aggregation functions on bounded lattices. Inf. Sci. **462**, 367–373 (2018)
11. Halaš, R., Mesiar, R., Pócs, J.: A new characterization of the discrete Sugeno integral. Inf. Fusion **29**, 84–86 (2016)
12. Halaš, R., Mesiar, R., Pócs, J.: Congruences and the discrete Sugeno integrals on bounded distributive lattices. Inf. Sci. **367–368**, 443–448 (2016)
13. Halaš, R., Mesiar, R., Pócs, J.: Generators of Aggregation Functions and Fuzzy Connectives. IEEE Trans. Fuzzy Syst. **24**(6), 1690–1694 (2016)
14. Halaš, R., Mesiar, R., Pócs, J.: On generating sets of the clone of aggregation functions on finite lattices. Inf. Sci. **476**, 38–47 (2019)
15. Halaš, R., Mesiar, R., Pócs, J., Torra, V.: A note on some algebraic properties of discrete Sugeno integrals. Fuzzy Sets Syst. **355**, 110–120 (2019)
16. Halaš, R., Pócs, J.: On lattices with a smallest set of aggregation functions. Inf. Sci. **325**, 316–323 (2015)
17. Halaš, R., Pócs, J.: On the clone of aggregation functions on bounded lattices. Inf. Sci. **329**, 381–389 (2016)
18. Marichal, J.-L.: Weighted lattice polynomials. Discrete Math. **309**(4), 814–820 (2009)
19. Mayor, G., Torrens, J.: Triangular norms on discrete settings. In: Klement, E.P., Mesiar, R. (eds.) Logical, Algebraic, Analytic and Probabilistic Aspects of Triangular Norms, pp. 189–230. Elsevier, Amsterdam (2005)

Mixture Functions Based on Deviation and Dissimilarity Functions

Jana Špirková[✉] and Pavol Kráľ

Faculty of Economics, Matej Bel University,
Tajovského 10, 975 90 Banská Bystrica, Slovakia
{jana.spirkova,pavol.kral}@umb.sk

Abstract. Mixture functions represent a special class of weighted averaging functions with weights determined by continuous weighting functions which depend on the input values. If they are monotone increasing, they also belong to the important class of aggregation functions. Their construction can be based on minimization of special (weighted) penalty functions using dissimilarity function or based on zero value of the special (weighted) strictly increasing function using deviation functions.

1 Introduction

Mixture functions represent a special class of weighted averaging functions with weights determined by continuous weighting functions which depend on the input values. They were introduced by Marques Pereira and Pasi in [9] and later studied by several other authors, see, e.g., papers [1,2,10–14,17]. Under certain conditions, mixture functions belong to the wide class of aggregation functions. Theoretical and applied research in the field of aggregation operators resulted in a number of recently published monographs, e.g., [1,5,8].

Aggregation of data into one representative value is an indispensable tool in many scientific domains, especially those dealing primarily with quantitative information. Naturally, in this context, an aggregation function is defined as a function assigning a single output to several input values. Moreover, we assume that the output value is not entirely arbitrary, but, in the end, it characterizes the input data in some aspect. Mixture functions can also be naturally obtained as solutions to an optimization problem using a local penalty function in the form of special dissimilarity or as zero function value of special strictly increasing function using deviation functions. Our contribution is directly linked to work related to deviation-based aggregation functions (see [6,7]), and penalty functions in data aggregation (see [3,4]).

The paper is organized as follows. Section 2 presents basic notations and definition of mixture functions and their generalizations. Moreover, this section contains definitions of local penalty function, weighted penalty function using continuous monotone weighting function and deviation and dissimilarity functions. Section 3 discusses a special dissimilarity function, the use of which, under certain conditions, leads directly to a deviation function. Additionally, we study

© Springer Nature Switzerland AG 2019
R. Halaš et al. (Eds.): AGOP 2019, AISC 981, pp. 255–266, 2019.
https://doi.org/10.1007/978-3-030-19494-9_24

there monotonicity of a special class of the mixture functions, so-called quasi-mixture functions. Section 4 gives some concluding remarks.

2   Preliminaries

The following notations will be used. Consider any closed non-empty interval $\mathbf{I} = [a, b] \subset \overline{\mathbf{R}} = [-\infty, \infty]$. Then $\mathbf{I}^n = \{\mathbf{x} = (x_1, \ldots, x_n) \mid x_i \in \mathbf{I}, i = 1, \ldots, n\}$ is the set of all input vectors $\mathbf{x}$. Considering $\mathbf{x}, \mathbf{y} \in \mathbf{I}^n$, $\mathbf{x} = (x_1, \ldots, x_n)$, $\mathbf{y} = (y_1, \ldots, y_n)$, we say that $\mathbf{x} \le \mathbf{y}$ if and only if $x_i \le y_i$ for each $i = 1, \ldots, n$. An increasing permutation of an input vector $(x_1, \ldots, x_n)$, i.e., $x_{(1)} \le x_{(2)} \le \cdots \le x_{(n)}$ will be denoted by $(x_{(1)}, \ldots, x_{(n)})$.

Definition 1 [2,8]. *A function $F : \mathbf{I}^n \to \mathbf{I}$ is monotone increasing if and only if, for all $\mathbf{x}, \mathbf{y} \in \mathbf{I}^n$, such that $\mathbf{x} \le \mathbf{y}$, it holds that $F(\mathbf{x}) \le F(\mathbf{y})$.*

Definition 2 [2,8]. *A function $F : \mathbf{I}^n \to \mathbf{I}$ is called an aggregation function if it is monotone increasing in each variable and satisfies the boundary conditions $F(\mathbf{a}) = a$, $F(\mathbf{b}) = b$, where $\mathbf{a} = (a, a, \ldots, a)$, $\mathbf{b} = (b, b, \ldots, b)$.*

2.1   Mixture Functions

In this part we recall basic definition of the mixture functions and their generalizations. A discussion about their properties can also be found in [9–13, 15].

Definition 3 [9]. *A function $M_g : \mathbf{I}^n \to \mathbf{I}$ given by*

$$M_g(x_1, \ldots, x_n) = \frac{\sum_{i=1}^{n} g(x_i) \cdot x_i}{\sum_{i=1}^{n} g(x_i)}, \tag{1}$$

where $g : \mathbf{I} \to]0, \infty[$ is a continuous weighting function, is called a mixture function.

Sometimes different continuous weighting functions are applied for different input values, what leads to a generalized mixture function (see [12–14]).

Definition 4 [13]. *A function $M_{\mathbf{g}} : \mathbf{I}^n \to \mathbf{I}$ given by*

$$M_{\mathbf{g}}(x_1, \ldots, x_n) = \frac{\sum_{i=1}^{n} g_i(x_i) \cdot x_i}{\sum_{i=1}^{n} g_i(x_i)}, \tag{2}$$

where $g_i : \mathbf{I} \to]0, \infty[$, $i = 1, 2, \ldots, n$, are continuous weighting functions of the weighting vector $\mathbf{g} = (g_1, g_2, \ldots, g_n)$, is called a generalized mixture function.

Moreover, we can modify the previous function by reordering the input values which leads to an ordered generalized mixture function $OM_{\mathbf{g}}$. The most essential step in calculation of the $OM_{\mathbf{g}}$ functions is to create an increasing permutation $(x_{(1)}, \ldots, x_{(n)})$ of an input vector $(x_1, \ldots, x_n)$.

Definition 5 [13]. *A function $OM_{\mathbf{g}} : \mathbf{I}^n \to \mathbf{I}$ given by*

$$OM_{\mathbf{g}}(x_1, \ldots, x_n) = \frac{\displaystyle\sum_{i=1}^{n} g_i(x_{(i)}) \cdot x_{(i)}}{\displaystyle\sum_{i=1}^{n} g_i(x_{(i)})}, \tag{3}$$

where $g_i : \mathbf{I} \to]0, \infty[$, $i = 1, 2, \ldots, n$, are continuous weighting functions of the weighting vector $\mathbf{g} = (g_1, g_2, \ldots, g_n)$, and $(x_{(1)}, \ldots, x_{(n)})$ is the vector obtained by the increasing ordination of $(x_1, \ldots, x_n)$, i.e., $x_{(1)} \leq x_{(2)} \leq \ldots \leq x_{(n)}$, is called an ordered generalized mixture function.

Generalized mixture functions $M_{\mathbf{g}}$ and also ordered generalized mixture functions $OM_{\mathbf{g}}$ are continuous and idempotent, i.e. $M_{\mathbf{g}}(OM_{\mathbf{g}})(x, \ldots, x) = x$.

In the following part, we introduce modifications of the previous functions based on a transformation using a strictly monotone continuous function $f : \mathbf{I} \to \mathbf{R}$.

Definition 6. *A function $M_g^f : \mathbf{I}^n \to \mathbf{I}$ given by*

$$M_g^f(x_1, \ldots, x_n) = f^{-1} \left(\frac{\displaystyle\sum_{i=1}^{n} g(x_i) \cdot f(x_i)}{\displaystyle\sum_{i=1}^{n} g(x_i)} \right), \tag{4}$$

where $g : \mathbf{I} \to]0, \infty[$ is a continuous weighting function and $f : \mathbf{I} \to \mathbf{R}$ is a strictly monotone continuous function, is called a quasi-mixture function.

If we assume different weighting functions for single input values, we can define a generalized quasi-mixture function as follows.

Definition 7. *A function $M_{\mathbf{g}}^f : \mathbf{I}^n \to \mathbf{I}$ given by*

$$M_{\mathbf{g}}^f(x_1, \ldots, x_n) = f^{-1} \left(\frac{\displaystyle\sum_{i=1}^{n} g_i(x_i) \cdot f(x_i)}{\displaystyle\sum_{i=1}^{n} g_i(x_i)} \right), \tag{5}$$

where $g_i : \mathbf{I} \to]0, \infty[$, $i = 1, 2, \ldots, n$, are continuous weighting functions of the weighting vector $\mathbf{g} = (g_1, \ldots, g_n)$ and $f : \mathbf{I} \to \mathbf{R}$ is a strictly monotone continuous function, is called a generalized quasi-mixture function.

An ordered generalized quasi-mixture function is defined as follows.

Definition 8. *A function $OM_{\mathbf{g}}^f : \mathbf{I}^n \to \mathbf{I}$ given by*

$$OM_{\mathbf{g}}^f(x_1,\ldots,x_n) = f^{-1}\left(\frac{\sum\limits_{i=1}^{n} g_i(x_{(i)}) \cdot f(x_{(i)})}{\sum\limits_{i=1}^{n} g_i(x_{(i)})}\right), \tag{6}$$

where $g_i : \mathbf{I} \to]0,\infty[$, $i = 1,2,\ldots,n$, are continuous weighting functions of the weighting vector $\mathbf{g} = (g_1,\ldots,g_n)$, $f : \mathbf{I} \to \mathbf{R}$ is strictly monotone continuous function, and $(x_{(1)},\ldots,x_{(n)})$ is the vector obtained by the increasing ordination of $(x_1,\ldots,x_n)$, i.e., $x_{(1)} \leq x_{(2)} \leq \ldots \leq x_{(n)}$, is called an ordered generalized quasi-mixture function.

2.2 Penalty Based Function

Based on [3,4,6,7] we recall basic definitions related to a penalty based function.

Definition 9. *A local penalty function is a function $LP : \mathbf{R}^2 \to \mathbf{R}^+$ satisfying for any u, v, $y \in \mathbf{I}$ the following requirements:*

(i) $LP(u,y) = 0$ if and only if $u = y$;
(ii) $LP(u,y) \geq LP(v,y)$ if $|u - y| \geq |v - y|$.

The notion of a local penalty function can be used for introducing a penalty function P on $\mathbf{I}^{n+1}$.

Definition 10. *A penalty function $P : \mathbf{I}^{n+1} \to \mathbf{R}^+$ is defined for any $\mathbf{x} \in \mathbf{I}^n$ and $y \in \mathbf{I}$ as*

$$P(\mathbf{x},y) = \sum_{i=1}^{n} LP(x_i,y), \tag{7}$$

where $y \in \mathbf{I}$ is called the fused value of $\mathbf{x} \in \mathbf{I}^n$.

Definition 11. *Let $LP : \mathbf{I}^2 \to \mathbf{R}^+$ be a local penalty function (see Definition 9). Then a weighted penalty function $gP : \mathbf{I}^{n+1} \to \mathbf{R}^+$ is defined, for any $\mathbf{x} \in \mathbf{I}^n$ and $y \in \mathbf{I}$ as follows:*

$$gP(\mathbf{x},y) = \sum_{i=1}^{n} g_i(x_i) \cdot LP(x_i,y), \tag{8}$$

where $g_i : \mathbf{I} \to]0,\infty[$ are continuous weighting functions associated to the input values x_i, $i = 1,2,\ldots,n$.

The best fused value of $\mathbf{x}$ is obtained as the value y^* such that

$$gP(\mathbf{x},y^*) = \arg\min_y gP(\mathbf{x},y), \tag{9}$$

where either y^* is the unique minimizer or y^* is the set of minimizers which is the interval $]a,b[$ ($[a,b]$) then $y^* = \frac{a+b}{2}$, and it is called a penalty based function.

2.3   Deviation and Dissimilarity Functions

On the basis of [4,6] and [7] we present the definition of a deviation function.

Definition 12. *A mapping* $D : \mathbf{I}^2 \to \mathbf{R}$, *which is a continuous function, strictly increasing with respect to the second argument, i.e., for all* $x \in \mathbf{I}$, $D : (x, \cdot) : \mathbf{I} \to \mathbf{R}$ *is continuous strictly increasing, and for all* $x \in \mathbf{I}$ *satisfying* $D(x, x) = 0$, *is called a deviation function.*

The notion of deviation function leads to the so called Daróczy mean (see [6] and [7]).

Definition 13. *Let* $D : \mathbf{I}^2 \to \mathbf{R}$ *be a deviation function and* $\mathbf{x} = (x_1, \ldots, x_n) \in \mathbf{I}^n$ *be an input vector. A unique solution* y^* *of the equation* $\sum_{i=1}^{n} D(x_i, y) = 0$ *is called the Daróczy mean.*

Differences between elements of $\mathbf{I}$ might be also expressed using so called dissimilarity functions defined as follows.

Definition 14. *Let* $K : \mathbf{R} \to \mathbf{R}^+$ *be a convex function with a unique minimum* $K(0) = 0$ *and let* $f : \mathbf{I} \to \mathbf{R}$ *be a continuous strictly monotone function. Then the function* $L : \mathbf{I}^2 \to \mathbf{R}^+$ *given by* $L(x, y) = K\left(f(x) - f(y)\right)$ *is called a dissimilarity function (on* $\mathbf{I}$).

3   Mixture Functions as Penalty Based Functions

We start this section by discussing properties of dissimilarity and deviation and relationship between them, mainly in the context of generating mixture functions using various penalty functions.

It is obvious that, we can derive a dissimilarity function from a deviation function and vice-versa.

Lemma 1. *Let* L *be a dissimilarity function. If the corresponding* f *is a continuous strictly decreasing function, then* $D(x, y) = f(x) - f(y)$ *is a deviation function. If the corresponding* f *is a continuous strictly increasing function, then* $D(x, y) = f(y) - f(x)$ *is a deviation function.*

Proof. Let L be a dissimilarity function with a continuous strictly decreasing function f. Let $D(x, y) = f(x) - f(y)$. Then, from the properties of f, it directly follows that, for all $x \in \mathbf{I}$, $D(x, x) = 0$, D is continuous, and, for each x, $D(x, \cdot) = f(x) - f(\cdot)$ is a continuous strictly increasing function, i.e., D is a dissimilarity function.

Analogously, for a continuous strictly increasing function f and $D(x, y) = f(y) - f(x)$. □

Lemma 2. *Let* D *be a deviation function and* $K : \mathbf{R} \to \mathbf{R}^+$ *be a convex function with a unique minimum* $K(0) = 0$. *Then* $L(x, y) = K(f(x) - f(y))$, *and for a fixed* $c \in \mathbf{I}$, $f(x) = D(c, x)$, *is a dissimilarity function.*

Proof. Straightforwardly, from properties of a deviation function, we get that, for a fixed $c \in \mathbf{I}$, $f(x) = D(c, x)$ is a continuous strictly increasing function. Then $L(x, y) = K(f(x) - f(y))$, where $K : \mathbf{R} \to \mathbf{R}^+$ is a convex function with a unique minimum $K(0) = 0$, is a dissimilarity function according to Definition 14. □

Both dissimilarity and deviation functions can be also connected to a local penalty function.

Proposition 1. *Let D be a deviation function such that $D(\cdot, y)$ is a decreasing function, for each $y \in \mathbf{I}$, and $|D(2y - x, y)| = |D(x, y)|$, for all $x, y \in \mathbf{I}$. Then $D^*(x, y) = |D(x, y)|$ is a local penalty function.*

Proof. It is obvious that $\text{Rng}(D^*) \subset \mathbf{R}^+$. From the boundary condition we have that for $x = y$, $D^*(x, y) = D^*(x, x) = 0$. Let as assume that there exist $x_0, y_0 \in \mathbf{I}$ such that $x_0 \neq y_0$ and $D(x_0, y_0) = 0$. As, for all $x \in \mathbf{I}$, $D(x, \cdot)$ is a strictly increasing function, it holds that $D^*(x_0, y_0) > 0$ which contradicts our assumptions. Therefore D^* satisfies condition (i) in Definition 9.

Let $u, v, y \in \mathbf{I}$, such that $|u - y| \geq |v - y|$. Let us assume the following cases:

1. $u \leq v \leq y$. From the properties of a deviance function, we get that $D(u, y) \geq 0, D(v, y) \geq 0$. As we restrict ourselves to deviance function decreasing in the first coordinate, it holds

$$D^*(u, y) = |D(u, v)| = D(u, v) \geq D(v, y) = |D(v, y)| = D^*(v, y).$$

2. $u \leq y \leq v \leq 2y - u$. Analogously to the case above, $D(2y - u, y) \leq 0, D(v, y) \leq 0$ and $D(2y - u, y) \leq D(v, y)$. Then

$$D^*(u, y) = |D(u, y)| = |D(2y - u, y)| \geq |D(v, y)| = D^*(v, y).$$

3. $y \leq v \leq u$. In this case, we get that $D(2y - u, y) \geq 0, D(v, y) \geq 0$ and $D(2y - u, y) \geq D(v, y)$. Then

$$D^*(u, y) = |D(u, y)| = |D(2y - u, y)| = D(2y - u, y) \geq D(v, y) = |D(v, y)| = D^*(v, y).$$

4. $2y - u \leq v \leq y \leq u$. Finally, for $2y - u \leq v \leq y \leq u$, $D(u, y) \leq 0, D(v, y) \leq 0$ and $D(u, y) \leq D(v, y)$. Then

$$D^*(u, y) = |D(u, y)| \geq |D(v, y)| = D^*(v, y).$$

Therefore condition (ii) in Definition 9 is fulfilled.   □

Proposition 2. *Let L be a dissimilarity function. Then L is a local penalty function.*

Proof. From the boundary condition and the fact that f used to define L is a strictly monotone function we get that L satisfies condition (i) required by a local penalty function. The fact that L satisfies condition (ii) follows directly from the strict monotonicity of f and convexity and a unique minimum of K. □

Let us assume penalty function (8) and penalty based function (9) with dissimilarity function $L(x,y) = (f(x) - f(y))^2$ as a local penalty function. Without any problems, we can rewrite weighted penalty function as follows: $g\,P(y) = \sum_{i=1}^{n} g_i(x_i) \cdot (f(y) - f(x_i))^2$. The function f may be differentiable on an interval $\mathbf{I}$ but not necessarily.

(a) If a function f is not differentiable on $\mathbf{I}$, we can rewrite weighted penalty function as $g\,P(y) = \sum_{i=1}^{n} g_i(x_i) \cdot (f^2(x_i) - 2f(x_i) \cdot f(y) + f^2(y))$, hence it is convex quadratic function and has just one minimum.

(b) If a function f is differentiable on $\mathbf{I}$, we can express minimum as a solution of a derivative equal to zero. So, we have $\sum_{i=1}^{n} g_i(x_i) \cdot 2 \cdot (f(y) - f(x)) \cdot f'(y) = 0$. After small modification we can separate the expression $\sum_{i=1}^{n} g_i(x_i) \cdot (f(y) - f(x))$, which expresses strictly increasing function given by deviation function in the form $f(y) - f(x)$. As a solution of equality $\sum_{i=1}^{n} g_i(x_i) \cdot (f(y) - f(x)) = 0$ we give mixture function.

Obviously, the above mentioned idea works similarly for more general dissimilarity functions in the form $L(x,y) = (f(y) - f(x))^{2k}$ $k \in N$, $k > 1$. However, from the point of view of applications, using of the second power is sufficient in majority of cases.

3.1 Mixture Function Based on Special Dissimilarity and Deviation Functions

In this section we extend our approach related to use of dissimilarity functions which was studied in [11]. Mixture functions can be naturally introduced as solutions of optimization problems, when for each input vector $(x_1, \ldots, x_n) \in \mathbf{I}^n$ and a fixed local penalty function evaluating distances between points in $\mathbf{I}^2$, we look for an output $y^* \in \mathbf{I}$ with a minimal difference from the given input values.

Definition 15 [11]. *Let $LP : \mathbf{I}^2 \to R^+$ be a given local penalty function and let $\mathbf{g}$ be a vector of continuous weighting functions. The function $A_{\mathbf{g},LP}$ is defined by*

$$A_{\mathbf{g},LP}(x_1, \ldots, x_n) = y^*,$$

where

$$\sum_{i=1}^{n} g_i(x_i) \cdot LP(x_i, y^*) = \arg\min_{y} \left\{ \sum_{i=1}^{n} g_i(x_i) \cdot LP(x_i, y) \mid y \in \mathbf{I} \right\} \qquad (10)$$

or,
if there are more minimization solutions which represent a closed subinterval

$$[\underline{y}^*, \overline{y^*}] \subset \mathbf{I}$$

$$[\underline{y}^*, \overline{y^*}] = \left\{ t^* \in I \mid \sum_{i=1}^{n} g_i(x_i) \cdot LP(x_i, t^*) = \arg\min_y \left\{ \sum_{i=1}^{n} g_i(x_i) \cdot LP(x_i, y) \mid y \in \mathbf{I} \right\} \right\},$$

then $y^* = \frac{(\underline{y}^* + \overline{y^*})}{2}$.

Theorem 1 [11,16]. *For a given interval* $\mathbf{I} \subset [0, \infty[$, *let the dissimilarity function* $L : \mathbf{I}^2 \to \mathbf{R}^+$ *be given by* $L(x, y) = (f(x) - f(y))^2$ *and* $f : \mathbf{I} \to \mathbf{R}$ *be a continuous strictly monotone function. Then for a vector* $\mathbf{g}$ *of weighting functions and for a local penalty function* $LP = L$ *we recognize in Definition 15 the function* $A_{\mathbf{g}, LP}$ *as:*

1. *a mixture function* M_g *(1) linked to* $f = id$ *and* $g_1 = \ldots = g_n = g$;
2. *a generalized mixture function* $M_{\mathbf{g}}$ *(2) (Losonczi mean) linked to* $f = id$ *and* $\mathbf{g}$;
3. *an ordered generalized mixture function* $OM_{\mathbf{g}}$ *(3) linked to* $f = id$, $\mathbf{g}$ *and* $(x_{(1)}, \ldots, x_{(n)})$;
4. *a quasi-mixture function* M_g^f *(4) linked to* f *and* $g_1 = \ldots = g_n = g$;
5. *a generalized quasi-mixture function* $M_{\mathbf{g}}^f$ *(5) (Bajraktarević mean) linked to* f *and* $\mathbf{g}$;
6. *an ordered generalized quasi-mixture function* $OM_{\mathbf{g}}^f$ *(6) linked to* f, $\mathbf{g}$ *and* $(x_{(1)}, \ldots, x_{(n)})$.

As discussed at the beginning of this section, on the basis of Definitions 12 and 13, we can get a mixture function using a deviation function.

Theorem 2. *Let the deviation function* $D : \mathbf{I}^2 \to \mathbf{R}$ *be given by* $D(x, y) = f(y) - f(x)$, *where* $f : \mathbf{I} \to \mathbf{R}$ *be a continuous strictly monotone function. Then for a vector* $\mathbf{g}$ *of weighting functions as a solution of equation* $\sum_{i=1}^{n} g_i(x_i) \cdot (f(y) - f(x_i)) = 0$ *we recognize:*

1. *a mixture function* M_g *(1) linked to* $f = id$ *and* $g_1 = g_2 = \ldots = g_n = g$;
2. *a generalized mixture function* $M_{\mathbf{g}}$ *(2) (Losonczi mean) linked to* $f = id$ *and* $\mathbf{g}$;
3. *an ordered generalized mixture function* $OM_{\mathbf{g}}$ *(3) linked to* $f = id$, $\mathbf{g}$ *and* $(x_{(1)}, \ldots, x_{(n)})$;
4. *a quasi-mixture function* M_g^f *(4) linked to* f *and* $g_1 = \ldots = g_n = g$;
5. *a generalized quasi-mixture function* $M_{\mathbf{g}}^f$ *(5) (Bajraktarević mean) linked to* f *and* $\mathbf{g}$;
6. *an ordered generalized quasi-mixture function* $OM_{\mathbf{g}}^f$ *(6) linked to* f, $\mathbf{g}$ *and* $(x_{(1)}, \ldots, x_{(n)})$.

Proof. By solution of $\sum_{i=1}^{n} g_i(x_i) \cdot (f(y) - f(x)) = 0$, we obtain function given by (4).

Proof for other generalizations is obvious. □

3.2 Properties of Selected Quasi-mixture Function

As we mentioned, mixture functions need not be monotone in general. If we want
to select mixture functions as a special class of aggregation functions, we need
to know conditions under which they are monotone increasing. These conditions
were studied, for instance, in [9–13, 15].

For illustration, we present here one selected generalization of mixture function, namely, quasi mixture function M_g^f which is given by (4). We discuss this
on the interval $[0, 1]$ which is generated by $g(x) = x + l; l \geq 0$ and $f(x) = \log x$.
Therefore, we define the mentioned function using limits as follows: let us assume
weighting functions $g(x) = x + l$, $g_\epsilon(x) = x + l + \epsilon$ and transformation functions
$f(x) = \log x$ and $f_\epsilon(x) = \log x + \epsilon$. Then M_g^f function is well defined in the
form

$$
M_g^f(\mathbf{x}) = \begin{cases} M_g^f(\mathbf{x}); & \mathbf{x} \in]0, 1]^n; \\ \lim_{\epsilon \to 0} M_{g_\epsilon}^{f_\epsilon}(\mathbf{x}); & \mathbf{x} = \mathbf{0}. \end{cases} \tag{11}
$$

Proposition 3. *Let $M_g^f : [0, 1]^n \to [0, 1]$ be a function defined by (4), with an
increasing weighting function $g : [0, 1] \to]0, \infty[$ and $f = \log x$. Then M_g^f is
monotone increasing for x satisfying condition*

$$
g'(x) \cdot \log x + \frac{g(x)}{x} \geq 0. \tag{12}
$$

Proof. If we apply in (4) the function $f = \log x$, we obtain weighted geometric

mean in the form $M_g^f = \left(\prod_{i=1}^n x_i^{g(x_i)} \right)^{\frac{1}{\sum_{i=1}^n g(x_i)}}$. Without loss of generality, we

can rewrite the function as follows: $M_g^f = x_1^{g(x_1)} \cdot \left(\prod_{i=2}^n x_i^{g(x_i)} \right)^{\frac{1}{g(x_1) + \sum_{i=2}^n g(x_i)}} =$

$\exp\left\{ \frac{1}{g(x_1) + \sum_{i=2}^n g(x_i)} \cdot \left[\log x_1^{g(x_1)} + \log \left(\prod_{i=2}^n x_i^{g(x_i)} \right) \right] \right\}$ and from the first derivative

of M_g^f according to x_1 we get the condition

$$
g'(x_1) \cdot \log x_1 + \frac{g(x_1)}{x_1} \geq \frac{\log \prod_{i=1}^n x_i^{g(x_i)}}{\sum_{i=1}^n g(x_i)} \cdot g'(x_1).
$$

When we modify right-hand side of previous inequality (without $g'(x_1)$), we see
that maximal value on $[0, 1]$ is 0, hence our condition is simplified to the form

$$
g'(x_1) \cdot \log x_1 + \frac{g(x_1)}{x_1} \geq 0,
$$

whence we get condition (12). $\square$

Proposition 4. *Let $M_g^f : [0,1]^n \to [0,1]$ be a M_g^f function defined by (3) with the weighting function $g(x) = x + l$; $l \geq 0$ and the transformation function $f(x) = \log x$. Then M_g^f is monotone increasing for*

$$l \geq \exp\{-2\}. \tag{13}$$

Proof. W.r.t. condition (12) and $g(x) = x+l$ we solve the inequality $\ln x + \frac{x+l}{x} \geq 0$. The first derivative of the function on the left-hand side is $\frac{x-l}{x^2}$, whence we see that the function has a minimum at $x = l$. We get immediately condition for l in the form (13). $\qquad\square$

For illustration see Figs. 1 and 2.

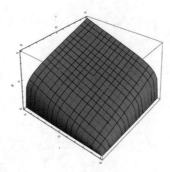

Fig. 1. *Quasi $- M_g$* (4) with $g(x) = x + 0.5$ and $f(x) = \log x$.

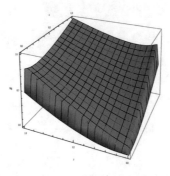

Fig. 2. *Quasi $- M_g$* (4), with $g(x) = x + 0.05$ and $f(x) = \log x$.

Remark 1. W.r.t. [15], let $M_g^f : [0,1]^n \to [0,1]$ be a quasi mixture function defined by (4). If M_g^f is monotone increasing, then M_{Bg}^f, with $B > 0$ is also monotone increasing.

Corollary 1. *Let $M_g^f : [0,1]^n \to [0,1]$ be a M_g^f function defined by (4) with the weighting function $g(x) = cx + 1 - c$; $c \in [0,1]$ and the transformation function $f(x) = \log x$. Then M_g^f is monotone increasing for*

$$0 \le c \le \frac{\exp\{2\}}{1 + \exp\{2\}}. \tag{14}$$

Proof. W.r.t. condition (12), Remark 1 and the proof of Proposition 4.   □

4   Conclusion

In the paper we summarized the overall view on the generating of mixture functions. Using the special deviation function and dissimilarity function in the weighted penalty function for solving an optimization problem, we can derive all generalizations of mixture functions. We also mentioned basic properties of mixture functions, especially the property of increasing monotonicity. In our future investigation, we plan to deeply study the relationships between deviation and dissimilarity functions.

Acknowledgement. Jana Špirková has been supported by the project VEGA no. 1/0093/17 Identification of risk factors and their impact on products of the insurance and savings schemes.

Pavol Kráľ has been supported by the project VEGA 1/0767/18 SMART model a decision support tool in management of enterprises.

References

1. Beliakov, G., Pradera, A., Calvo, T.: Aggregation Functions: A Guide for Practitioners. Springer, Berlin (2007)
2. Beliakov, G., Bustince Sola, H., Calvo Sánchez, T.: A Practical Guide to Averaging Functions. Springer, Berlin (2016)
3. Bustince, H., Beliakov, G., Dimuro, G.P., Bedregal, B., Mesiar, R.: On the definition of penalty functions in data aggregation. Fuzzy Sets Syst. **323**, 1–18 (2017)
4. Calvo, T., Beliakov, G.: Aggregation functions based on penalties. Fuzzy Sets Syst. **12**(1), 1420–1436 (2010)
5. Calvo, T., Kolesárová, A., Komorníková, M., Mesiar, R.: Aggregation operators: properties, classes and construction methods. In: Calvo, T., Mayor, G., Mesiar, R. (eds.) Aggregation Operators: New Trends and Applications, pp. 3–104. Physica, Heidelberg (2002)
6. Decký, M., Mesiar, R., Stupňanová, A.: Deviation-based aggregation functions. Fuzzy Sets Syst. **332**, 29–36 (2018)
7. Decký, M., Mesiar, R., Stupňanová, A.: Aggregation Functions Based on Deviation. In: Medina, J., et al. (eds.) IPMU 2018. CCIS, vol. 853, pp. 151–159. Springer, Heidelberg (2018)
8. Grabisch, J., Marichal, J., Mesiar, R., Pap, E.: Aggregation Functions. Cambridge University Press, Cambridge (2009)

9. Marques Pereira, R. A., Pasi, G.: On non-monotonic aggregation: mixture operators. In: Proceedings of the 4th Meeting of the EURO Working Group on Fuzzy Sets (EUROFUSE 1999) and 2nd International Conference on Soft and Intelligent Computing (SIC 1999), Budapest, Hungary, pp. 513–517 (1999)
10. Mesiar, R., Špirková, J.: Weighted means and weighting functions. Kybernetika **42**(2), 151–160 (2006)
11. Mesiar, R., Špirková, J., Vavríková, L.: Weighted aggregation operators based on minimization. Inf. Sci. **17**(4), 1133–1140 (2008)
12. Ribeiro, R.A., Marques Pereira, R.A.: Generalized mixture operators using weighting functions: a comparative study with WA and OWA. Eur. J. Oper. Res. **145**, 329–342 (2003)
13. Ribeiro, R.A., Marques Pereira, R.A.: Aggregation with generalized mixture operators using weighting functions. Fuzzy Sets Syst. **137**, 43–58 (2003)
14. Ribeiro, R.A., Marques Pereira, R.A.: Weights as functions of attribute satisfaction values. In: Proceedings of the Workshop on Preference Modelling and Applications (EUROFUSE), Granada, Spain, pp. 131–137 (2001)
15. Špirková, J., Beliakov, G., Bustince, H., Fernandez, J.: Mixture functions and their monotonicity. Inf. Sci. **481**, 520–549 (2019)
16. Špirková, J.: Weighted operators based on dissimilarity function. Inf. Sci. **281**, 172–181 (2014)
17. Wilkin, T., Beliakov, G.: Weakly monotonic averaging functions. Int. J. Intell. Syst. **30**(2), 144–169 (2014)

F-transform and Dimensionality Reduction: Common and Different

Jiří Janeček[1] and Irina Perfilieva[2(✉)]

[1] Department of Mathematics, University of Ostrava,
30. dubna 22, 701 03 Ostrava 1, Czech Republic
`jiri.janecek@osu.cz`
[2] IRAFM, University of Ostrava,
30. dubna 22, 701 03 Ostrava 1, Czech Republic
`irina.perfilieva@osu.cz`

Abstract. This contribution is focused on the connection between two theories: dimensionality reduction and the F-transform. We show how the graph Laplacian can be extracted from a fuzzy partition and how the Laplacian eigenmaps can be associated with the F-transform components in their functional forms. We analyse the spectrum of the graph Laplacian and its dependence on basic functions in a fuzzy partition. We support our theoretical results by numerical experiments.

Keywords: F-transform · Graph Laplacian · Dimensionality reduction

1 Introduction

One of the major focus of the current machine learning theory is the processing of big data. Usually, the data consists of a high number of measurements on a large group of objects. Further analysis requires simplifying of the data into a lower-dimensional representation. Let us briefly describe the general dimensionality reduction problem assuming that all variables are real.

Suppose that the points $x_1, \ldots, x_{l_k} \in \mathbb{R}^l$ (the input data, $l > 1$) are given. The aim of the dimensionality reduction problem is to map them onto points $y_1, \ldots, y_{l_k} \in \mathbb{R}^m$ ($l > m$) in a way that preserves the most important features.

In general, the aim of the dimensionality reduction problem is to find the images of high-dimensional data points (after linear or nonlinear transformation) in a lower-dimensional representation in order to preprocess (simplify) original data before further analysis (e.g. regression, classification, clustering) or visualization. The reduction can be performed in many ways, e.g., in a form of a projection on a lower-dimensional subspace (e.g. manifold) embedded in the original space. The projection can be made in different directions so that we are finding the most suitable one such that the resulting reconstruction error (information loss) is minimal.

There are many techniques that can be used, because there is no unique way how to reduce the dimensionality and because there can be different requirements

© Springer Nature Switzerland AG 2019
R. Halaš et al. (Eds.): AGOP 2019, AISC 981, pp. 267–278, 2019.
https://doi.org/10.1007/978-3-030-19494-9_25

on the result. The dimensionality reduction methods include clustering, classification, segmentation, discriminant analysis, graph-based approaches, PCA, factor analysis, autoencoders etc.

The approach proposed in [1] emphasizes the importance of the neighborhood relations and is aimed at preserving the local information rather than trying to include everything. On the basis of the mutual distances of the initial data points the adjacency graph is created. The eigenvectors of the corresponding Laplacian matrix are then used for the embedding (mapping on the lower-dimensional subspace).

The concept of the dimensionality reduction in the soft computing sense contains a large spectrum of methods and usages (e.g. clustering of the data or fuzzy IF-THEN rules creation) including the fuzzy transform (F-transform) which also agrees with the machine learning approach (granulation of the data).

The F-transform is based on a fuzzy partition formed by units that characterize closeness among the data points. Each partition unit generates one F-transform component. The components are weighted projections on polynomials with weights given by the units.

The main contribution of this paper consists in providing technical verification of some non-supported claims in [1,4]. The definition of principal notions and explanation of the problem is taken from [1,4].

2   Preliminaries

This section summarizes important properties of the graph-based approach to the dimensionality reduction and of F-transform.

2.1   Laplacian Eigenmaps for Dimensionality Reduction

In this section, we remind some important details regarding the weighted graph-based approach to the dimensionality reduction, originated from [1].

Firstly, the mentioned approach establishes which data points ($\boldsymbol{x}_1, \ldots, \boldsymbol{x}_k \in \mathbb{R}^l$) are close enough to be connected in a weighted graph $G = (V, E)$. Each data point corresponds to one vertex, so that $|V| = k$. The edges between vertices corresponding to close data points are assigned positive weights, whereas others are assigned zeros (the corresponding vertices are not connected). This setting ensures the locality-preserving character of the graph representation. The lower-dimensional images ($\boldsymbol{y}_1, \ldots, \boldsymbol{y}_k \in \mathbb{R}^m$, $m \ll l$) of the original data points will be determined by the generalized eigenvectors of the graph Laplacian.

We assume that the graph G is connected (consists of one connected component), otherwise we treat each component in the same manner separately.

For simplicity, we consider the reduction mapping onto $\mathbb{R}^1$, i.e., the case $m = 1$. The algorithm [1,4] runs as follows:

Step 1 (constructing the adjacency graph). We put an edge between nodes i and j, if $\boldsymbol{x}_i$ and $\boldsymbol{x}_j$ are "close".

Step 2 (choosing the weights). In [1], the two options for weighting the edges were considered:

(a) *Heat kernel.* If vertices i and j are connected, then

$$w_{ij} = \mathrm{e}^{-\frac{\|\boldsymbol{x}_i - \boldsymbol{x}_j\|^2}{t}},$$

otherwise $w_{ij} = 0$. Parameter $t \in \mathbb{R}$ should be specified beforehand.

(b) *Simple-minded assignment.* $w_{ij} = 1$, if vertices i and j are connected by an edge, and $w_{ij} = 0$, otherwise.

Step 3 (eigenmaps). Compute generalized eigenvalues and eigenvectors for the Laplacian matrix L in accordance with

$$L\boldsymbol{f} = \lambda D \boldsymbol{f}, \tag{1}$$

where D is the diagonal matrix such that

$$d_{ii} = \sum_{j=1}^{k} w_{ij},$$

and $L = D - W$. Below, we will show that the matrix L is symmetric and positive semidefinite.

Let $\boldsymbol{f}_0, \ldots, \boldsymbol{f}_{k-1}$ be solutions of Eq. (1), ordered according to their eigenvalues, that is $\lambda_0 \leq \lambda_1 \leq \ldots \leq \lambda_{k-1}$. If we leave out the eigenvector $\boldsymbol{f}_0$ corresponding to the eigenvalue 0 and use the eigenvector $\boldsymbol{f}_1$ corresponding to the second smallest eigenvalue, then the embedding in 1-dimensional Euclidean space is given by

$$\boldsymbol{x}_i \mapsto \boldsymbol{f}_1(i).$$

We remark that by preserving local information in the embedding, the algorithm naturally emphasizes the data clusters.

In order to justify the proposed algorithm, we show that the matrix L is symmetric and positive semidefinite. The first property is obvious, and to prove the second one, we choose an arbitrary vector $\boldsymbol{y} = (y_1, \ldots, y_k)^\top$ and show that

$$\boldsymbol{y}^\top L \boldsymbol{y} = \frac{1}{2} \sum_{i,j=1}^{k} (y_i - y_j)^2 w_{ij}. \tag{2}$$

Proof.

$$\forall \boldsymbol{y} \in \mathbb{R}^k : \boldsymbol{y}^\top L \boldsymbol{y} = \boldsymbol{y}^\top (D - W) \boldsymbol{y} = \boldsymbol{y}^\top D \boldsymbol{y} - \boldsymbol{y}^\top W \boldsymbol{y}$$

$$= \frac{1}{2}(2\boldsymbol{y}^\top D \boldsymbol{y} - 2\boldsymbol{y}^\top W \boldsymbol{y}) = \frac{1}{2}\left(\sum_{i=1}^{k} y_i^2 d_{ii} + \sum_{j=1}^{k} y_j^2 d_{jj} - 2\sum_{i,j=1}^{k} y_i w_{ij} y_j\right)$$

$$= \frac{1}{2}\left[\sum_{i=1}^{k}\left(y_i^2 \sum_{j=1}^{k} w_{ij}\right) + \sum_{j=1}^{k}\left(y_j^2 \sum_{i=1}^{k} w_{ij}\right) - 2\sum_{i,j=1}^{k} y_i y_j w_{ij}\right]$$

$$= \frac{1}{2}\sum_{i,j=1}^{k}\left(y_i^2 w_{ij} + y_j^2 w_{ij} - 2y_i y_j w_{ij}\right) = \frac{1}{2}\sum_{i,j=1}^{k}\left[(y_i^2 + y_j^2 - 2y_i y_j)w_{ij}\right]$$

$$= \frac{1}{2}\sum_{i,j=1}^{k}\left(y_i^2 w_{ij} + y_j^2 w_{ij} - 2y_i y_j w_{ij}\right) = \frac{1}{2}\sum_{i,j=1}^{k}\left[(y_i - y_j)^2 w_{ij}\right] \geq 0.$$

$\square$

It is known that vector $\boldsymbol{y}$ that minimizes $\boldsymbol{y}^\top L \boldsymbol{y}$ corresponds to the minimal eigenvalue solution to the generalized eigenvalue problem (1). The latter follows from

$$L\boldsymbol{y} = \lambda D \boldsymbol{y} \Rightarrow \boldsymbol{y}^\top L \boldsymbol{y} = \lambda \boldsymbol{y}^\top D \boldsymbol{y} \Rightarrow \min_{\boldsymbol{y}(\lambda)} \boldsymbol{y}^\top L \boldsymbol{y} = \min_{\boldsymbol{y}(\lambda)} \lambda \boldsymbol{y}^\top D \boldsymbol{y}.$$

Because any eigenvector is determined up to any nonzero multiple, the normalization requirement $\boldsymbol{y}^\top D \boldsymbol{y} = 1$ was proposed in [1]. Finally, the problem reduces to

$$\min_{\boldsymbol{y}^\top D \boldsymbol{y} = 1} \boldsymbol{y}^\top L \boldsymbol{y},$$

and the solution is determined by the minimal (normalized) generalized eigenvalue.

By (2), this minimum eigenvalue solution minimizes the following objective function:

$$\sum_{i,j=1}^{k} (y_i - y_j)^2 w_{ij}. \tag{3}$$

This leads to the expected property of the solution: the local closeness of initial data points is kept.

By [1], the minimal eigenvalue is always 0, and if the graph G is connected, then $\boldsymbol{y} = \boldsymbol{1}$ is the only normalized eigenvector for $\lambda = 0$.

In the case $m > 1$ (embedding into a higher-dimensional subspace) the algorithm remains the same and the locality-preserving map of $\boldsymbol{x_1}, \ldots, \boldsymbol{x_k} \in \mathbb{R}^l$ to the m-dimensional Euclidean space gives a matrix $Y \in \mathbb{R}^{k \times m}$, such that each row represents the image of the corresponding initial data point.

2.2   F-transforms

The substantial characterization of the technique of F-transform (originally, *fuzzy transform*, [2]) is that it is an integral transform that uses a *fuzzy partition* of a universe of discourse (usually, a bounded interval of reals $[a, b]$). The F-transform has two phases: direct and inverse. The direct F-transform (FzT) is applied to functions from $L^2([a, b])$ and maps them linearly onto (finite) sequences of numeric/functional components. Each component is a weighted orthogonal projection of a given function on a certain linear subspace of $L^2([a, b])$. Dimensionality reduction by the F-transform is thought as a projection of original data onto the set of the F-transform components, see Sect. 3 for the details.

Below, we recall some definitions from [3].

Fuzzy Partition. Let $[a, b]$ be an interval on the real line $\mathbb{R}$. Fuzzy sets on $[a, b]$ are identified by their membership functions; i.e., they are mappings from $[a, b]$ into $[0, 1]$.

Definition 1. *Let $[a, b]$ be an interval on $\mathbb{R}$, $n \geq 2$, and let $x_0, x_1, \ldots, x_n, x_{n+1}$ be nodes such that $a = x_0 \leq x_1 < \ldots < x_n \leq x_{n+1} = b$. We say that fuzzy sets $A_1, \ldots, A_n : [a, b] \to [0, 1]$, which are identified with their membership functions, constitute a* fuzzy partition *of $[a, b]$ if for $k = 1, \ldots, n$, if they fulfill the following conditions:*

1. *(locality) - $A_k(x) - 0$ if $x \in [a, x_{k-1}] \cup [x_{k+1}, b]$,*
2. *(continuity) - $A_k(x)$ is continuous,*
3. *(covering) - $A_k(x) > 0$ if $x \in (x_{k-1}, x_{k+1})$.*

The membership functions $A_1, \ldots, A_n$ are called basic functions.

Hilbert Space with Weighted Inner Product. Let us fix $[a, b]$ and its fuzzy partition $A_1, \ldots, A_n$, $n \geq 2$. Let k be a fixed integer from $\{1, \ldots, n\}$ and let $L^2([a, b])$ denote a set of square-integrable functions on $[a, b]$. Let us denote

$$s_k = \int_{x_{k-1}}^{x_{k+1}} A_k(x)dx,$$

and consider $\frac{1}{s_k} \int_{x_{k-1}}^{x_{k+1}} f(x)g(x)A_k(x)dx$ as Lebesgue integral $\int_{x_{k-1}}^{x_{k+1}} f(x)g(x)d\mu_k$, where $d\mu = A_k(x)dx/s_k$ and the measure μ_k on $[x_{k-1}, x_{k+1}]$ is defined as follows:

$$\mu_k(E) = \frac{\int_E A_k(x)dx}{\int_{x_{k-1}}^{x_{k+1}} A_k(x)dx}.$$

Let $L^2(A_k)$ be a set of square-integrable functions on $[x_{k-1}, x_{k+1}]$ with the weighted inner product

$$\langle f, g \rangle_k = \int_{x_{k-1}}^{x_{k+1}} f(x)g(x)A_k(x)dx. \tag{4}$$

The functions $f, g \in L^2(A_k)$ are *orthogonal* in $L^2(A_k)$ if $\langle f, g \rangle_k = 0$. The function $f \in L^2(A_k)$ is orthogonal to a subspace B of $L^2(A_k)$ if $\langle f, g \rangle_k = 0$ for all $g \in B$.

Let us denote by $L_2^m(A_k)$ a linear subspace of $L^2(A_k)$ with the basis given by orthogonal functions $P_k^0, P_k^1, P_k^2 \ldots, P_k^m$ where $P_k^0(x) = 1$, $x \in [x_{k-1}, x_{k+1}]$.

The following lemma gives an analytic representation of the orthogonal projection on the subspace $L_2^m(A_k)$.

Lemma 1. *([3]) Let function F_k^m be the orthogonal projection of $f \in L^2(A_k)$ on $L_2^m(A_k)$. Then,*

$$F_k^m = c_{k,0} P_k^0 + c_{k,1} P_k^1 + \cdots + c_{k,m} P_k^m, \tag{5}$$

where for all $i = 0, 1, \ldots, m$,

$$c_{k,i} = \frac{\langle f, P_k^i \rangle_k}{\langle P_k^i, P_k^i \rangle_k} = \frac{\int_{x_{k-1}}^{x_{k+1}} f(x) P_k^i(x) A_k(x) dx}{\int_{x_{k-1}}^{x_{k+1}} P_k^i(x) P_k^i(x) A_k(x) dx}. \tag{6}$$

The n-tuple $(F_1^m, \ldots, F_n^m)$ is an F^m-transform of f with respect to $A_1, \ldots, A_n$, or formally,

$$F^m[f] = (F_1^m, \ldots, F_n^m).$$

F_k^m is called the kth F^m-transform component of f.

In particular, let us consider the case where the basis of $L_2^m(A_k)$ is given by orthogonal polynomials $P_k^0, P_k^1, P_k^2, \ldots, P_k^m$ and P_k^0 is a constant function with the value 1. Then, the F^0-transform of f or simply, the F-transform of f with respect to the partition $A_1, \ldots, A_n$ is given by the n-tuple $(c_{1,0}, \ldots, c_{n,0})$ of constant functions (0-degree polynomials) where for $k = 1, \ldots, n$,

$$c_{k,0} = \frac{\langle f, 1 \rangle_k}{\langle 1, 1 \rangle_k} = \frac{\int_{x_{k-1}}^{x_{k+1}} f(x) A_k(x) dx}{\int_{x_{k-1}}^{x_{k+1}} A_k(x) dx}. \tag{7}$$

The F^1-transform of f with respect to $A_1, \ldots, A_n$ is given by the n-tuple $(c_{1,0} + c_{1,1}(x - x_1), \ldots, c_{n,0} + c_{n,1}(x - x_n))$ of linear functions (1-degree polynomials). The latter are fully represented by their 2D coefficients $((c_{1,0}, c_{1,1}), \ldots, (c_{n,0}, c_{n,1}))$, which in addition to (7), have the following particular representation:

$$c_{k,1} = \frac{\langle f, x - x_k \rangle_k}{\langle (x - x_k), (x - x_k) \rangle_k} = \frac{\int_{x_{k-1}}^{x_{k+1}} f(x)(x - x_k) A_k(x) dx}{\int_{x_{k-1}}^{x_{k+1}} (x - x_k)^2 A_k(x) dx}. \tag{8}$$

The *inverse F^m-transform* of function f with respect to the partition $A_1, \ldots, A_n$ is a function represented by the following *inversion formula*:

$$f_{F,n}^m(x) = \sum_{k=1}^{n} F_k^m A_k(x). \tag{9}$$

The following results demonstrate approximation properties of the direct and inverse F^m-transforms.

Lemma 2. *([3]) Let $m \geq 0$, and let functions F_k^m and F_k^{m+1} be the k-th F^m- and F^{m+1}-transform components of f, respectively. Then,*

$$\|f - F_k^{m+1}\| \leq \|f - F_k^m\|.$$

Theorem 1. *([2]) Let f be a continuous function on $[a,b]$. For every $\varepsilon > 0$, there exist an integer n_ε and the related fuzzy partition $A_1, \ldots, A_{n_\varepsilon}$ of $[a,b]$ such that for all $x \in [a,b]$,*

$$|f(x) - f_{F,n_\varepsilon}(x)| < \varepsilon,$$

where f_{F,n_ε} is the inverse F-transform of f with respect to $A_1, \ldots, A_{n_\varepsilon}$.

The discrete F-transforms were introduced in [2].

3 Dimensionality Reduction by the F-transform

In this section, we show that the F-transform technique can be reinterpreted within the framework of dimensionality-reduction based on intrinsic geometry of data. In more detail, we will show that the F^0- and the F^1-transform are solutions of Eq. (1), provided that the Laplacian matrix L is properly constructed. This fact confirms that these F-transforms provide with a low-dimensional representation of a given dataset.

At first, we specify a dataset that will be characterized by the F-transform-based low-dimensional representation in the form of embedding maps and finally, by components. This dataset will be connected with a discrete representation of a function, say f on some domain (subset of the Euclidean $\mathbb{R}^l$). This function can be a signal, time series, image, etc. For simplicity, we assume that the domain is an interval $[a,b]$ of the real line, and the function f is given on a discrete set, say P of points where $P \subseteq [a,b]$. Then, we assume that $[a,b]$ is partitioned into the collection of fuzzy sets $A_1, \ldots, A_n$ as it is described in Definition 1. Moreover, we assume that for every k, $1 \leq k \leq n$, there is one point $x_k \in P$ (we call it *node*) such that $A_k(x_k) = 1$ and $A_j(x_k) = 0$, $j \neq k$. Further on, we distinguish between points (denoted by using letter p) and nodes (denoted by using letter x).

The chosen partition determines a geometry of the set P, given by the following directed weighted graph $D = (V(D), A(D))$. Each vertex from $V(D)$ corresponds to one point in P, and if a point $p_i \in P$ is *covered by basic function* A_k, i.e. $A_k(p_i) > 0$, then the vertex i corresponding to p_i is connected by the directed edge (k,i) with the vertex k, corresponding to the node x_k. Let us emphasize that there is no connection between "pure" points, if there is no node among them. The weight of the directed edge (k,i) is equal to $A_k(p_i)$. It is easy to see that graph D is covered by n weakly connected components $D_1, \ldots, D_n$, each has a "star" shape. In more detail, $D_k = (V(D_k), A(D_k))$ is a subgraph of D associated with a "source" vertex k. Then, the vertex set $V(D_k)$ contains vertex k and moreover, vertex i belongs to $V(D_k)$, if $(k,i) \in A(D_k)$. Therefore, the set of directed edges $A(D_k)$ consists of the edges (k,i) where $i \in V(D_k)$

containing also the loop (k, k). We will continue with each connected component D_k separately, and construct the low-dimensional representation of the part of the dataset corresponding to it (covered by the A_k).

Let a finite set of points $\{p_1, \ldots, p_{l_k}\}$ be covered by A_k (a fuzzy partition unit). Let us remark that the node x_k is among these points. Let $f_i = f(p_i)$, $1 \le i \le l_k$. The set X_k of data points allocated for the low-dimensional representation is $\{(p_1, f_1), \ldots, (p_{l_k}, f_{l_k})\}$. Let D_k be the corresponding directed subgraph of D and $G_k = (V(G_k), E(G_k))$ be the corresponding ordinary weighted graph derived from D_k after removing edge directions. Therefore, $V(G_k) = V(D_k)$, and if $(k, i) \in A(D_k)$, then $\{k, i\} \in E(G_k)$. In the weight matrix W corresponding to G_k, each edge, connecting i and k, is represented by the two symmetrical elements $w_{ik} = w_{ki} = A_k(p_i)$, the loop is represented by one diagonal entry $w_{kk} = A_k(x_k) = 1$. Therefore, W is a symmetrical $l_k \times l_k$ matrix. Then, we proceed with *Step 3* of the algorithm described in Sect. 2.

Proposition 1. *If the adjacency graph is constructed in the way proposed above, then the multiplicity of the generalized eigenvalue 1 of the graph Laplacian is* $l_k - 2$.

Proof. By the assumption

$$
W = \begin{bmatrix}
0 & 0 & \cdots & 0 & A_k(p_1) & 0 & \cdots & 0 \\
0 & 0 & \cdots & 0 & A_k(p_2) & 0 & \cdots & 0 \\
\vdots & \vdots & & \vdots & \vdots & \vdots & & \vdots \\
0 & 0 & \cdots & 0 & A_k(p_{k-1}) & 0 & \cdots & 0 \\
A_k(p_1) & A_k(p_2) & \cdots & A_k(p_{k-1}) & 1 & A_k(p_{k+1}) & \cdots & A_k(p_{l_k}) \\
0 & 0 & \cdots & 0 & A_k(p_{k+1}) & 0 & \cdots & 0 \\
\vdots & \vdots & & \vdots & \vdots & \vdots & & \vdots \\
0 & 0 & \cdots & 0 & A_k(p_{l_k}) & 0 & \cdots & 0
\end{bmatrix},
$$

$$
L = D - W = \begin{bmatrix}
A_k(p_1) & 0 & 0 \cdots 0 & -A_k(p_1) & 0 \cdots 0 & 0 \\
0 & A_k(p_2) & 0 \cdots 0 & -A_k(p_2) & 0 \cdots 0 & 0 \\
& & & \vdots & & \\
-A_k(p_1) & -A_k(p_2) & \cdots & \sum_{j=1}^{l_k} A_k(p_j) - 1 & \cdots & -A_k(p_{l_k}) \\
& & & \vdots & & \\
0 & 0 & 0 \cdots 0 & -A_k(p_{l_k}) & 0 \cdots 0 & A_k(p_{l_k})
\end{bmatrix}
$$

and

$$
L\boldsymbol{y} = \lambda D\boldsymbol{y} \Rightarrow (L - \lambda D)\boldsymbol{y} = 0 \Rightarrow \det(L - \lambda D) = 0.
$$

$$L - \lambda D =$$

$$= \begin{bmatrix} (1-\lambda)A_k(p_1) & 0 & 0 \ldots 0 & -A_k(p_1) & 0 \ldots 0 & 0 \\ 0 & (1-\lambda)A_k(p_2) & 0 \ldots 0 & -A_k(p_2) & 0 \ldots 0 & 0 \\ & & & \vdots & & \\ -A_k(p_1) & -A_k(p_2) & \ldots & (1-\lambda)s_k - 1 & \ldots & -A_k(p_{l_k}) \\ & & & \vdots & & \\ 0 & 0 & 0 \ldots 0 & -A_k(p_{l_k}) & 0 \ldots 0 & (1-\lambda)A_k(p_{l_k}) \end{bmatrix},$$

where

$$s_k = \sum_{j=1}^{l_k} A_k(p_j).$$

Let $M_{0,\ldots,h;0,\ldots,h}$ denote the minor of the matrix $L - \lambda D$ (the determinant of its square submatrix such that the columns and rows $0,\ldots,h$ are deleted). Then, expanding the determinant along the first rows, for all $h = 0,\ldots,k-2$, we have:

$$M_{0,\ldots,h;0,\ldots,h} = (1 - \lambda)A_k(p_{h+1})$$

$$\cdot \left[M_{0,\ldots,h+1;0,\ldots,h+1} - (1-\lambda)^{l_k-3-h} A_k(p_{h+1}) \prod_{\substack{j=h+2, \\ j \neq k}}^{l_k} A_k(p_j) \right]$$

and

$$M_{0,\ldots,k-1;0,\ldots,k-1} = [(1-\lambda)s_k - 1](1-\lambda)^{l_k-k} \prod_{j=k+1}^{l_k} A_k(p_j)$$

$$- (1-\lambda)^{l_k-k-1} \sum_{i=k+1}^{l_k} A_k^2(p_i) \prod_{\substack{j=k+1, \\ j \neq i}}^{l_k} A_k(p_j),$$

where

$$s_k = \sum_{j=1}^{l_k} A_k(p_j).$$

Therefore, $\det(L - \lambda D)$ can be rewritten in the form

$$(1-\lambda)^{l_k-2} \cdot R(\lambda), \tag{10}$$

which implies that there are $l_k - 2$ generalized eigenvalues equal to 1.   □

The symmetry of the matrix L ensures that the geometric multiplicity of the eigenvalue 1 is also $l_k - 2$. Assuming $l_k > 3$, there is no unique generalized eigenvector corresponding to the second smallest generalized eigenvalue. The Laplacian matrix of a connected graph has also one generalized eigenvalue equal to 0, so besides 0 and 1, there is only one positive generalized eigenvalue left.

The formula (10) shows that $l_k - 1$ generalized eigenvalues are independent of the choice of the basic function A_k.

The minimal eigenvalue solution to the generalized eigenvalue problem (1) is a constant vector that corresponds to the zero eigenvalue. This solution minimizes the objective function $y^\top L y$ (see explanations in Sect. 2.1). To have a non-trivial specification of the eigenvector corresponding to 0, we put additional constraint to this minimization problem and consider

$$f_0 = \operatorname*{argmin}_{\langle f, 1\rangle_k = (Dy)_k} y^\top L y, \tag{11}$$

where $f = (f_1, \ldots, f_{l_k})$, $\langle \cdot, \cdot \rangle_k$ is the weighted inner product in $l^2(A_k)$ (discrete case of (4)) and $(\cdot)_k$ is the k-th vector component. Then, f_0 is a constant vector, whose components are equal to

$$(f_0)_1 = \ldots (f_0)_{l_k} = \frac{\sum_{j=1}^{l_k} A_k(p_j) f_j}{\sum_{j=1}^{l_k} A_k(p_j)}. \tag{12}$$

The proof of (12) is below.

$$\langle f, 1\rangle_k = \sum_{j=1}^{l_k} f(p_j) \cdot 1 \cdot A_k(p_j)$$

$$(Dy)_k = d_{kk} \cdot y_k = \sum_{j=1}^{l_k} w_{kj} \cdot y_k = \sum_{j=1}^{l_k} A_k(p_j) \cdot y_k$$

$$\sum_{j=1}^{l_k} f(p_j) A_k(p_j) = \sum_{j=1}^{l_k} A_k(p_j) \cdot y_k \Rightarrow y_k = \frac{\sum_{j=1}^{l_k} A_k(p_j) f(p_j)}{\sum_{j=1}^{l_k} A_k(p_j)}$$

This means that the k-th coordinate of the vector y is equal to the claimed value. It remains to show that all other coordinates are the same. By

$$\lambda_0 = 0 \Rightarrow \lambda_0 D y' = 0 \Rightarrow L y' = 0 \Rightarrow 0 = y'^\top L y' \leq y^\top L y \quad \forall y \in \mathbb{R}^{l_k},$$

and because L is positive semi-definite, we deduce that there exists vector $y' \neq 0$ minimizing the quadratic form with L. The connectedness of the subgraph corresponding to A_k ensures that the minimal generalized eigenvalue $\lambda_0 = 0$ has the multiplicity one. Moreover, the equalities $Ly = \lambda_0 Dy = 0$ and $L1 = 0$ (because L is singular and $L1$ is equal to the sum of columns of L) show that the vector $\mathbf{1}$ is a solution to the generalized eigenvalue problem. Hence, all coordinates of $\mathbf{f}_0$ are the same and

$$\mathbf{f}_0 = \mathbf{y}' \cdot \frac{\sum\limits_{j=1}^{l_k} A_k(p_j) f(p_j)}{\|\mathbf{y}'\| \sum\limits_{j=1}^{l_k} A_k(p_j)}.$$

$\square$

This solution agrees with the discrete version of the k-th F^0-transform component in (7). Moreover, it can be interpreted as a weighted projection of $\mathbf{f}$ on the constant vector $\mathbf{1}$.

To obtain a non-constant vector solution to the minimization of $\mathbf{y}^\top L \mathbf{y}$, let us put the constraint of orthogonality and consider

$$\mathbf{y}_1 = \operatorname*{argmin}_{\langle \mathbf{y}, \mathbf{1} \rangle_k = 0} \mathbf{y}^\top L \mathbf{y}. \tag{13}$$

Then, the k-th F^1-transform component $\mathbf{f}_1$ has the following representation

$$\mathbf{f}_1 = \mathbf{f}_0 + \frac{\langle \mathbf{y}_1, \mathbf{f} \rangle_k}{\langle \mathbf{y}_1, \mathbf{y}_1 \rangle_k} \mathbf{y}_1.$$

This representation agrees with (5) and (6).

To conclude, the vectorial F^0- and F^1-transform components, respectively, $\mathbf{f}_0$ and $\mathbf{f}_1$, provide with a low-dimensional representation of the considered data set X_k. In particular, every (x_i, f_i) from X_k is represented by $(\mathbf{f}_{0,i}, \mathbf{f}_{1,i})$.

3.1 Numerical Experiments

In order to support and verify the presented results, we created small datasets and computed the generalized eigenvalues and eigenvectors of the corresponding Laplacian matrices. In those computations we used h-uniform fuzzy partitions with generating functions $A_{01}(x) = 1 - |x|$ and $A_{02}(x) = 1 - x^2$. The basic functions, being shifted and scaled copies, satisfy

$$A_k(x) = A_{0i}\left(\frac{x - x_k}{h}\right),$$

where $h = x_{k+1} - x_k$ is the uniform distance between nodes. For both types of datasets (with symmetric and non-symmetric distributions of the original data points) the spectrum consists of three values: 0, 1 and $\lambda > 1$. The multiplicities of the generalized eigenvalues are equal to those presented in Proposition 1. For example, if the data points $\left\{-\frac{2}{3}, -\frac{1}{3}, 0, \frac{1}{3}, \frac{2}{3}\right\}$ are covered by the basic function $A_{02}(x) = 1 - x^2$ (with no shift or scaling), the generalized eigenvalues of resulting Laplacian matrix

$$L = \begin{bmatrix} \frac{5}{9} & 0 & -\frac{5}{9} & 0 & 0 \\ 0 & \frac{8}{9} & -\frac{8}{9} & 0 & 0 \\ -\frac{5}{9} & -\frac{8}{9} & \frac{26}{9} & -\frac{8}{9} & -\frac{5}{9} \\ 0 & 0 & -\frac{8}{9} & \frac{8}{9} & 0 \\ 0 & 0 & -\frac{5}{9} & 0 & \frac{5}{9} \end{bmatrix},$$

are 0, 1, 1, 1 and 1.743 (after rounding up) which supports our claim.

4   Conclusion

In this contribution, we summarized the techniques of weighted graph-based dimensionality reduction [1] and F-transforms [4] and their distinctive properties. We showed the differences and similarities between these two reduction techniques. Moreover, we gave theoretical justification to some claims in [4].

Acknowledgement. The work of Irina Perfilieva has been partially supported by the project "LQ1602 IT4Innovations excellence in science" and by the Grant Agency of the Czech Republic (project No. 18-06915S).

References

1. Belkin, M., Niyogi, P.: Laplacian eigenmaps for dimensionality reduction and data representation. Neural Comput. **15**(6), 1373–1396 (2003)
2. Perfilieva, I.: Fuzzy transform: theory and application. Fuzzy Sets Syst. **157**, 993–1023 (2006)
3. Perfilieva, I., Daňková, M., Bede, B.: Towards a higher degree F-transform. Fuzzy Sets Syst. **180**, 3–19 (2011)
4. Perfilieva, I.: Dimensionality reduction by fuzzy transforms with applications to mathematical finance. In: Anh, L., Dong, L., Kreinovich, V., Thach, N. (eds.) Econometrics for Financial Applications: ECONVN 2018. Studies in Computational Intelligence, vol. 760. Springer, Cham (2018)

Note on Aggregation Functions and Concept Forming Operators

Peter Butka[1]([⊠]), Jozef Pócs[2,3], and Jana Pócsová[4]

[1] Department of Cybernetics and Artificial Intelligence,
Faculty of Electrical Engineering and Informatics, Technical University of Košice,
Letná 9, 04200 Košice, Slovakia
peter.butka@tuke.sk
[2] Department of Algebra and Geometry, Faculty of Science,
Palacký University Olomouc, 17. listopadu 12, 771 46 Olomouc, Czech Republic
[3] Mathematical Institute, Slovak Academy of Sciences,
Grešákova 6, 040 01 Košice, Slovakia
pocs@saske.sk
[4] Institute of Control and Informatization of Production Processes, BERG Faculty,
Technical University of Košice, Boženy Němcovej 3, 043 84 Košice, Slovakia
jana.pocsova@tuke.sk

Abstract. A certain connection between the theory of aggregation functions and the theory of concept lattices is discussed. We describe a generalization of residuated mappings, convenient for creating a monotone analogue of antitone concept lattices. Examples of such mappings are also presented.

Keywords: Residuated mappings · Concept lattice ·
Closure operator · Interior operator · Formal Concept Analysis

1 Introduction

Nowadays there is an evident emergence of usage of various mathematical tools in many applied areas of information sciences such as data mining, information retrieval, knowledge discovery, decision making, etc. Among several theoretical tools applied within the mentioned areas, a very important role is played by aggregation functions. The aggregation functions were originally introduced to act on real intervals, however recently the aggregation on posets, and in particular on lattices, has became a rapidly growing topic, especially due to the fact that these algebraic structures are involved in representation of various data structures.

Such example of application represent the so-called Formal Concept Analysis (FCA), where information contained in data table is represented in the form of a concept lattice. Mathematically see [10], FCA is based on the notion of a formal context which is represented by two sets, objects and attributes, and by a binary relation between the set of objects and the set of attributes representing

© Springer Nature Switzerland AG 2019
R. Halaš et al. (Eds.): AGOP 2019, AISC 981, pp. 279–288, 2019.
https://doi.org/10.1007/978-3-030-19494-9_26

the relationship between them. From a formal context all conceptual abstractions (concepts) combining sets of individuals (objects) with the sets of shared properties (attributes) are extracted using concept forming operators. The set of all concepts with respect to a hierarchical order forms a complete lattice, commonly known as a concept lattice.

Classical FCA method is suitable for hierarchical analysis of crisp binary data tables. However, in practice, there are natural examples od object-attribute models where the usage of fuzzy values is more appropriate. Therefore, several attempts to fuzzify FCA have been proposed. Briefly, these attempts can be characterized as a search for concept forming operators whether in antitone or monotone setting, e.g., [1–4, 6, 13–18].

The main aim of this paper is to point out on a possibility to include aggregation functions into a definition of concept forming operators. As aggregation functions are monotone, we consider a generalization of monotone fuzzy concept lattices which are defined via residuated pairs of mappings. For this we introduce the so-called w-residuated mappings and show that their fixed points form a complete lattice. Also we provide an example of such w-residuated mappings, which are composed of residuated mappings and interior and closure operator respectively.

2   Aggregation Functions, Residuated Mappings, Concept Lattices

We assume that the reader is familiar with the basic notions concerning partially ordered sets and lattices. First, we briefly recall the definition of aggregation functions widely accepted and used as the definition in the case of bounded partially ordered sets.

Let $(P, \leq, 0, 1)$ be a bounded partially ordered set (poset for short), let $n \in \mathbb{N}$ be a positive integer. A mapping $A: P^n \to P$ is called an (n-ary) aggregation function on P if it is monotone (nondecreasing), i.e., for any $\mathbf{x}, \mathbf{y} \in P^n$:

$$A(\mathbf{x}) \leq A(\mathbf{y}) \quad \text{whenever} \quad \mathbf{x} \leq \mathbf{y},$$

and it satisfies two boundary conditions

$$A(0, \ldots, 0) = 0 \quad \text{and} \quad A(1, \ldots, 1) = 1.$$

Further, we give basic facts concerning the so-called residuated mappings, cf. [9]. Let $(P, \leq)$ and $(Q, \leq)$ be two posets and let $\varphi : P \to Q$ and $\psi : Q \to P$ be mappings between these posets. A pair (φ, ψ) is called *residuated mappings* or *an isotone Galois connection* between P and Q provided for all $p \in P$ and $q \in Q$

$$\varphi(p) \leq q \quad \text{if and only if} \quad p \leq \psi(q). \tag{1}$$

In such a case the mapping φ is referred to as *residuated (lower adjoint)*, while ψ is called *residual (upper adjoint)*. From the condition (1) it follows that for a given residuated mapping φ there is unique residual map ψ and vice versa. Let us note that residuated mappings are equivalently defined as monotone mappings fulfilling for all $p \in P$, $q \in Q$ the properties

$$p \leq \psi\varphi(p) \quad \text{and} \quad \varphi\psi(q) \leq q. \tag{2}$$

One of the important property of residuated mappings is their "mutual stability" under the composition of function, i.e., $\varphi\psi\varphi(p) = \varphi(p)$ for all $p \in P$ and similarly $\psi\varphi\psi(q) = \psi(q)$ for all $q \in Q$. If residuated mappings are given between complete lattices, there is the well-known important characterization of such mappings.

Proposition 1. *Let L and M be complete lattices. A pair (φ, ψ) of mappings between L and M is residuated if and only if f is $\bigvee$-preserving and g is $\bigwedge$-preserving mapping.*

From another point of view, residuated mappings between complete lattices are closely related to the notion of *closure operator (closure system)* and *interior operator (interior system)*, respectively. Let L be a complete lattice. By a closure operator in L we understand a mapping $c\colon L \to L$ which satisfies:

(c1) $x \leq c(x)$ for all $x \in L$,
(c2) $c(x_1) \leq c(x_2)$ for $x_1 \leq x_2$,
(c3) $c(c(x)) = c(x)$ for all $x \in L$, (i.e. c is idempotent).

Dually, by an interior operator in L we understand a mapping $i\colon L \to L$ satisfying:

(i1) $i(x) \leq x$ for all $x \in L$,
(i2) $i(x_1) \leq i(x_2)$ for $x_1 \leq x_2$,
(i3) $i(i(x)) = i(x)$ for all $x \in L$.

A subset X of a complete lattice L is called a closure system in L if X is closed under arbitrary infima and it is called an interior system if it is closed with respects to arbitrary suprema. Let us note that there is a bijective correspondence between closure (interior) operators on the one side, and closure (interior) systems on the other side.

From the properties of residuated mappings it can be deduced that if L, M are complete lattices and (φ, ψ) is a pair of residuated mappings, then the composition $\varphi\psi\colon L \to L$ forms a closure operator in L and the composition $\psi\varphi\colon M \to M$ forms an interior operator in M. Moreover the corresponding interior system and closure system, respectively, are isomorphic as complete lattices. Dually, any pair of isomorphic closure system in L and interior systems in M give rise to a residuated mappings between the lattices L and M.

The properties of residuated mappings allow to construct a monotone analogue of classical concept lattices, which are originally defined in antitone setting, cf. [10]. Formally, let (φ, ψ) be a pair of residuated mappings between complete lattices L and M. Denote by $\mathsf{CL}_{\varphi,\psi}$ a subset of $L \times M$ consisting of all pairs (x, y) with $\varphi(x) = y$ and $\psi(y) = x$. Define a partial order on $\mathsf{CL}_{\varphi,\psi}$ as follows:

$$(x_1, y_1) \leq (x_2, y_2) \quad \text{if} \quad x_1 \leq x_2, \; y_1 \leq y_2. \tag{3}$$

The following proposition describing the structure of $\mathsf{CL}_{\varphi,\psi}$ can be easily deduced from the basic fact, that for a residuated mappings the corresponding ranges form isomorphic closure and interior system respectively.

Proposition 2. *Let* (φ, ψ) *be a pair of residuated mappings between complete lattices L and M. Then* $(\mathsf{CL}_{\varphi,\psi}, \leq)$ *forms a complete lattice, where*

$$\bigwedge_{i \in I}(x_i, y_i) = \left(\bigwedge_{i \in I} x_i, \varphi\psi\left(\bigwedge_{i \in I} y_i \right) \right) , \quad \bigvee_{i \in I}(x_i, y_i) = \left(\psi\varphi\left(\bigvee_{i \in I} x_i \right), \bigvee_{i \in I} y_i \right)$$

for each family $(x_i, y_i)_{i \in I}$ *of elements from* $\mathsf{CL}_{\varphi,\psi}$.

3   General Concept Structures

We have seen that monotone concept lattices, defined via residuated pair of mappings, correspond to isomorphic closure and interior systems respectively. In order to provide a generalization of monotone concept lattices, we relax the condition on isomorphic order substructures of given complete lattices. For this, we recall the notions of order retraction and retract respectively.

Let P be a poset. A mapping $\mu \colon P \to P$ is an *order retraction* of P if μ is an monotone and idempotent operator in P, i.e., it satisfies the following two properties:

(i) $x_1 \leq x_2$ implies $\mu(x_1) \leq \mu(x_2)$,
(ii) $\mu(\mu(x))$.

Note, that any retraction is an idempotent order endomorphism of P.

By a *retract* of a poset P we understand the range of any retraction in P. Hence a subset $X \subseteq P$ is a retract of P if $X = \mu(P)$ for some retraction $\mu \colon P \to P$.

In what follows we will focus on complete lattices. Note, that $\bigvee$ and $\bigwedge$ denote the lattice operations of an underlying lattice.

Lemma 1. *Let L be a complete lattice and $S = \mu(L)$, for some order retraction $\mu \colon P \to P$ be a retract of L. Then $(S, \leq)$ forms a complete lattice, where the corresponding lattice operations are given by*

$$\inf_S\{x_i : i \in I\} = \mu\left(\bigwedge_{i \in I} x_i \right) \quad \text{and} \quad \sup_S\{x_i : i \in I\} = \mu\left(\bigvee_{i \in I} x_i \right).$$

Proof. Let $X = \{x_i : i \in I\}$ be a subset of S. Since $\bigwedge_{i\in I} x_i \leq x_i$ for all $i \in I$, due to monotonicity of μ we obtain $\mu(\bigwedge_{i\in I} x_i) \leq \mu(x_i) = x_i$ for all $i \in I$, i.e., $\mu(\bigwedge_{i\in I} x_i)$ is a lower bound of the subset X. Now, let $x \in S$ be such that $x \leq x_i$ for each $i \in I$. Then $x \leq \bigwedge_{i\in I} x_i$, hence $x = \mu(x) \leq \mu(\bigwedge_{i\in I} x_i)$. This yields that $\mu(\bigwedge_{i\in I} x_i)$ is the greatest lower bound of X, i.e., $\mu(\bigwedge_{i\in I} x_i) = \inf_S\{x_i : i \in I\}$.
Similarly, we obtain $\mu(\bigvee_{i\in I} x_i) = \sup_S\{x_i : i \in I\}$. □

Lemma 2. *Let L be a complete lattice and S be a subset of L forming a complete lattice with respect to the inherited order from L. Then S is a retract of L.*

Proof. We define an order retraction $\mu: L \to L$. For each $x \in L$ put $\mu(x) = \inf_S\{u \geq x : u \in S\}$. If $x_1 \leq x_2$ then $\{u \geq x_1 : u \in S\} \supseteq \{u \geq x_2 : u \in S\}$, thus $\mu(x_1) \leq \mu(x_2)$.
Since $\mu(x)$ is an element of S, we obtain

$$\mu(\mu(x)) = \inf_S\{u \geq \mu(x) : u \in S\} = \mu(x),$$

i.e., μ is idempotent. □

Further, we define a notion of w–residuated mappings and describe its relationship to the isomorphic retracts of complete lattices.

Let P and Q be partially ordered sets. We say that a pair of mappings (φ, ψ), $\varphi: P \to Q$, $\psi: Q \to P$ is *w-residuated*[1] between P and Q if it satisfies the following properties for all $p, p_1, p_2 \in P$ and for all $q, q_1, q_2 \in Q$:

(i) $p_1 \leq p_2$ implies $\varphi(p_1) \leq \varphi(p_2)$,
(ii) $q_1 \leq q_2$ implies $\psi(q_1) \leq \psi(q_2)$,
(iii) $\psi\varphi(p)) = \psi\varphi\psi\varphi(p)$ and $\varphi\psi(q) = \varphi\psi\varphi\psi(q)$.

It can be easily seen that these properties imply that two compositions $\psi\varphi: P \to P$ and $\varphi\psi: Q \to Q$ are order retractions of P and Q respectively. Note, that the importance of the third property can be also stressed by the algorithmic aspect, since fixed points of w-residuated pairs can be identified using simple two step iteration.

Observe, that each residuated pair of mappings is w-residuated. However, contrary to the case of residuated mappings, particular components of a w-residuated pair need not be determined uniquely.

Theorem 1. *Let L and M be complete lattices and (φ, ψ) be a w-residuated pair of mappings between them. Then $\psi\varphi(L)$ and $\varphi\psi(M)$ are isomorphic retracts of L and M respectively.*

Converly, if $\mu_L: L \to L$ and $\mu_M: M \to M$ are order retractions such that $\mu_L(L)$ being isomorphic to $\mu_M(M)$ via order isomorphism f, then the pair of mappings $(f\mu_L, f^{-1}\mu_M)$ is w-residuated.

---

[1] The authors are aware that this notion can be defined in other contexts under another name.

Proof. It is obvious that $S_L = \psi\varphi(L)$ is a retract of L and $S_M = \varphi\psi(M)$ is a retract of M. We show that $\varphi\lceil_{S_L} : S_L \to S_M$, i.e., the mapping φ restricted to the set S_L, is an order isomorphism with inverse $\psi\lceil_{S_M} : S_M \to S_L$.

To see this, assume that $x \in S_L$, i.e., $x = \psi\varphi(x_1)$ for some $x_1 \in L$. In this case $\varphi(x) \in S_M$ as $\varphi\psi(\varphi\psi\varphi(x_1)) = \varphi\psi\varphi(x_1)$ moreover $\psi\varphi(x) = x$. Similarly for any $y \in S_M$ we obtain $\psi(y) \in S_L$ and $\varphi\psi(y) = y$. Hence the mappings $\varphi\lceil_{S_L}$ and $\psi\lceil_{S_M}$ are monotone mutually inverse.

Further, let $\mu_L : L \to L$ and $\mu_M : M \to M$ be order retractions and f an isomorphism between the retracts $\mu_L(L)$ and $\mu_M(M)$. Since all four mappings are monotone, the same holds for compositions $\varphi = f\mu_L$ and $\psi = f^{-1}\mu_M$ as well. As $\varphi(x) = f\mu_L(x) \in \mu_M(M)$ for all $x \in L$, it follows that $\psi\varphi(x) = f^{-1}\mu_M(f\mu_L(x)) = f^{-1}(f\mu_L(x)) = \mu_L(x)$, which yields

$$\psi\varphi(\psi\varphi(x)) = \psi\varphi(\mu_L(x)) = \mu_L(\mu_L(x)) = \mu_L(x) = \psi\varphi(x).$$

In a similar way, it can be shown $\varphi\psi(\varphi\psi(y)) = \varphi\psi(y)$ for all $y \in M$.   □

Similarly as in the case of residuated mappings, we can define the following structure of "concepts".

Let L and M be complete lattices and (φ, ψ) be a w-residuated mappings between L and M. Denote by $\mathsf{wCL}_{\varphi,\psi}$ a subset of $L \times M$ such that

$$\mathsf{wCL}_{\varphi,\psi} = \{(x, y) : y = \varphi(x) \text{ and } x = \psi(y)\}.$$

Evidently, the set of fixed points $\mathsf{wCL}_{\varphi,\psi}$ can be ordered by the relation $\leq$ given by (3). The following theorem describes the structure of the set $\mathsf{wCL}_{\varphi,\psi}$.

Theorem 2. *The poset* $(\mathsf{wCL}_{\varphi,\psi}, \leq)$ *forms a complete lattice, where*

$$\inf\{(x_i, y_i) : i \in I\} = \left(\psi\varphi\left(\bigwedge_{i\in I} x_i\right), \varphi\psi\left(\bigwedge_{i\in I} y_i\right)\right),$$

$$\sup\{(x_i, y_i) : i \in I\} = \left(\psi\varphi\left(\bigvee_{i\in I} x_i\right), \varphi\psi\left(\bigvee_{i\in I} y_i\right)\right).$$

Proof. We show the formula for the infimum. The formula for the supremum can be proved analogously.

Denote $x = \psi\varphi(\bigwedge_{i\in I} x_i)$ and $y = \varphi\psi(\bigwedge_{i\in I} y_i)$. Since the compositions $\psi\varphi$ and $\varphi\psi$ form order retractions, from Lemma 1 we obtain that the element (x, y) represents the infimum of the indexed family $((x_i, y_i))_{i\in I}$ in the lattice $\psi\varphi(L) \times \varphi\psi(M)$. Hence to show that $(x, y) \in \mathsf{wCL}_{\varphi,\psi}$, it suffices to show that $\varphi(x) = y$ and $\psi(y) = x$. However this follows easily from the facts that $\varphi\lceil_{\psi\varphi(L)}$ as well as $\psi\lceil_{\varphi\psi(M)}$ are inverse order isomorphisms, and every order isomorphism preserves the existing infima and suprema.   □

Now we describe a possible method how to obtain certain concepts from data tables via w-residuated pair of mappings. First we recall the well-know and widely recognized concept forming operators, which form residuated mappings and are used in monotone analogue of FCA.

Let L be a fixed complete lattice, B be a set of objects and A be a set of attributes used for characterization of particular objects. Further, we assume that a data table is given in the form of a many-valued binary relation $R\colon B \times A \to V$, where V is a set of some values.

Let us note that there is only one essential way how to induce concept forming operators forming a residuated pair between the structures of L-valued fuzzy subsets L^B and L^A, see e.g., [12]. Assume there is given a mapping from the set V into the set of all residuated mappings, where $f_a\colon L \to L$, $g_a\colon L \to L$ denote a residuated pair of mappings associated to an element $a \in V$. Consequently, we obtain the pair of residuated mappings $F\colon L^B \to L^A$ and $G\colon L^A \to L^B$ given by

$$F(\mathbf{x})(a) = \bigvee_{b \in B} f_{R(b,a)}(\mathbf{x}(b)), \quad \text{for all } \mathbf{x} \in L^B, \tag{4}$$

$$G(\mathbf{y})(b) = \bigwedge_{a \in A} g_{R(b,a)}(\mathbf{y}(a)), \quad \text{for all } \mathbf{y} \in L^A. \tag{5}$$

To mention an example of such fuzzy concept forming operators, consider the so-called monotone $\mathbf{L}$-Galois connections, introduced e.g., in [11]. In this case, $\mathbf{L} = (L, \wedge, \vee, \otimes, \to, 0, 1)$ is a complete commutative residuated lattice and $R\colon B \times A \to L$ is a binary $\mathbf{L}$-fuzzy relation. In each residuated lattice, the two fuzzy connectives $\otimes$ and $\to$ are related by the so-called adjointness property

$$x \otimes a \leq y \quad \text{iff} \quad x \leq a \to y.$$

Hence, this property gives rise for an arbitrary element $a \in L$ a lower adjoint $f_a\colon L \to L$, $f_a(x) = x \otimes a$ for all $x \in L$, with the corresponding upper adjoint given by $g_a(y) = a \to y$ for all $y \in L$. Observe that each such pair fulfills the condition (1).

Consequently, the concept forming operators $\uparrow\colon L^B \to L^A$ and $\downarrow\colon L^A \to L^B$ are given, in accordance with formulas (4) and (5) respectively, by the rule

$$\uparrow(\mathbf{x})(a) = \bigvee_{b \in B} \mathbf{x}(b) \otimes R(b, a),$$

$$\downarrow(\mathbf{y})(b) = \bigwedge_{a \in A} R(b, a) \to \mathbf{y}(a).$$

To introduce an example of some w-residuated mappings derived from a data table we show the following lemma.

Lemma 3. *Let L and M be complete lattices, (φ, ψ) be a residuated pair of mappings, $i\colon L \to L$ be an interior operator on L and $c\colon M \to M$ be a closure operator on M. Define two mappings $F\colon L \to M$ and $G\colon M \to L$ for all $x \in L$ and $y \in M$ by*

$$F(x) = \varphi(i(x)) \quad \text{and} \quad G(y) = \psi(c(y)). \tag{6}$$

Then (F, G) forms a w-residuated pair of mappings between L and M.

Proof. Obviously $x_1 \leq x_2$ implies $i(x_1) \leq i(x_2)$, thus due to monotonicity of φ we obtain $F(x_1) = \varphi(i(x_1)) \leq \varphi(i(x_2)) = F(x_2)$. Similarly, from $y_1 \leq y_2$ we obtain $c(y_1) \leq c(y_2)$ and $G(y_1) = \psi(c(y_1)) \leq \psi(c(y_2)) = G(y_2)$.

Further we show the equalities $GF(GF(x)) = GF(x)$ and $FG(FG(y)) = FG(y)$ for all $x \in L$ and $y \in M$. First, from (2) we obtain $i(x) \leq \psi\varphi(i(x)) \leq \psi(c(\varphi(i(x)))) = GF(x)$ and similarly $c(y) \geq \varphi\psi(c(y)) \geq \varphi(i(\psi(c(y)))) = FG(y)$.

Also $F(x) = \varphi(i(x)) = \varphi(ii(x)) = F(i(x))$ as well as $G(y) = \psi(c(y)) = \psi(cc(y)) = G(c(y))$.

Further $i(x) \leq GF(x)$ yields $F(i(x)) \leq FGF(x)$ and $i(G(y)) \leq GF(G(y))$. Similarly $c(y) \geq FG(y)$ gives $G(c(y)) \geq GFG(y)$ and $c(F(x))) \geq FG(F(x))$, hence we obtain

$$F(x) = F(i(x)) \leq FGF(x) \leq c(F(x)), \; G(y) = G(c(y)) \geq GFG(y) \geq i(G(y)).$$

Finally, from the previous inequalities we have

$$GF(x) \leq GFGF(x) \leq G(c(F(x))) = GF(x),$$

$$FG(y) \geq FGFG(y) \geq F(i(G(y))) = FG(y).$$

$\square$

Let L be a complete lattice and assume that a data table is given in the form of relation $R: B \times A \to V$, where B is a nonempty set of objects, A nonempty set of attributes and V a set of values which are used for characterization of particular objects by attributes. Also there is a mapping from the set V into the set of all residuated mappings on L, where $f_a: L \to L$, $g_a: L \to L$ denote a residuated pair of mappings associated to an element $a \in V$. Further there are two operators $i: L^B \to L^B$ an interior operator and $c: L^A \to L^A$ a closure operator. These operators can be used for effective reduction of the underlying complete lattices corresponding to the fuzzy subsets L^B and L^A respectively. In fact, instead of taking L^B as the domain of a concept forming operator, the set $\mathsf{rng}(i) \subseteq L^B$ is used, which forms an interior operator in L^B. Similarly, in the case of the closure operator c, the! range $\mathsf{rng}(c)$ forming a closure system is used as the domain of a concept forming operator. This fact can be effectively used for a parametrized reduction of the resulting concept lattice and also it can be used for speeding up an algorithm for building a concept lattice. The basic type of concept forming operators can be defined as follows:

$$\uparrow(\mathbf{x})(a) = \bigvee_{b \in B} f_{R(b,a)}\big(i(\mathbf{x})(b)\big), \quad \text{for all } \mathbf{x} \in L^B, \tag{7}$$

$$\downarrow(\mathbf{y})(b) = \bigwedge_{a \in A} g_{R(b,a)}\big(c(\mathbf{y})(a)\big), \quad \text{for all } \mathbf{y} \in L^A. \tag{8}$$

Let us notice that $i(\mathbf{x})(b)$ denotes the b-th projection of the element $i(\mathbf{x}) \in L^B$ as well as $c(\mathbf{y})(a)$ denotes the a-th projection of $c(\mathbf{y}) \in L^A$. Evidently $\uparrow = Fi$ and $\downarrow = Gc$ where F and G form a residuated pair of mappings given by (4) and (5) respectively. Hence due to Lemma 3, the pair $(\uparrow, \downarrow)$ forms a w-residuated pair.

As we can see, for all $a \in A$ the particular values $f_{R(b,a)}\big(i(\mathbf{x})(b)\big)$ in formula (7) are aggregated using the operation of suprema. However, it is also possible for an attribute $a \in A$ to aggregate values $f_{R(b,a)}\big(i(\mathbf{x})(b)\big)$ by means of $|B|$-ary supremum preserving aggregation function, which can be more appropriate for an evaluation connected with the considered attribute. In this case, the concept forming operator $\uparrow\colon L^B \to L^A$ has the following form:

$$\uparrow(\mathbf{x})(a) = F_a\big(\ldots, f_{R(b,a)}\big(i(\mathbf{x})(b)\big), \ldots\big), \quad \text{for all } \mathbf{x} \in L^B,$$

As F_a for all $a \in A$ is supremum preserving, the same is valid for the resulting operator $\uparrow$. However, to find the corresponding functions G_b, $b \in B$ one must find a decomposition of the mapping F_a on $|B|$ unary components $(\bar{f}_{b,a})_{b \in B}$, where $\bar{f}_{b,a}(x) = F_a(0, \ldots, x, \ldots, 0)$ for all $x \in L$ see [12] for more details. Consequently, for a fixed $b \in B$ and $\mathbf{x} \in L^A$ the mapping is given by $G_b(\mathbf{x}) = \bigvee_{a \in A} \bar{g}_{b,a}(x_a)$, where $\bar{g}_{b,a}$ represents the upper adjoint to $\bar{f}_{b,a}$. Consequently, we obtain the following formula for the operator $\downarrow$.

$$\downarrow(\mathbf{y})(b) = G_b\big(\ldots, g_{R(b,a)}\big(c(\mathbf{y})(a)\big), \ldots\big), \quad \text{for all } \mathbf{y} \in L^A.$$

Again it can be easily seen that the pair $(\uparrow, \downarrow)$ fulfills the assumptions of Lemma 3, it is a composition of residuated mappings and interior and closure operators, thus it forms a w-residuated pair.

Considering other types of aggregation functions to aggregate data from tables, can be useful in order to incorporate other types of information usually available for the considered data, but not explicitly contained in the data table. Such view allows to consider about concept lattice based clustering methods in the realm of various theories, e.g., possibility theory, data fusion, decision making or optimalization.

4   Conclusion

In this paper we have discussed some application of certain aggregation function within the theory of concept lattices. Based on the results in [16] we have introduced w-residuated mappings, which can be used as a framework for inclusion of monotone aggregation functions to the definition of concept forming operators. Formal concepts can be studied and interpreted within many different theories. Beside the many-valued framework, we think that the study of links between aggregation functions and concept forming operators can bring better understanding of information occurring in the resulting concept lattices. In our future work we would like to describe and to investigate wider family of w-residuated mappings defined within different framework, e.g., residuated lattices, quantales or ordered quasigroups cf. [5,7,8].

Acknowledgments. The first author was supported by the Slovak Research and Development Agency under the contract no. APVV-16-0213. The second author was supported by the project of Grant Agency of the Czech Republic (GAČR) no. 18-06915S and by the Slovak Research and Development Agency under the contract no. APVV-16-0073. The third author was supported by the Slovak VEGA Grant 1/0365/19.

References

1. Antoni, L., Krajči, S., Krídlo, O., Macek, B., Pisková, L.: On heterogeneous formal contexts. Fuzzy Set. Syst. **234**, 22–33 (2014)
2. Antoni, L., Krajči, S., Krídlo, O.: On fuzzy generalizations of concept lattices. Stud. Comput. Intell. **758**, 79–103 (2018)
3. Bartl, E., Konecny, J.: L-concept analysis with positive and negative attributes. Inf. Sci. **360**, 96–111 (2016)
4. Bělohlávek, R., Konecny, J.: Concept lattices of isotone vs. antitone Galois connections in graded setting: mutual reducibility revisited. Inf. Sci. **199**, 133–137 (2012)
5. Brajerčík, J., Demko, M.: On sheaf spaces of partially ordered quasigroups. Quasigroups Relat. Syst. **22**(1), 51–58 (2014)
6. Butka, P., Pócs, J., Pócsová, J.: On equivalence of conceptual scaling and generalized one-sided concept lattices. Inf. Sci. **259**, 57–70 (2014)
7. Demko, M.: On congruences and ideals of partially ordered quasigroups. Czechoslovak Math. J. **58**(3), 637–650 (2008)
8. Demko, M.: Lexicographic product decompositions of half linearly ordered loops. Czechoslovak Math. J. **57**(2), 607–629 (2007)
9. Derderian, J.-C.: Residuated mappings. Pac. J. Math. **20**(1), 35–43 (1967)
10. Ganter, B., Wille, R.: Formal Concept Analysis: Mathematical Foundations. Springer, Berlin (1999)
11. Georgescu, G., Popescu, A.: Non-dual fuzzy connections. Arch. Math. Log. **43**(8), 1009–1039 (2004)
12. Halaš, R., Mesiar, R., Pócs, J.: Description of sup- and inf-preserving aggregation functions via families of clusters in data tables. Inf. Sci. **400401**, 173–183 (2017)
13. Medina, J., Ojeda-Aciego, M., Ruiz-Calviño, J.: Formal concept analysis via multi-adjoint concept lattices. Fuzzy Set. Syst. **160**, 130–144 (2009)
14. Konecny, J.: Isotone fuzzy Galois connections with hedges. Inf. Sci. **181**(10), 1804–1817 (2011)
15. Konecny, J., Osicka, P.: Triadic concept lattices in the framework of aggregation structures. Inf. Sci. **279**, 512–527 (2014)
16. Pócs, J.: On possible generalization of fuzzy concept lattices using dually isomorphic retracts. Inf. Sci. **210**, 89–98 (2012)
17. Rodriguez-Jimenez, J.M., Cordero, P., Enciso, M., Mora, A.: A generalized framework to consider positive and negative attributes in formal concept analysis. In: 11th International Conference on Concept Lattices and Their Applications (CLA 2014), vol. 1252, pp. 267–278. CEUR Workshop Proceedings (2014)
18. Rodriguez-Jimenez, J.M., Cordero, P., Enciso, M., Mora, A.: Negative attributes and implications in formal concept analysis. Procedia Comput. Sci. **31**, 758–765 (2014)

Ordinal Sums of t-norms and t-conorms on Bounded Lattices

Antonín Dvořák[(✉)] and Michal Holčapek

Institute for Research and Applications of Fuzzy Modeling, CE IT4Innovations,
University of Ostrava, 30. dubna 22, 701 03 Ostrava, Czech Republic
{Antonin.Dvorak,Michal.Holcapek}@osu.cz

Abstract. This contribution extends a recently proposed novel approach to ordinal sum constructions of t-norms and t-conorms on bounded
lattices that are determined by interior and closure operators. The extension lies in a possibility to consider also infinite sets of indices.

Keywords: t-norms · t-conorms · Bounded lattices · Ordinal sum

The operations of t-norms and t-conorms on the unit interval [12,18] are
popular operations used in fuzzy set theory and fuzzy logic. They serve as natural interpretations of operations of conjunction and disjunction, respectively.[1]
Because fuzzy logic started to use more general structures of truth values, following the seminal work of Goguen [9], and these structures fall under the concept of
bounded lattices, it was quite natural to begin to study t-norms and t-conorms
on bounded lattices [1,5,20].

Ordinal sums in the sense of Clifford [4] are very important constructions of
t-norms and t-conorms on the unit interval. They also provide a basis of a well-
known representation theorem for continuous t-norms (t-conorms) as ordinal
sums of isomorphic images of Łukasiewicz and product t-norms (t-conorms) [12,
14]. Generalizations of ordinal sum constructions for t-norms and t-conorms on
bounded lattices have been intensively studied [3,8,13,16,17].

Recently [7] we proposed a new and more general definition of an ordinal
sum of t-norms and t-conorms on bounded lattices and show that ordinal sums
proposed in [3,8] are special cases of our definition. We used the concept of
lattice interior and closure operators [15], which allow us to pick a sublattice (or
sublattices) of a given bounded lattice appropriate for our construction.

The aim of this paper is to extend further the approach proposed in [7]. Unlike
that paper, we consider here the possibility to use an infinite set of indices in

---

[1] Naturally, the role of t-norms and t-conorms is very important in applications of
fuzzy logic, e.g., in multicriteria decision-making, fuzzy control, image processing,
etc.

This research was partially supported from the ERDF/ESF project AI-Met4AI (No.
CZ.02.1.01/0.0/0.0/17_049/0008414). The additional support was also provided by the
Czech Science Foundation through the project of No. 18-06915S.

© Springer Nature Switzerland AG 2019
R. Halaš et al. (Eds.): AGOP 2019, AISC 981, pp. 289–301, 2019.
https://doi.org/10.1007/978-3-030-19494-9_27

ordinal sum constructions and show examples of ordinal sums of t-norms and t-conorms on products of bounded lattices. Moreover, we analyze the ordinal sums of t-representable t-norms and t-conorms.

The paper is structured as follows. In Sect. 1 we recall notions of bounded lattice, t-norms and t-conorms on bounded lattices and also interior and closure operators. Section 2 contains the main results: a new construction of t-norms (t-conorms) on bounded lattices using interior (closure) operators, called h-ordinal sum (g-ordinal sum) (Theorems 5 and 6). In Sect. 3 we investigate ordinal sums of t-norms and t-conorms on product lattices. We are interested in t-representable t-norms (t-conorms) on a sublattice of a product lattice that are determined by t-norms (t-conorms) on its coordinates. Section 4 contains conclusions and directions of further research.

1   Preliminaries

Let $L = (L, \wedge, \vee)$ denote a lattice and let $\leqslant$ denote the lattice ordering on L, which is determined by $x \leqslant y$ if $x \wedge y = x$ for any $x, y \in L$. A lattice L is *bounded* if there exist two elements $0_L, 1_L \in L$ such that for all $x \in L$ it holds that $0_L \leqslant x \leqslant 1_L$. We call 0_L and 1_L the zero and the unit element, respectively, and write this bounded lattice as $(L, \wedge, \vee, 0_L, 1_L)$. A dual lattice of L is a lattice $L^{\mathrm{d}} = (L, \wedge^{\mathrm{d}}, \vee^{\mathrm{d}})$, where $x \wedge^{\mathrm{d}} y = x \vee y$ and $x \vee^{\mathrm{d}} y = x \wedge y$ hold for any $x, y \in L$. One can see that if L is bounded, then L^{d} is also bounded with $0_{L^{\mathrm{d}}} = 1_L$ and $1_{L^{\mathrm{d}}} = 0_L$. A lattice L is said to be self-dual if there exists an isomorphism of lattices $\varphi : L \to L^{\mathrm{d}}$. Note that L is self-dual if there exists a bijective map $\varphi : L \to L$ such that $\varphi(x \wedge y) = \varphi(x) \vee \varphi(y)$ and $\varphi(x \vee y) = \varphi(x) \wedge \varphi(y)$ for any $x, y \in L$ (see, e.g., [19]). Let P be a non-empty set of indices, and let L_p be a lattice for any $p \in P$. Then $\prod_{p \in P} L_P$ denotes the product of lattices and $\pi_k : \prod_{p \in P} L_p \to L_k$ denotes the k-th projection for $k \in P$. An element of $\prod_{p \in P} L_p$ will be denoted as $(x_p)_{p \in P}$ or simply (x_p) if no confusion can appear. Let $a, b \in L$ be such that $a \leqslant b$. The closed subinterval $[a, b]$ of L is the sublattice

$$[a, b] = \{x \in L \mid a \leqslant x \leqslant b\}.$$

Similarly, the open subinterval (a, b) of L is defined as $(a, b) = \{x \in L \mid a < x < b\}$. Definitions of semi-open intervals $(a, b]$ and $[a, b)$ are obvious. For more information about bounded lattices, we refer to [10, 11, 19].

The definition of a t-norm and a t-conorm on a bounded lattice has been proposed by Saminger in [16, Definition 3.1].

Definition 1. *An operation $T : L^2 \to L$ ($S : L^2 \to L$) on a bounded lattice $(L, \wedge, \vee, 0_L, 1_L)$ is a t-norm (t-conorm) if it is commutative, associative, non-decreasing with respect to both variables and 1_L (0_L) is its neutral element.*

Obviously, an operation T on a bounded lattice L is a t-norm if and only if T is a t-conorm on its dual L^{d}, and similarly for a t-conorm S on L. Let T and S be a t-norm and a t-conorm on a self-dual bounded lattice L, respectively. We say that S is a φ-dual t-conorm to T on L if $\varphi(T(x,y)) = S(\varphi(x), \varphi(y))$ holds for $\varphi : L \cong L^d$. Similarly, one can define a φ-dual t-norm to a given t-conorm.

An important concept in our analysis of ordinal sum of t-norms (t-conorms) on bounded lattices is interior (closure) operator on these lattices. Let us recall their definitions.

Definition 2. *Let L be a bounded lattice. A map $h : L \to L$ is said to be an interior operator on L if*

1. $h(1_L) = 1_L$,
2. $h(h(x)) = h(x)$ for any $x \in L$,
3. $h(x \wedge y) = h(x) \wedge h(y)$ for any $x, y \in L$,
4. $x \geqslant h(x)$.

Definition 3. *Let L be a bounded lattice. A map $g : L \to L$ is said to be a closure operator on L if*

1. $g(0_L) = 0_L$,
2. $g(g(x)) = g(x)$ for any $x \in L$,
3. $g(x \vee y) = g(x) \vee g(y)$ for any $x, y \in L$,
4. $x \leqslant g(x)$.

It is easy to see that the identity map id_L is an interior as well as a closure operator on L. Moreover, if $h(L) = g(L) = L$, then $h = g = id_L$. Further, if h is an interior operator on L, then h is a closure operator on L^d and vice versa. If L is self dual and $\varphi : L \cong L^d$, then $\varphi^{-1} \circ h \circ \varphi$ is a closure operator on L for any interior operator h on L and vice versa.

The following two examples show a simple construction of the interior and closure operators that are used later in the constructions of t-norms and t-conorms, respectively.

Example 1. Let L be a bounded lattice, and let $h \in L \backslash \{0_L, 1_L\}$ be arbitrary. Then, the maps $h_b, g_b : L \to L$ defined by

$$h_b(x) = \begin{cases} x, & x \geqslant b, \\ x \wedge b, & \text{otherwise,} \end{cases} \qquad g_b(x) = \begin{cases} x, & x \leqslant b, \\ x \vee b, & \text{otherwise,} \end{cases} \qquad (1)$$

for any $x \in L$, are an interior and a closure operators on L, respectively.

Example 2. Let L be a bounded lattice, and let $b \in L \backslash \{0_L, 1_L\}$ be arbitrary. Then, the maps $h_b, g_b : L \to L$ defined by

$$h_b(x) = \begin{cases} x, & x \geqslant b, \\ 0_L, & \text{otherwise,} \end{cases} \qquad g_b(x) = \begin{cases} x, & x \leqslant b, \\ 1_L, & \text{otherwise,} \end{cases} \qquad (2)$$

for any $x \in L$, are an interior and a closure operator on L, respectively.

One can simply verify that the composition of the interior (closure) operators on a bounded lattice L from the previous two examples is commutative, more precisely, $h_a \circ h_b = h_b \circ h_a$ holds for any $a, b \in L \backslash \{0_L, 1_L\}$ with $a \leqslant b$. Obviously, not all interior (closure) operators commute. And moreover, their composition is again an interior (closure) operator. This motivates us to introduce the following definition.

Definition 4. *Let $h, g : L \to L$ be maps on a bounded lattice L. We say that h and g commute on L provided that $h \circ g = g \circ h$.*

The commutativity of interior (closure) operators is a sufficient condition to ensure that their composition is again an interior (closure) operator as the following lemma states.

Lemma 1. *Let h, g be interior (closure) operators on a bounded lattice L that commute on L. Then $h \circ g$ is an interior (closure) operator on L.*

As a simple consequence of the previous lemma we obtain the following theorem which is formulated only for interior operators, but an analogous theorem can be stated also for closure operators.

Theorem 1. *Let H be a set of mutually commutative interior operators on L, i.e., $h \circ g = g \circ h$ for any $h, g \in H$, and $id_L \in H$ and let $G = \{h_1 \circ \cdots \circ h_n \mid h_1, \ldots, h_n \in H\}$, then $(G, \circ)$ is a monoid of interior operators on L.*

Example 3. Let $L = ([0,1], \wedge, \vee, 0, 1)$ be the bounded lattice on the unit interval determined by the common linear ordering of reals, and let $M = L \times L$ be the product, where

$$(a, b) \wedge (c, d) = (a \wedge c, b \wedge d) \quad \text{and} \quad (a, b) \vee (c, d) = (a \vee c, b \vee d)$$

holds for any $(a, b), (c, d) \in L$. Consider the interior operator $h_{(a,1)}$ on M defined by (1) for any $a \in [0, 1)$ and put $h_{(1,1)} = id_M$. Obviously, $h_{(a,1)} \circ h_{(b,1)} = h_{(a \wedge b, 1)}$, hence the composition of interior operators $\circ$ is closed on $H = \{h_{(a,1)} \mid a \in [0,1]\}$ and $h_{(1,1)}$ is the neutral element. Since $\circ$ is also associative, we get that $(H, \circ)$ is a monoid of interior operators on M.

2   Ordinal sum of t-norms and t-conorms

For a discussion of the concept of ordinal sum in the context of partially ordered sets (ordinal sum in the sense of Birkhoff [2]) and in the context of semigroup theory (ordinal sum in the sense of Clifford [4]) as well as for an overview of ordinal sums of t-norms on the unit interval see [16, Sect. 2]. In the following part, we extend the results in [7].

The following theorem provides a construction of t-norms (t-conorms) on bounded lattices with the help of an interior (a closure) operator. Note that this construction forms a core of the ordinal sum of t-norms (t-conorms) as will be demonstrated later. The proof of theorem can be found in [7].

Theorem 2. *Let L be a bounded lattice, and let $h : L \to L$ be an interior operator on L and $g : L \to L$ be a closure operator. Let H and G denote the images of L under h and g, i.e. $h(L) = H$ and $g(L) = G$, respectively. Then,*

(i) if V is a t-norm on H, then there exists its extension to a t-norm T on L as follows:

$$T(x,y) = \begin{cases} V(h(x),h(y)), & x,y \in L\setminus\{1_L\}; \\ x \wedge y, & otherwise, \end{cases}$$

(ii) if W is a t-conorm on G, then there exists its extension to a t-conorm S on L as follows:

$$S(x,y) = \begin{cases} W(g(x),g(y)), & x,y \in L\setminus\{0_L\}; \\ x \vee y, & otherwise. \end{cases}$$

To introduce the ordinal sum of t-norms and t-conorms, we assume that the set of indices K is linearly ordered with the least element 0_K and the greatest element 1_K. We say that a subset $\{b_k \mid k \in K\} \subseteq L$ is a *complete ascending chain* in L if $b_k < b_j$ when $k < j$ and

$$b_k^+ = \bigwedge_{j \in K, j > k} b_j > b_k \tag{3}$$

exists in L for any $k \in K$. Similarly, a subset $\{b_k \mid k \in K\} \subseteq L$ is called a *complete descending chain* in L if $b_k < b_j$ when $k > j$ and

$$b_k^- = \bigvee_{j \in K, j > k} b_j < b_k \tag{4}$$

exists in L for any $k \in K$. One can see that each finite chain $b_1 < \cdots < b_n$ such that $b_n < 1_L$ is a complete ascending chain in L and $b_n < \cdots < b_1$ such that $0_L < b_n$ in L is a complete descending chain in L.

The following two theorems introduce the ordinal sum of t-norms and t-conorms on bounded lattices that can be partitioned into a chain of lattice subintervals.

Theorem 3. *Let L be a bounded lattice, and let $\{b_k \mid k \in K\}$ be a complete ascending chain in L such that $L = \bigcup_{k \in K}[b_k, b_k^+]$. If V_k is a t-norm on $[b_k, b_k^+]$ for any $k \in K$, then the ordinal sum of t-norms $\{V_k \mid k \in K\}$ defined as follows:*

$$T(x,y) = \begin{cases} V_k(x,y), & x,y \in [b_k, b_k^+), \\ x \wedge y, & otherwise; \end{cases} \tag{5}$$

is a t-norm on L.

Theorem 4. *Let L be a bounded lattice, and let $\{b_k \mid k \in K\}$ be a complete descending chain in L such that $L = \bigcup_{k \in K}[b_k^-, b_k]$. If W_k is a t-conorm on $[b_k^-, b_k]$ for any $k \in K$, then the* ordinal sum of t-conorms $\{W_k \mid k \in K\}$ *defined as follows:*

$$S(x,y) = \begin{cases} W_k(x,y), & x,y \in (b_k^-, b_k], \\ x \vee y, & \text{otherwise;} \end{cases} \tag{6}$$

is a t-conorm on L.

The proof of the previous theorems can be designed in the same way as the proof of Theorem 3.2. in [7], where, moreover, one need to show that the axioms of t-norm and t-conorm are also fulfilled when the elements of $F^+ = \{b_k^+ \mid k \in K\}$ and $F^- = \{b_k^- \mid k \in K\}$ are considered. The ordinal sums of t-norms $\{V_k \mid k \in K\}$ and t-conorms $\{W_k \mid k \in K\}$ will be denoted as

$$\bigoplus_{k \in K} V_k \quad \text{and} \quad \boxplus_{k \in K} W_k, \tag{7}$$

respectively.

Example 4. Let $L = ([0,1], \wedge, \vee, 0, 1)$ be the bounded lattice on the unit interval determined by the common linear ordering of reals, and let $M = L \times_{\text{lex}} L$ be the lexicographic product, where

$$(a,b) \wedge_{\text{lex}} (c,d) = \begin{cases} (a, b \wedge d), & a = c, \\ (a,b), & a < c, \\ (c,d), & a > c, \end{cases} \quad (a,b) \vee_{\text{lex}} (c,d) = \begin{cases} (a, b \vee d), & a = c, \\ (c,d), & a < c, \\ (a,b), & a > c, \end{cases}$$

hold for any $(a,b), (c,d) \in M$. One can see that $(M, \wedge_{\text{lex}}, \vee_{\text{lex}}, (0,0), (1,1))$ is a bounded linearly ordered lattice. Let $K = [0,1]$ with the standard order $\leqslant$. Consider $B = \{b_k \mid k \in K\} \subset M$, where $b_k = (k,0)$. Since $b_k^+ = (k,1) \in M$, the set B is a complete ascending chain in M such that $M = \bigcup_{k \in K}[b_k, b_k^+]$. Let T_k be a t-norm on $[0,1]$ for any $k \in K$. Obviously, T_k can be naturally transformed to a t-norm V_k defined on $[b_k, b_k^+] = [(k,0), (k,1)]$ by

$$V_k((k,x), (k,y)) = (k, T_k(x,y)), \quad x,y \in [0,1].$$

The ordinal sum $\bigoplus_{k \in K} V_k$ introduces a t-norm on M. If $T_k = \wedge$ for any $k \in K$, then the ordinal sum of $\{V_k \mid k \in K\}$ defines $\wedge_{\text{lex}}$ on M. Similarly, the set $C = \{b_k \mid k \in K\}$ with $b_k = (1-k, 1)$ defines a complete descending chain in M such that $b_k^- = (k,0)$ for any $k \in K$ and $M = \bigcup_{k \in K}[b_k^-, b_k]$. For any $k \in K$, let S_k be a t-conorm on $[0,1]$, and let W_k denote its natural transformation to $[b_k^-, b_k]$. Then $\boxplus_{k \in K} W_k$ is a t-conorm on M. If $S_k = \vee$ for any $k \in K$, then the ordinal sum of $\{W_k \mid k \in K\}$ defines $\vee_{\text{lex}}$ on M.

The previous theorem shows a construction of t-norms (t-conorms) on a lattice L by the ordinal sum of t-norms (t-conorms) on subintervals $[b_k, b_k^+]$ ($[b_k^-, b_k]$) that cover L. In the following part, we extend this construction to a more general situation, where the union of subintervals can also partially cover the lattice L, i.e.,

$$\{0_L, 1_L\} \subset M = \bigcup_{k \in K} [b_k, b_k^+] \subseteq L \quad \text{or} \quad \{0_L, 1_L\} \subset M = \bigcup_{k \in K} [b_k^-, b_k] \subseteq L,$$

in other words, M is generally a sublattice of L that have at least three elements.

Theorem 5. *Let L be a bounded lattice, and let $\{b_k \mid k \in K\}$ be a complete ascending chain in L such that $\{0_L, 1_L\} \subset M = \bigcup_{k \in K}[b_k, b_k^+] \subseteq L$. If V_k is a t-norm on $[b_k, b_{k+}]$ for any $k \in K$ and h is an interior operator on L such that b_k and b_k^+ are fixed points for h for any $k \in K$, $h(L) \subseteq M$ and V_k restricted to $J_{k+} = h(L) \cap [b_k, b_k^+]$ is a t-norm on J_{k+} for any $k \in K$, then*

$$T(x,y) = \begin{cases} V_k(h(x), h(y)), & (h(x), h(y)) \in J_k^2, \\ h(x) \wedge h(y), & (h(x), h(y)) \in J_k \times J_j, \text{ for } k \neq j, \\ x \wedge y, & \text{otherwise;} \end{cases} \tag{8}$$

for any $x, y \in L$, where $J_k = h(L) \cap [b_k, b_k^+)$, is a t-norm on L, which is called the h-ordinal sum of t-norms $\{V_k \mid k \in K\}$.

Theorem 6. *Let L be a bounded lattice, and let $\{b_k \mid k \in K\}$ be a complete descending chain in L such that $\{0_L, 1_L\} \subset M = \bigcup_{k \in K}[b_k^-, b_k] \subset L$. If W_k is a t-conorm on $[b_k^-, b_k]$ for any $k \in K$ and g is a closure operator on L such that b_k and b_k^- are fixed points for g for any $k \in K$, $g(L) \subseteq M$ and W_k restricted to $J_{k+} = g(L) \cap [b_k^-, b_k]$ is a t-conorm on J_{k+} for any $k \in K$, then*

$$S(x,y) = \begin{cases} W_k(g(x), g(y)), & (g(x), g(y)) \in J_k^2, \\ g(x) \vee y(y), & (g(x), g(y)) \in J_k \times J_j, \text{ for } k \neq j, \\ x \vee y, & \text{otherwise;} \end{cases} \tag{9}$$

for any $x, y \in L$, where $J_k = g(L) \cap (b_k^-, b_k]$, is a t-conorm on L, which is called the g-ordinal sum of t-conorms $\{W_k \mid k \in K\}$.

One can simply verify that if we admit that $h(L) = M = L$, then $h = id_L$ and the h-ordinal sum of t-norms defined in (8) coincides with the standard ordinal sum of t-norms introduced in Theorem 3, and similarly for the g-ordinal sum of t-conorms. The h-ordinal sum of t-norms $\{V_k \mid k \in K\}$ and the g-ordinal sum of t-conorms $\{W_k \mid k \in K\}$ will be denoted as

$$h \text{-} \bigoplus_{k \in K} V_k \quad \text{and} \quad g \text{-} \boxplus_{k \in K} W_k, \tag{10}$$

respectively. As a simple consequence of Theorems 3 and 5, we obtain

$$h - \bigoplus_{k \in K} V_k(x,y) = \bigoplus_{k \in K} V_k(h(x),h(y)) \tag{11}$$

for any $x,y \in L \backslash \{1_L\}$, and a similar equation holds for g-ordinal sum of t-conorms.

For self-dual lattices, we can provide a correspondence between h-ordinal sums of t-norms and g-ordinal sums of t-conorms, which is a simple consequence of Theorems 5 and 6 and the self-duality of lattices.

Corollary 1. *Let L be a self-dual lattice, let $\varphi : L \cong L^d$, and let $\{b_k \mid k \in K\}$ be a complete ascending chain in L. If $T = h - \bigoplus_{k \in K} V_k$ on L, where V_k is a t-norm on $[b_k, b_k^+]$, then*

$$W_k(x,y) = \varphi(V_k(\varphi^{-1}(x), \varphi^{-1}(y))) \tag{12}$$

defines a t-conorm on $[\varphi(b_k^+), \varphi(b_k)]$ for any $k \in K$ and $S = g - \boxplus_{k \in K} W_k$ is φ-dual t-conorm to T on L, where $g = \varphi \circ h \circ \varphi^{-1}$ is a closure operator on L.

In the following part, we will demonstrate the ordinal sum only for t-norms, the analogous construction of ordinal sums of t-conorms is left for a reader. The first example shows a general construction of a t-norm based on an h-ordinal sum of t-norms for a finite ascending chain, where h is defined in Example 1.

Example 5. Let L be a bounded lattice, and let $b_1, \ldots, b_n \in L$ such that $0 = b_1 < \cdots < b_n < 1_L$. It is easy to see that $b_1 < \cdots < b_n$ is a complete ascending chain in L with $b_k^+ = b_{k+1}$ for any $k = 1, \ldots, n-1$ and $b_n^+ = 1_L$. Assume that V_k is a t-norm on $[b_k, b_k^+]$ for any $k = 1, \ldots, n$ and put $h_k = h_{b_k}$ for any $k = 2, \ldots, n$, where h_{b_k} is defined in Example 1. According to the same example, the map h_k is an interior operator on L for any $k = 2, \ldots, n$, and as we have noted $h_k \circ h_j = h_j \circ h_k$ for any $k, j = 2, \ldots, n$. As a consequence of Theorem 3, we obtain that $h = h_2 \circ \cdots \circ h_n$ is an interior operator on L. By a straightforward verification, one can show that $h(L) = M$, where $M = \bigcup_{k=1}^{n} [b_k, b_k^+]$. From Theorem 5, the h-ordinal sum of t-norms $V_1, \ldots, V_n$ on L can be expressed as:

$$T(x,y) = \begin{cases} V_k(x \wedge b_k^+, y \wedge b_k^+), & (x \wedge b_k^+, y \wedge b_k^+) \in [b_k, b_k^+)^2, \\ x \wedge y \wedge b_k^+ \wedge b_j^+, & (x \wedge b_k^+, y \wedge b_j^+) \in [b_k, b_k^+) \times [b_j, b_j^+), \\ & \text{for } k \neq j, \\ x \wedge y, & \text{otherwise.} \end{cases} \tag{13}$$

Note that $h(x) = x$ for $x \in [b_n, 1_L)$, nevertheless, we consider $h(x) = x \wedge b_n^+ = x \wedge 1_L$ in (13) to get a more compact formula for the ordinal sum of t-norms. Similarly, one can introduce the ordinal sum of t-conorms for a complete descending chain $0_L < b_1 < \cdots b_n = 1_L$.

As a special case of the previous construction one can obtain the ordinal sum of t-norms on L provided by Ertuğrul et al. in [8].

Example 6. Let L be a bounded lattice, and let $b \in L\backslash\{0_L, 1_L\}$. Let V be a t-norm on $[b, 1_L]$, let $\wedge$ be a t-norm on $[0_L, b]$, and let $h = h_b$, where h_b is defined in Example 1. Then h_b-ordinal sum of t-norms V and $\wedge$ on L is a t-norm expressed as

$$T(x, y) = \begin{cases} V(x, y), & (x, y) \in [b, 1_L]^2, \\ x \wedge y & x = 1_L \text{ or } y = 1_L, \\ x \wedge y \wedge b, & \text{otherwise.} \end{cases} \tag{14}$$

The following example shows that a class of new t-norms on a bounded lattice L recently proposed by Çayli in [3] is a special case of the h-ordinal sum.

Example 7. Let L be a bounded lattice, and let $b \in L\backslash\{0_L, 1_L\}$. Let V be a t-norm on $[b, 1_L]$, let $\wedge$ be a t-norm on $[0_L, b]$, and let $h = h_b$, where h_b is defined in Example 2. Then the h_b-ordinal sum of t-norms V and $\wedge$ on L is a t-norm expressed as

$$T(x, y) = \begin{cases} V(x, y), & (x, y) \in [b, 1_L]^2, \\ x \wedge y & x = 1_L \text{ or } y = 1_L, \\ 0_L, & \text{otherwise.} \end{cases} \tag{15}$$

Note that $h_b(x) = 0_L$ for any $x \notin [b, 1_L]$. Hence, $h(L) = \{0_L, b\} \cup [b, 1_L] \subset M = [0_L, b] \cup [b, 1_L]$. Since the restriction of $\wedge$ to $\{0_L, b\}$ is a t-norm on $\{0_L, b\}$, the assumptions of Theorem 5 are satisfied and (15) can be simply derived from (8).

3 Ordinal Sums of t-representable t-norms and t-conorms

Let P be a non-empty set of indices, and let $M \subseteq L = \prod_{p \in P} L_p$ be a sublattice of the product of bounded lattices L_p. Note that M is bounded, $0_M = 0_L$ and $1_M = 1_L$. In what follows, we use M_p to denote the image of the p-th projection of M to L_p, i.e., $M_p = \pi_p(M)$. A broad class of t-norms and t-conorms on a sublattice M assumes that these t-norms and t-conorms are determined by t-norms and t-conorms on its coordinates, respectively. This type of t-norms and t-conorms are called t-representable. Note that a mixture of t-norms and t-conorms is also admissible, e.g., fuzzy intuitionistic t-norms are determined by t-norms in the first component and t-conorms in the second component [6]. As we have mentioned in Preliminaries, each t-norm on L becomes a t-conorm on L^d, and similarly for t-conorms on L. Hence, if a t-norm (t-conorm) on $M \subseteq \prod_{p \in P} L_p$ is t-representable and it is determined by t-norms and t-conorms on its coordinates in the same time, then it is sufficient to replace all lattices L_p on which t-conorms (t-norms) are considered by the dual lattices L_p^d to get a t-representable t-norm (t-conorm) determined only with t-norms (t-conorms) on its coordinates. Using this property, we can restrict our considerations to t-representable t-norms (t-conorms) that are determined by t-norms (t-conorms) on their coordinates.

In what follows, we assume that M is a sublattice of $L = \prod_{p \in P} L_p$ such that $\pi_p(M) = M_p = L_p$ for any $p \in P$. Note that if $M_p \subset L_p$, then we can replace L_p by M_p in the product to ensure the previous assumption.

Definition 5. *A t-norm T or a t-conorm S on M is said to be t-representable if for any $p \in P$ there exists a t-norm T_p or a t-conorm S_p on L_p such that the following diagram commutes*

$$
\begin{array}{ccc}
M \times M & \xrightarrow{\ T\ } & M \\
\pi_p \times \pi_p \downarrow & & \downarrow \pi_p \\
L_p \times L_p & \xrightarrow{\ T_p\ } & L_p
\end{array}
\qquad or \qquad
\begin{array}{ccc}
M \times M & \xrightarrow{\ S\ } & M \\
\pi_p \times \pi_p \downarrow & & \downarrow \pi_p \\
L_p \times L_p & \xrightarrow{\ S_p\ } & L_p,
\end{array}
$$

respectively.

Note that if a t-norm T (t-conorm S) is t-representable, then this representation is unique, i.e., there exists only one system of t-norms T_p (t-conorms S_p), $p \in P$, from which the t-norm T (t-conorm S) is determined. We say that a map $\xi_p : L_p \to M$ is a right inversion of π_p if $\pi_p \circ \xi_p = id_{L_p}$. Note that for our analysis of t-representable t-norms (t-conorms) we need not assume that ξ_p is a lattice homomorphism. Since $\pi_p(M) = L_p$, a right inversion of π_p always exists, but it is not unique. The product of two maps ξ_p is a map $\xi_p \times \xi_p : L_p \times L_p \to M \times M$ defined by $\xi_p \times \xi_p(x, y) = (\xi_p(x), \xi_p(y))$ for any $(x, y) \in L_p \times L_p$. Now, a simple consequence of the definition of t-representable t-norms on M is the following useful lemma.

Lemma 2. *A t-norm T on M is t-representable if and only if $T_p = \pi_p \circ T \circ \xi_p \times \xi_p$ is a t-norm on L_p for any $p \in P$, where ξ_p is an arbitrary right inversion of π_p. Moreover, it holds that*

$$
T((x_p)_{p \in P}, (y_p)_{p \in P}) = (T_p(x_p, y_p))_{p \in P}, \quad (x_p)_{p \in P}, (y_p)_{p \in P} \in M. \tag{16}
$$

If we replace t-norms by t-conorms in the previous lemma, we obtain an analogous statement for t-representable t-conorms. The following theorem shows a condition under which a t-representable t-norm on M which is an h-ordinal sum of other t-norms can be determined by ordinal sums ($p \in P$) of t-norms on L_p.

Theorem 7. *Let T be a t-representable t-norm on M. If T is the h-ordinal sum of t-norms $\{V_k \mid k \in K\}$ such that $\pi_p \circ h(M) = L_p$ for any $p \in P$, where $\pi_p \circ h(M)$ denotes the image of the support of lattice M under the map $\pi_p \circ h$ and L_p denotes the support of the lattice L_p, then V_k, $k \in K$, is t-representable and*

$$
T((x_p), (y_p)) = \begin{cases} \left(\bigoplus_{k \in K} V_{k,p}(\pi_p(h((x_p))), \pi_p(h((y_p)))) \right)_p, & (x_p) \neq 1_M \neq (y_p), \\ \left(\bigoplus_{k \in K} V_{k,p}(\pi_p((x_p)), \pi_p((y_p))) \right)_p, & otherwise, \end{cases}
$$

holds for any $(x_p), (y_p) \in M$ with $V_{k,p} = \pi_p \circ V_k \circ \xi_p \times \xi_p$ for any $p \in P$, where ξ_p is an arbitrary right inversion of π_p on L_p.

Note that if the t-norm V_k from the previous theorem is defined on $[b_k, b_k^+]$, then the restriction of V_k on $B_k = h(M) \cap [b_k, b_k^+]$ has to be also a t-norm, which follows from the definition of h-ordinal sum. Since $h((x_p)) = (x_p)$ for any $(x_p) \in B_k$, we find that $T((x_p),(y_p)) = V_k((x_p),(y_p))$ for any $(x_p),(y_p) \in B_k$. Moreover, from the t-representability of T, we get $T_p(x_p, y_p) = \pi_p(V_k((x_p),(y_p)))$. Since the inequality holds for any $(x_p),(y_p) \in B_k$ and $\pi_p(B_k) = [b_{kp}, b_{kp}^+]$, where $b_{kp} = \pi_p(b_k)$, $b_{kp}^+ = \pi_p(b_k^+)$ and $\bigcup_{k \in K}[b_{kp}, b_{kp}^+] = L_p$ by assumption $\pi_p \circ h(M) = L_p$, one can see that V_k is t-representable by a system of t-norms T_p, $p \in P$, restricted to $[b_{kp}, b_{kp}^+]$. Similarly to the previous statement, if we replace t-norms by t-conorms in the previous theorem, we obtain an analogous statement for t-representable t-conorms.

Example 8. Let $L_1 = ([0,1], \wedge, \vee, 0, 1)$, $L_2 = L_1^d$, and let

$$M = \{(x,y) \mid x, y \in [0,1], \, x + y \leqslant 1\} \subset L_1 \times L_2,$$

where $\wedge$ and $\vee$ on M are defined as follows:

$$(x_1, y_1) \wedge (x_2, y_2) = (x_1 \wedge x_2, y_1 \wedge^d y_2) = (x_1 \wedge x_2, y_1 \vee y_2),$$
$$(x_1, y_1) \vee (x_2, y_2) = (x_1 \vee x_2, y_1 \vee^d y_2) = (x_1 \vee x_2, y_1 \wedge y_2).$$

The partial order on M determined by the meet is defined by $(x_1, y_1) \leqslant (x_2, y_2)$ if $x_1 \leqslant x_2$ and $y_1 \geqslant y_2$. Obviously, $0_M = (0,1)$ and $1_M = (1,0)$, and $M = (M, \wedge, \vee, 0_M, 1_M)$ is a bounded lattice. By [6], a necessary and sufficient condition for a t-representable (fuzzy intuitionistic) t-norm T on M is to be determined by a pair of a t-norm T on L_1 and a t-conorm S on L_1 (i.e., a t-norm S on L_2) such that $T \leqslant S^*$, where S^* denote a φ-dual t-norm to S on L_1 with $\varphi : L_1 \cong L_2 = L_1^d$ defined by $\varphi(x) = 1 - x$ for any $x \in [0,1]$. Let $b \in (0,1)$ and consider an interior operator $h_{(b,1-b)} : M \to M$ defined by (1) in Example 1, i.e.,

$$h_{(b,1-b)}(x,y) = \begin{cases} (x,y), & (b, 1-b) \leqslant (x,y), \\ (x,y) \wedge (b, 1-b), & \text{otherwise,} \end{cases}$$

for any $(x,y) \in M$. Put $h = h_{(b,1-b)}$. One can see easily that $\pi_1 \circ h(M) = \pi_2 \circ h(M) = [0,1]$, where $[0,1]$ is the support of lattices L_1 and L_2. Put $b_1 = 0_M$ and $b_2 = (b, 1-b)$, and let T be a t-representable t-norm on M and $T = h - \bigoplus_{k=1}^2 V_k$ for suitable t-norms V_1 on $[b_1, b_1^+]$ and V_2 on $[b_2, b_2^+]$. Note that $b_1^+ = b_2$ and $b_2^+ = 1_M$. Assume that T is determined by a pair (T, S) for which $T \leqslant S^*$. As was shown above, V_k on $[b_k, b_k^+]$, $k = 1, 2$, is t-representable by a (V_k, W_k), where V_k (W_k) is the restriction of t-norm T (t-conorm S) on $[b_{k1}, b_{k1}^+]$ $([b_{k2}, b_{k2}^+])$ with $b_{kp} = \pi_p(b_k)$ and $b_{kp}^+ = \pi_p(b_k^+)$ for $p = 1, 2$. By Theorem 7, the h-ordinal sum of t-norms V_1 and V_2 on M can be derived from the ordinal sum of t-norms V_1 and V_2 on L_1 and the ordinal sum of t-conorms of W_1 and W_2 on L_1 (or t-norms if L_2 is considered). To demonstrate equality in Theorem 7, for example, let $b_2^+ > (x_1, y_1) > b_2$ and $x_2 < b$ and $y_2 < 1 - b$, i.e., $(x_2, y_2) \parallel b_2$. Then $h(x_1, y_1) = (x_1, y_1)$ and $h(x_2, y_2) = (x_2, y_2) \wedge (b, 1-b) = (x_2, 1-b)$. Since

$((x_1, y_1), (x_2, 1 - b)) \in [b_2, b_2^+) \times [b_1, b_1^+)$, then

$$h - \bigoplus_{k=1}^{2} \mathcal{V}_k((x_1, y_1), (x_2, y_2)) = (x_1, y_1) \wedge (x_2, 1 - b) = (x_2, 1 - b).$$

On the other side, we have

$$(\bigoplus_{k=1}^{2} V_k(x_1, x_2), \bigoplus_{k=1}^{2} W_k(y_1, 1 - b)) = (x_1 \wedge x_2, W_2(y_1, 1 - b)) = (x_2, 1 - b),$$

where we used $(x_1, x_2) \in [b_{21}, b_{21}^+) \times [b_{11}, b_{11}^+)$ and $(y_1, 1 - b) \in [b_{22}, b_{22}^+)^2$.

4   Conclusion

In this paper we extended a new construction of t-norms (t-conorms) on bounded lattices [7], called the h-ordinal sum (g-ordinal sum), which generalized the previous ordinal sum approaches using the concept of lattice interior operator (closure operator). The extension lies in the possibility to use general (possibly infinite) sets of indices in ordinal sum constructions. We also investigated t-representable t-norms and t-conorms on product lattices. In further research, we plan to study similar constructions for other classes of aggregation operators.

References

1. De Baets, B., Mesiar, R.: Triangular norms on product lattices. Fuzzy Sets Syst. **104**, 61–75 (1999)
2. Birkhoff, G.: Lattice Theory. American Mathematical Society, Providence (1973)
3. Çayli, G.D.: On a new class of t-norms and t-conorms on bounded lattices. Fuzzy Sets Syst. **332**, 129–143 (2018)
4. Clifford, A.H.: Naturally totally ordered commutative semigroups. Am. J. Math. **76**, 631–646 (1954)
5. De Cooman, G., Kerre, E.E.: Order norms on bounded partially ordered sets. J. Fuzzy Math. **2**, 281–310 (1994)
6. Deschrijver, G., Cornelis, C., Kerre, E.E.: On the representation of intuitionistic fuzzy t-norms and t-conorms. IEEE Trans. Fuzzy Syst. **12**, 45–61 (2004)
7. Dvořák, A., Holčapek, M.: New construction of an ordinal sum of t-norms and t-conorms on bounded lattices. Inf. Sci. (2019, submitted)
8. Ertuğrul, U., Karaçal, F., Mesiar, R.: Modified ordinal sums of triangular norms and triangular conorms on bounded lattices. Int. J. Intell. Syst. **30**, 807–817 (2015)
9. Goguen, J.A.: L-fuzzy sets. J. Math. Anal. Appl. **18**, 145–174 (1967)
10. Grätzer, G.: General Lattice Theory. Academic Press, New York, London (1978)
11. Grätzer, G.: Lattice Theory: Foundation. Birkhäuser, Basel (2011)
12. Klement, E.P., Mesiar, R., Pap, E.: Triangular Norms, Trends in Logic, vol. 8. Kluwer, Dordrecht (2000)
13. Medina, J.: Characterizing when an ordinal sum of t-norms is a t-norm on bounded lattices. Fuzzy Sets Syst. **202**, 75–88 (2012)

14. Mostert, P.S., Shields, A.L.: On the structure of semigroups on a compact manifold with boundary. Ann. Math. Second Ser. **65**, 117–143 (1957)
15. Rutherford, D.E.: Introduction to Lattice Theory. Oliver & Boyd, Edinburgh and London (1965)
16. Saminger, S.: On ordinal sums of triangular norms on bounded lattices. Fuzzy Sets Syst. **157**, 1403–1416 (2006)
17. Saminger-Platz, S., Klement, E.P., Mesiar, R.: On extensions of triangular norms on bounded lattices. Indag. Math.-New Ser. **19**, 135–150 (2008)
18. Schweizer, B., Sklar, A.: Probabilistic Metric Spaces. North-Holland, New York (1983)
19. Szász, G.: Introduction to Lattice Theory. Academic Press, New York and London (1963)
20. Zhang, D.: Triangular norms on partially ordered sets. Fuzzy Sets Syst. **153**, 195–209 (2005)

Aggregation of Fuzzy Conformances

Miroslav Hudec[1,3]([✉]) and Miljan Vučetić[2]

[1] Faculty of Economic Informatics, University of Economics in Bratislava,
Dolnozemská cesta 1, Bratislava, Slovakia
miroslav.hudec@euba.sk
[2] Vlatacom Institute of High Technologies,
5 Milutina Milankovića Blvd, Belgrade, Serbia
miljan.vucetic@vlatacom.com
[3] Faculty of Organizational Sciences, University of Belgrade,
Jove Ilića 154, Belgrade, Serbia
miroslav.hudec@fon.bg.ac.rs

Abstract. Retrieving the most suitable items and sorting them downwards from the best face many challenges. The conformance measures are able to efficiently calculate similarities between the desired value and values of considered items' attribute regardless of different data types. These measures should be suitably aggregated, because the users usually provide different preferences among atomic conformances and therefore various aggregation functions should be considered. In this paper, we examine conjunctive functions (including non t-norms) as well as averaging and hybrid ones. In the hybrid aggregation, uninorms and ordinal sums of conjunctive and disjunctive functions have shown their perspectives in aggregating conformance measures. Diverse tasks require functions of desired behaviour and properly assigned weights or parameters. Thus, the perspectives for merging aggregation functions with the machine learning to the mutual benefits are outlined.

Keywords: Conformance measure ·
Conjunctive and disjunctive functions · Averaging functions ·
Hybrid functions · Data and function fitting

1   Introduction

Generally, an item (product, customer, service...) is expressed as a vector of attributes (e.g., colour, volume, weight, price, opinion ...). Retrieving the most suitable items and sorting them downwards from the best should closely follow users' requirements [18]. The users may have various requirements in mind (desired values of items' attributes and preferences among attributes), which the best matches should meet. The main problems in such data retrieval and aggregation tasks are: (i) heterogeneity of data types, and (ii) diverse preferences among attributes [1, 2, 21].

© Springer Nature Switzerland AG 2019
R. Halaš et al. (Eds.): AGOP 2019, AISC 981, pp. 302–314, 2019.
https://doi.org/10.1007/978-3-030-19494-9_28

The former covers cases when values of one attribute might be stored for some records as real numbers, whereas for other as fuzzy number or categorical data for instance. This fact is known as the variety feature of big–data [10]. Further, preferences expressed by users and data type for the considered attribute might not collide, i.e. a user may explain the desired distance to the nearest train stop as *very short*, or *short*; whereas the distance attribute is recorded as a real number greater than 0, or vice versa. The conformance measure based on the fuzzified attributes' domains and proximity measure among fuzzy partitions and linguistic terms is a robust solution for this problem, e.g. [21, 22]. The conformance between the desired value raised by user and item's value of respective attribute assumes value from the [0, 1] interval. This creates a space for diverse aggregation functions, which belong to the four main categories [7]: conjunctive, disjunctive, averaging and mixed.

This work focuses on the second problem: aggregating preferences among atomic conformances in order to cover diverse preferences raised by users. For instance, the user might declare that high conformance values should be emphasised and low values attenuated; when adding a piece of positive evidence the belief that a product is suitable is higher; weight of two conformances is significant only when both appear together and the like.

Another motivation is the possible synergy between the machine learning and aggregation functions. Generally, the machine learning methods are data hungry [17], i.e. they require a large input–output dataset for a simpler evaluation (e.g., classification). But, domain experts are usually able to explain the desired aggregation linguistically. From such explanation a small subclass of aggregation functions could be recognized and therefore a smaller set of input–output data may suffice. Recognizing the most suitable fuzzy partitions of attributes' domains and proximities among partitions is the further perspective for machine learning.

The reminder of this paper is organized as follows. Section 2 briefly introduces conformance measures. Section 3 is dedicated to various aggregation functions and reflections upon their applicability. Section 4 is focused on discussing findings and challenges, while Sect. 5 concludes the paper.

2 The Conformance Measures in Brief

When calculating similarity among the attribute's values, the fuzzy conformance–based approach has been shown as suitable for matching complex user requirements with the records in a dataset when heterogeneous (mixed type) attributes are considered [21].

In retrieving the most relevant items, the conformance measure expresses the similarity between the existing attribute's value with the desired value. In this sense, the value of conformance assumes value from the unit interval which is suitable for measuring how the user's requirements and the considered item's attribute match (1 is the perfect match, whereas 0 means absolutely no conformance). Therefore, amongst many methods, this approach provides a universal way for comparing given crisp, categorical and fuzzy data appearing in user preferences and attributes' values. Although data may be heterogeneous, we are able to straightforwardly measure the similarity between user requirements and item's features by the conformance equation presented in [22], which is based on the fuzzy sets and proximity relation:

$$C(A_i[t_u, t_j]) = \min(\mu_{tu}(A_i), \mu_{tj}(A_i), s(t_u(A_i), t_j(A_i))) \tag{1}$$

where C is a fuzzy conformance of attribute A_i defined on the domain D_i between the user requirement t_u and an item (or tuple in the database terminology) t_j, s is a proximity relation and $\mu_{tu}(A_i)$ and $\mu_{tj}(A_i)$ are membership degrees of user preferred value and attribute value in a dataset, respectively. In order to calculate membership degrees, domains of attributes are fuzzified into the fuzzy partitions as in [15]. The proximity relation explains how similar on the unit scale two labels (partitions) are. The literature offers several approaches, e.g. [19–21,23] to calculate the intensity of compatibility between the desired value and values of each item in a dataset.

The proximity relation expresses the degree of similarity or closeness among domain elements (or among flexible subdomains formalized by fuzzy sets) obtained from the domain expert [16] or learned from the data. The proximity relation on a domain D is a mapping $s : D \times D \mapsto [0, 1]$, where for each p, q holds $s(p, p) = 1$ and $s(p, q) = s(q, p)$. An example is shown in Table 1 for the attribute *walking distance to point P*. The number of terms and values of proximities depend on the particularities of considered task and users requirements.

Table 1. The proximity relation among subdomains of the attribute walking distance.

s	Low	Medium	High	Beyond walking distance
Low	1	0.75	0.25	0
Medium		1	0.6	0
High			1	0.2
Beyond wd				1

When the attributes' values are expressed by the discrete linguistic terms in the database model based on the similarity [14] the conformance of attribute A_i defined on domain D_u for t_u and domain D_j for t_j presented in the relation instance r and denoted by $C(A_i[t_u, t_j])$ is given as [19]:

$$C(A_i[t_u, t_j]) = \min\{ \min_{x \in D_u} \{ \max_{y \in D_j} [s(x, y)] \}, \{ \min_{x \in D_j} \{ \max_{y \in D_u} [s(x, y)] \} \} \tag{2}$$

where $s(x, y)$ is a similarity relation for values x and y. For instance, the user might say that desired value for A_i is {very low, low}, whereas value of attribute A_i for item j is stored as {low, more low than medium}. Buckles and Petry's model is extended in [16] by replacing similarity relation with proximity relation.

The significant benefit is in querying binary data and data containing multiple values. In a classical or fuzzy query item either fully satisfies or does not satisfy the binary requirement. For instance, an attribute *presence of balcony* assumes either *yes* or *no*. On the other hand, the proximity relation between two opposite binary values might belong to the (0, 1) interval. For instance, the proximity value 0.2 indicates that a flat without balcony is not explicitly rejected, but weakly contributes to the aggregation of all conformance measures. The high value of proximity between the opposite terms indicates that the presence of balcony is not very relevant, but might influence the solution.

The next section examines aggregation of conformances to calculate the overall matching degree considering various aspects of preferences among attributes.

3   The Aggregation of Atomic Conformances

Aggregation is a demanding task, because it should compute overall evaluation considering various influences among attributes.

3.1   Conjunctive Aggregation

When all atomic conformances should be at least partially met, the choice is usually a conjunctive function which meets t-norm's axioms due to desirable properties (monotonicity, associativity, symmetry, the presence of a neutral element) [8].

The minimum t-norm adjusted to the vector of conformances (1, 2) for the record t_j is expressed as [21]

$$T_{min}(\mathbf{t_j}) = \min_{i=1...n} \; C(A_i[t_u, t_j]) \tag{3}$$

where n is the number of atomic conformances. This t-norm provides the highest matching degree among all t-norms, but it lacks the compensation effect [2, 12], i.e. conformances higher than the minimal one are simply ignored. For the clarity throughout this paper we adopt the notation $x_i = C(A_i[t_u, t_j])$. Hence, we express (3) as

$$T_{min}(\mathbf{t_j}) = \min_{i=1...n} \; (x_i) \tag{4}$$

The other t-norms have the property of downward reinforcement [2], e.g. the product t-norm, adjusted to conformances (1, 2) is:

$$T_{prod}(\mathbf{t_j}) = \prod_{i=1}^{n}(x_i) \tag{5}$$

This aggregation produces the solution which is significantly lower than the lowest conformance, especially when we have higher number of atomic conformances. Therefore, the user might conclude that such low value indicates the low conformance of the considered item to his expectations. On the other hand, strict monotonicity ensures that all atomic conformances contribute to the aggregated matching degree. A hypothetical possibility is the transforming the aggregated matching degrees for t-norms having the limit property, i.e. all items assume the overall degree from the $[a, b]$ interval, except the rejected ones. This interval can be transformed into the $[c, d]$, where $c = 0$ or close to 0, and $d = 1$ or close to 1. In this way, the best matches are emphasized.

We should not forget conjunctive functions which does not meet all axioms of t-norms. For instance, function

$$f(\mathbf{t_j}) = (x_1) \cdot (x_2)^a \tag{6}$$

where $a > 1$ is a conjunctive asymmetric function (the result is bounded between 0 and MIN, which is required for conjunction [7]). In this function, the second conformance is not as relevant as the first one. This function can be extended to the $n$ary function, where each aggregated conformance might get its own power parameter a. This parameter should be greater or equal 1, otherwise we got the weighted geometric mean. The solution is obviously lower than or equal to the product t-norm, see Table 2. An $n$ary case is

$$f(\mathbf{t_j}) = \prod_{i=1}^{n} (x_i)^{a_i} \tag{7}$$

where $a_i \geq 1$, $i = 1 \dots n$. Value $a_i = 0$ for some i indicates an irrelevant conformance, i.e. we got the neutral element for this conjunction. When all a_i are equal to 1, we got the product t-norm (5), whereas for $\sum_{i=1}^{n} a_i = 1$ we got the weighted geometric mean with the absorbing element 0. Hence, for different values of parameter a_i we have different behaviour and therefore are able to cover various requirements. The same discussion related to the transformation $[a, b] \mapsto [c, d]$ holds here. Table 3 illustrates application of functions from Table 2 for aggregating five atomic conformances.

Table 2. The aggregation for three conjunctive functions.

Item	x_1	x_2	$a = 1.2$ in (6)	Product (5)	Min (4)
t1	0.2	0.8	0.1530	0.16	0.20
t2	0.8	0.2	0.1159	0.16	0.20
t3	0.9	0.8	0.6886	0.72	0.80
t4	0.2	0.95	0.1881	0.19	0.20
t5	0.21	0.3	0.0495	0.06	0.21

3.2 Disjunctive Aggregation

This kind of aggregation is suitable when conformances push each other up. The overall degree is not lower than the maximum of atomic conformances. If adding pth conformance should push the total degree up in comparison with the degree

Table 3. Fuzzy conformances of five attributes between the ideal values t_u and records t_1 to t_5, and overall matching degree for three conjunctive functions.

a in (7)	1	1.1	1.2	1	1.1	Conjunctive (7)	Product (5)	Minimum (4)
Item	x_1	x_2	x_3	x_4	x_5			
t1	0.85	0.85	0.85	0.4	0.85	0.1957	0.2088	0.4
t2	0.5	0.25	0.26	1	0.29	0.0055	0.0094	0.25
t3	0.1	0.65	0.46	1	0.41	0.0092	0.0123	0.1
t4	0	0.95	0.88	0.4	0.9	0	0	0
t5	0.45	0.25	0.65	0.4	0.25	0.0051	0.0073	0.25

x_i represents conformance $C(A_i[t_u, t_j])$, $i = 1, \ldots, 5$

calculated from the $p-1$ conformances, we should search for a disjunctive function which has properties of monotonicity with respect to argument cardinality and upward reinforcement [2]. This observation immediately excludes some disjunctive functions like MAX function. This aggregation is less restrictive and suitable for e.g., searching preferable books or movies. If a conformance related to the genre is high and further conformance related to the year of release is also high, then the overall degree increases. Due to duality with conjunction this type of aggregation is not further examined.

3.3 The Hybrid Aggregation

This type of aggregation usually covers conjunctive, disjunctive and averaging behaviour in a single function. Two well–known families are uninorms and null-norms [2]. The former attenuates values lower than the neutral element (conjunctive behaviour), emphasizes values higher than the neutral element (disjunctive behaviour) and calculates averages for lower values of some conformances and higher values for the reminder of the considered conformances [7,25]. The latter emphasises low values and attenuates high values (considering the absorbing element).

The well-known uninorm is $3 - \prod$ function [25]

$$u_{3 \prod}(\mathbf{t_j}) = \frac{\prod_{i=1}^{n} x_i}{\prod_{i=1}^{n} x_i + \prod_{i=1}^{n}(1 - x_i)} \tag{8}$$

where $x_i = C(A_i[r_u, r_j]$ (a conformance defined in (1, 2)).

The product in numerator ensures that only the items that are at least partially conformant by all examined attributes are considered, i.e. value 0 is an absorbing element. Applying (8) on data in Table 3, results are shown in Table 4. Items t_2 and t_3 fully meet the aggregation, whereas t_4 is excluded. The consequence of this aggregation function is that value 1 is also absorbing element when other conformances are significantly high. In our example, it makes indistinguishableness between items t_2 and t_3.

Table 4. Fuzzy conformances of five attributes between the ideal values t_u and records t_1 to t_6, and overall matching degree calculated by $3 - \prod$ function.

Item	x_1	x_2	x_3	x_4	x_5	$3 - \prod$ (8)
t1	0.85	0.85	0.85	0.4	0.85	0.9985
t2	0.5	0.25	0.26	1	0.29	1
t3	0.1	0.65	0.46	1	0.41	1
t4	0	0.95	0.88	0.4	0.9	0
t5	0.45	0.25	0.65	0.4	0.25	0.1012

The opposite behaviour we can find in nullnorms introduced by Calvo et al. [4]. A limiting case of idempotent nullnorms is the a–median. Its $n$ary representation for conformances is as follows:

$$\mathrm{med}_a(\mathbf{r_j}) = \mathrm{med}(x_1, a, x_2, a, ..., a, x_n) \tag{9}$$

where $a \in [0, 1]$.

When in a vector of conformances all elements have lower value than a, the solution is maximal value of conformances, whereas for a vector of conformances consisting higher values than a, the solution is minimal value of conformances. The latter ensures that minimal value limits the solution for high conformances, which is desired property related to emphasizing conjunctive behaviours (satisfying all conformances). However, when all conformances have low value, then the item is less suitable for user. Hence, uninorms are better fitted for the tasks of retrieving and ranking items according to the conformances of desired and existing values. Moreover, in real–world tasks it is highly probable that at least one conformance is in $[0, a]$ and at least one in $[a, 1]$, i.e. we can hardly expect that all elements of n-dimensional input vector are lower, or greater than a, see Table 5. Equation (9) can be expressed as

$$\mathrm{med}_a(\mathbf{r_j}) = \begin{cases} \max_{i=1...n} (x_i) & \text{for } (x_1, x_2, ..., x_n) \in [0, a]^n \\ \min_{i=1...n} (x_i) & \text{for } (x_1, x_2, ..., x_n) \in [a, 1]^n \\ a & \text{otherwise} \end{cases} \tag{10}$$

Table 5. The aggregation of conformances by a–median nullnorm.

Item	x_1	x_2	x_3	x_4	x_5	$a = 0.2$	$a = 0.5$	$a = 0.89$
t1	0.85	0.85	0.85	0.4	0.85	0.4	0.5	0.85
t2	0.5	0.25	0.26	1	0.29	0.25	0.5	0.89
t3	0.1	0.65	0.46	1	0.41	0.2	0.5	0.89
t4	0	0.95	0.88	0.4	0.9	0.2	0.5	0.89
t5	0.45	0.25	0.65	0.4	0.25	0.25	0.5	0.65

On the other hand, medians are applicable to inputs given on the categorical scale when the ordering is relevant, not the numbers. It is suitable for aggregating inputs like labels, such as *very high, high, rather high than medium, medium, rather low than medium, low* and *very low.* Thus, a–medians are perspective when all evaluations, including calculation of conformances are realized by words. Generally, both categories of mixed functions have their benefits and drawbacks. The future research should focus on fusion into the data intense applications.

The full reinforcement as illustrated in Table 4 is not sufficiently flexible for the noble reinforcement [24]. For instance, the reinforcement is activated when at least k conformances are high, whereas at most m conformances have low values.

3.4   The Ordinal Sums of Aggregated Conformances

Uninorms and nullnorms divide $[0, 1]^n$ into the disjunctive, conjunctive and averaging subdomains according to the assigned values to neutral and absorbing element, respectively. On the other hand, ordinal sums divide this interval into the higher number of subdomains [13], where we can apply different aggregations. The ordinal sums of a t-norm family is defined in the following way [2]

$$T_{ord}(x,y) = \begin{cases} a_i + (b_i - a_i)T_i(\frac{x-a_i}{b_i-a_i}, \frac{y-a_i}{b_i-a_i}) & \text{for } (x,y) \in [a_i, b_i]^2 \\ \min(x,y) & \text{otherwise} \end{cases} \tag{11}$$

where T_i is a t-norm. Analogously are defined ordinal sums of a t-conorm family.

Although, this way is effective and can cover different requirements for conjunctive aggregation in subdomains: strictness (e.g., t-norms generated by the additive generator with property $g(0) = \infty$), nilpotency (e.g., t-norms generated by the additive generator with property $g(0) < \infty$) and idempotency (minimum t-norm), the solution still has only conjunctive behaviour. The dual observation holds for the ordinal sums or t-conorms.

The solution could be the ordered sum of conjunctive and disjunctive functions proposed by de Baets and Mesiar [6]. Let us have $a \in]0, 1[$, then aggregation on $[0, a]$ can be conjunctive, whereas aggregation on $[a, 1]$ can be disjunctive. This is similar to uninorms, but a is any element, not only a neutral one. Further, we have freedom to select desired conjunctive and disjunctive functions. Formally, this aggregation is expressed as

$$Ag_{ord}(\mathbf{r_j}) = Con(\min(x_1, a), \ldots, \min(x_n, a)) + Dis(\max(x_1, a), \ldots, \max(x_n, a)) - a \tag{12}$$

where Con is a conjunctive function and Dis is a disjunctive function. An example for conformances in Tables 3, 4 and 5 is shown in Table 6. To reduce the length of table, conformances of considered five attributes are omitted.

In this example, functions MIN and MAX have shown greater variability than product t-norm and probabilistic sum t-conorm in Con and Dis, respectively. In comparison to Table 4 items t_2 and t_3 are distinguishable, but items

Table 6. The aggregation of conformances from Table 3 by the ordinal sum of conjunction and disjunction (12).

Item	$Con = \min,$ $Dis = \max,$ $a = 0.3$	$Con = $ product, $Dis = $ prob. sum, $a = 0.3$	$Con = \min,$ $Dis = \max,$ $a = 0.8$	$Con = $ product, $Dis = $ prob. sum, $a = 0.8$
t1	0.85	0.7021	0.45	0.3637
t2	0.95	0.7017	0.45	0.2075
t3	0.8	0.7008	0.3	0.2098
t4	0.65	0.6997	0.15	0.2
t5	0.6	0.6451	0.25	0.2070

t_1 and t_2 are not distinguishable for $a = 0.8$. Further, the ideal matching degree does not appear, because not a single item fully satisfies all conformances. Those items are emphasized without assigning them value 1. On the other hand, t_4 has quite high value when a assumes low values. When a is higher, then degree of t_4 is very low but not 0, because 0 is not the absorbing element.

The ordinal sum (12) has promising behaviour, but further research is recommended. The flexibility is high: Con and Dis could be mutually dual or two independent functions to meet specific users requirements for low and high values. Moreover, they can be conjunctive and disjunctive functions, which are not necessary t-norms or t-conorms, respectively, i.e. function (6) for conjunction.

3.5 Averaging Aggregation Considering Groups of Conformances

To complete this section, the aggregation when weights are assigned to the conformances as well as groups or coalitions of conformances is discussed. For instance, the requirement that *the orientation of flat to the west* and *the presence of balcony* are not important individually, but when both appear the weight is significant. The averaging function, a Choquet integral–based aggregation [5] is the solution for this situation. An expression of the Choquet integral is as

$$C_v(\mathbf{r_j}) = \sum_{i=1}^{n} [x_{(i)} - x_{(i-1)}] v(H_i) \qquad (13)$$

where $x_{(0)} = 0$ by convention, $x_{(i)}$ is a non-decreasing permutation of conformances for tuple t_j and v is a fuzzy measure of $H_i = \{(i), \ldots, (n)\}$. The fuzzy measure v is a set function $v : 2^{\mathcal{N}} \mapsto [0, 1]$ which is monotonic and satisfies $v(\emptyset) = 0$ and $v(\mathcal{N}) = 1$, where $\mathcal{N} = \{1, 2, \ldots, n\}$.

In our example of five conformances, user should assign 5^2 weights, which is a quite demanding task. Generally, a larger number of weights should be fitted to data and users' needs. This is also a possible future topic for machine learning.

4   Discussion

In our data intense society, data explaining the same attribute might be of different types. In addition, users' preferences might not be expressed by the same data type as a type used to express considered attribute. Due to their convexity and normality fuzzy numbers can be easily defuzzified [3] and therefore compared with classical numbers. But, when attribute is expressed by the subset of categorical data, proximity distribution and fuzzy data, the task becomes more complex. Further, the users might expect that a degree of proximity exists between two mutually excluded binary values. Such query cannot be realized by the usual database query engines. The conformance measure reveals how user requirements and items (records) in a dataset are similar regardless of data types.

The aggregation functions should cover variety of preferences among atomic conformances demanded by users. Section 3 discussed the most expected cases. When a higher number of conformances is included, where all of them should be at least partially met the solution is a t-norm, or a conjunction which does not necessary meet all t-norms axioms, e.g. Eqs. (6, 7). These functions (except minimum t-norm (4)) have the downward reinforcement property, causing that the high values also influence the solution. Moreover, in (7) higher values of parameter a indicate even stronger reinforcement when considered attribute has a low value.

When the best matches should be emphasized (upward reinforcement), and the weak matches attenuated (downward reinforcement), the solution could be reached by the uninorm function, e.g. 3 − || function (8). The upward reinforcement causes that item ideally meets requirements regardless several lower conformances. On the other hand, idempotent nullnorms, e.g. a–median (9) which reinforces in the opposite direction are not suitable. A very perspective option is ordinal sums consisting of conjunctive and disjunctive function (12). This function cover higher flexibility in assigning parameter a and selecting suitable conjunctive and disjunctive functions, which might not be necessarily mutually dual, and which might not meet all axioms of t-norms and t-conorms, respectively.

When atomic conditions create so–called coalitions or groups, the aggregation by the discrete Choquet integral (13) is the solution. This is an averaging aggregation, i.e. not excluding items which does not meet all atomic conformances.

We should be aware when associating properties of functions with the users' requirements. The perspective topic in this direction is fitting function to data [11] (when historical input-output data, or data evaluated by domain experts are available), as well as fitting to users preferences usually expressed linguistically. Generally speaking, the fitting task can be formalized as

$$\min \sum_{k=1}^{K} [\sum_{i=1}^{n} f(x_{ik}) - y_k]^2$$
subject to
f satisfies $P_1, \dots, P_n$
w_i satisfies D
$$\tag{14}$$

where the goal is minimizing the norm of residuals, but satisfying required properties P_i, $i = 1, \ldots, n$ for functions and demands D for weights or importance vector. More about the fitting tasks is in, e.g. [2, 11].

This task is a perspective topic for machine learning, especially nowadays, when the focus in deep neural networks is on the explainable solutions, or "explainable artificial intelligence" [9]. Calculating conformances between heterogeneous data and users preferences might be also a promising topic for the machine learning. The recent tendency is: *operations research meets machine learning*. We could say that a tendency: *aggregation functions meet machine learning* is also promising. Machine learning might help in suggesting the suitable aggregation functions and their respective parameters. In addition, such machine learning solution is tractable and explainable, and therefore applicable for other similar tasks. The same holds for the possibility of learning how to divide attributes domains into the fuzzy partitions and calculate proximities and conformances among these partitions.

5 Conclusion

The theory of aggregation functions is rich, but in practice it is not an easy task to recognize suitable functions for a given situation. In this paper, we shed a light on aggregating conformances among attributes for evaluating the most suitable items from heterogeneous data sources.

The conjunction is usually expressed by t-norms, but we have illustrated that the conjunctive functions which are not t-norms have also their benefits. The disjunctive functions are not suitable, because users are usually interested in items which (fully or partially) meet as much as possible atomic conformances, or all of them should be at least partially met. The most suitable are uninorms and ordinal sums of conjunctive and disjunctive functions due to their full reinforcement (emphasizing high and attenuating low values). The ordinal sums can be created by various combinations of conjunctive and disjunctive functions and therefore not necessary by t-norms and t-conorms. The future research should be focused on deeper examination of ordinal sums and conjunctive functions other than t-norms.

Diverse tasks require functions of desired behaviour and properly assigned weights or parameters. Therefore, it opens the perspective for complementing aggregation functions with machine learning. In the ideal case, a desired function is recognized and its parameters learned, but such solution should be re–tractable and explainable.

Acknowledgments. This paper was partially supported by the project: VEGA No. 1/0373/18 entitled "Big data analytics as a tool for increasing the competitiveness of enterprises and supporting informed decisions" supported by the Ministry of Education, Science, Research and Sport of the Slovak Republic.

References

1. Bashon, Y., Neagu, D., Ridley, M.J.: A framework for comparing heterogeneous objects: on the similarity measurements for fuzzy, numerical and categorical attributes. Soft Comput. **17**(9), 1595–1615 (2013)
2. Beliakov, G., Pradera, A., Calvo Sánchez, T.: Aggregation Functions: A Guide for Practitioners. Springer, Heidelberg (2007)
3. Bojadziev, G., Bojadziev, M.: Fuzzy Logic for Business, Finance and Management, 2nd edn. World Scientific Publishing, London (2007)
4. Calvo Sánchez, T., De Baets, B., Fodor, J.: The functional equations of Frank and Alsina for uninorms and nullnorms. Fuzzy Sets Syst. **120**, 385–394 (2001)
5. Choquet, G.: Theory of capacities. Ann. Inst. Fourier **5**, 1953–1954 (1954)
6. De Baets, B., Mesiar, R.: Ordinal sums of aggregation operators. In: 8th International Conference on Information Processing and Management of Uncertainty, IPMU 2000, Madrid (2000)
7. Dubois, D., Prade, H.: On the use of aggregation operations in information fusion processes. Fuzzy Sets Syst. **142**, 143–161 (2004)
8. Dubois, D., Prade, H.: Fuzzy Sets and Systems: Theory and Applications. Academic Press, New York (1980)
9. Goebel, R., Chander, A., Holzinger, K., Lecue, F., Akata, Z., Stumpf, S., Kieseberg, P., Holzinger, A.: Explainable AI: the new 42? In: Holzinger, A., Kieseberg, P., Tjoa, A., Weippl, E. (eds.) Machine Learning and Knowledge Extraction. Lecture Notes in Computer Science, LNCS, vol. 11015, pp. 295–303. Springer, Cham (2018)
10. Ishwarappa, J., Anuradha, J.: A brief introduction on big data 5Vs characteristics and hadoop technology. Procedia Comput. Sci. **48**, 319–324 (2015)
11. Joe, H.: Dependence Modeling with Copulas. Monographs on Statistics and Applied probability, vol. 134. CRC Press, Boca Raton (2015)
12. Klement, E.P., Mesiar, R., Pap, E.: Triangular Norms. Kluwer, Dordrecht (2000)
13. Ling, C.M.: Representation of associative functions. Publ. Math. Debrecen **12**, 189–212 (1965)
14. Petry, F.E.: Fuzzy Databases: Principles and Applications. Kluwer, Boston (1996)
15. Ruspini, E.H.: A new approach to clustering. Inf. Control **15**, 22–32 (1969)
16. Shenoi, S., Melton, A.: Proximity relations in the fuzzy relational database model. Fuzzy Sets Syst. **100**, 51–62 (1989)
17. Singh, S., Ribeiro, M., Guestrin, C.: Programs as black–box explanations. In: Workshop on Interpretable Machine Learning in Complex Systems, NIPS 2016, Barcelona (2016)
18. Snasel, V., Kromer, P., Musilek, P., Nyongesa, H.O., Husek, D.: Fuzzy modeling of user needs for improvement of web search queries. In: Annual Meeting of the North American Fuzzy Information Processing Society, NAFIPS 2007, San Diego (2007)
19. Sözat, M., Yazici, A.: A complete axiomatization for fuzzy functional and multi-valued dependencies in fuzzy database relations. Fuzzy Sets Syst. **117**(2), 161–181 (2001)
20. Tung, A.K.H., Zhang, R., Koudas, N., Ooi, B.C.: Similarity search: a matching based approach. In: 32nd International Conference on Very Large Data Bases, Seoul (2006)
21. Vučetić, M., Hudec, M.: A flexibile approach to matching user preferences with records in datasets based on the conformance measure and aggregation functions. In: 10th International Joint Conference on Computational Intelligence, IJCCI 2018, Seville (2018)

22. Vucetic, M., Hudec, M., Vujošević, M.: A new method for computing fuzzy functional dependencies in relational database systems. Expert Syst. Appl. **40**, 2738–2745 (2013)
23. Vučetić, M., Vujošević, M.: A literature overview of functional dependencies in fuzzy relational database models. Technics Technol. Educ. Manag. **7**, 1593–1604 (2012)
24. Yager, R.R.: Noble reinforcement in disjunctive aggregation operators. IEEE Trans. Fuzzy Syst. **11**, 754–767 (2003)
25. Yager, R., Rybalov, A.: Uninorm aggregation operators. Fuzzy Sets Syst. **80**, 111–120 (1996)

Construction of Fuzzy Implication Functions Based on F-chains

Radko Mesiar[1]($\boxtimes$) and Anna Kolesárová[2]

[1] Faculty of Civil Engineering, Slovak University of Technology,
Radlinského 11, 810 05 Bratislava, Slovakia
radko.mesiar@stuba.sk
[2] Faculty of Chemical and Food Technology, Slovak University of Technology,
Radlinského 9, 812 37 Bratislava, Slovakia
anna.kolesarova@stuba.sk

Abstract. In this paper, we present a construction method for fuzzy implication functions based on F-chains. Using this method, which comes from the theory of aggregation functions, one can construct from any given fuzzy implication function a new one. We discuss some properties of fuzzy implication functions which are preserved by this construction, and also provide several examples. Moreover, we introduce for fuzzy implication functions ordinal sums based on F-chains in the case of a single fuzzy implication function as well as in the case of n fuzzy implication functions.

Keywords: Aggregation function · Fuzzy implication function ·
F-chain · Ordinal sum of fuzzy implication functions

1 Introduction

Fuzzy implication functions belong to essential operators in fuzzy logic and approximate reasoning, in particular when dealing with inferences and fuzzy relations. Fuzzy implication functions also play an important role in many other fields, such as image processing, integration theory based on monotone measures, etc. For more details we recommend [4,5,16].

Such great variety of applications has led to the systematical investigation of fuzzy implication functions from a purely theoretical point of view (see, e.g., [3, 4]), and, in particular, to searching for new construction methods. Among known construction methods, there are some methodologies enabling one to construct a new fuzzy implication function from one or two given ones, for instance,

- classical methods, such as conjugation, reciprocation, the upper, lower and medium contrapositivisations, or
- more recent construction methods that have been introduced and studied by many authors, e.g., some new types of contrapositivisations [1], horizontal and vertical threshold generation methods [17–19], the FNI-method [2,24], the ⊛-composition [25,26], etc. For more details, see also the monograph [3], and specially, the chapter [20], and

© Springer Nature Switzerland AG 2019
R. Halaš et al. (Eds.): AGOP 2019, AISC 981, pp. 315–326, 2019.
https://doi.org/10.1007/978-3-030-19494-9_29

– as one of the newest methods enabling one to construct a new fuzzy implication function from a given one, we recall the construction based on quadratic polynomials of 3 variables initiated in [21], and presented in more details in [15].

Observe that several results dealing with fuzzy implication functions were obtained due to duality between these functions and binary aggregation functions with annihilator $a \in \{0, 1\}$, see, e.g., [22,23]. Even, if in some cases new construction methods introduced in aggregation theory cannot be directly rewritten for fuzzy implication functions, they can serve as motivation for the study of similar usable methods. As an example, recall a recent proposal of constructing aggregation functions, and, in particular, binary copulas, by means of ternary quadratic polynomials proposed and discussed in [13], see also [14]. This approach has served as a background for the above mentioned construction method for fuzzy implication functions based on quadratic polynomials studied in [15]. This contribution to construction methods of fuzzy implication functions was also inspired by a recent construction method for aggregation functions, namely by construction considering the so-called F-chains [11], i.e., by an approach applicable both in the framework of aggregation functions as well as in the framework of fuzzy implication functions.

The paper is organized as follows. In the next section, some necessary preliminaries concerning fuzzy implication functions are given. In Sect. 3, F-chains based construction of fuzzy implication functions, starting from a given fuzzy implication function I, is introduced and studied, including several examples. Moreover, in this section, some properties of fuzzy implication functions preserved by this construction are also discussed. Section 4 is devoted to the proposal of ordinal sums of fuzzy implication functions based on F-chains related to a given weighted arithmetic mean. Finally, some concluding remarks are added.

2   Preliminaries

We start by recalling some basic concepts that will be used throughout the paper. First, we give the definition of fuzzy implication function.

Definition 2.1. *A binary operation* $I : [0,1]^2 \to [0,1]$ *is called a* fuzzy implication function *if it satisfies:*

(I1) $I(x, z) \geq I(y, z)$ *when* $x \leq y$, *for all* $z \in [0, 1]$;
(I2) $I(x, y) \leq I(x, z)$ *when* $y \leq z$, *for all* $x \in [0, 1]$;
(I3) $I(0, 0) = I(1, 1) = 1$ *and* $I(1, 0) = 0$.

Let us denote by $\mathcal{I}$ the class of all fuzzy implication functions. Note that from the definition, it can be deduced that for all $I \in \mathcal{I}$, $I(0, x) = 1$ and $I(x, 1) = 1$ for all $x \in [0, 1]$, while the values $I(x, 0)$ and $I(1, x)$ for $x \in]0, 1[$ are not determined from the definition.

We recall some additional properties of fuzzy implication functions which will be used in this work:

- The *left neutrality principle*:

$$I(1, y) = y, \quad y \in [0, 1]. \tag{NP}$$

- The *identity principle*:

$$I(x, x) = 1, \quad x \in [0, 1]. \tag{IP}$$

- The *ordering property*:

$$x \leq y \iff I(x, y) = 1, \quad x, y \in [0, 1]. \tag{OP}$$

- The *law of contraposition* with respect to the standard fuzzy negation:

$$I(1 - y, 1 - x) = I(x, y), \quad x, y \in [0, 1]. \tag{CP}$$

Recall that the standard fuzzy negation $N_s \colon [0, 1] \to [0, 1]$ is given by $N_s(x) = 1 - x$.

Definition 2.2. *Let I be a fuzzy implication function. The function N_I, defined by $N_I(x) = I(x, 0)$ for all $x \in [0, 1]$, is called the* natural negation *of I.*

Fuzzy implication functions are closely related to aggregation functions [6–8, 10].

Definition 2.3. *Let $n \in \mathbb{N}$. A monotone function $A \colon [0, 1]^n \to [0, 1]$ is called an n-ary aggregation function whenever it satisfies the boundary conditions $A(0, \ldots, 0) = 0$ and $A(1, \ldots, 1) = 1$. The class of all n-ary aggregation functions on $[0, 1]$ will be denoted by $\mathcal{A}_n$.*

Obviously, from the boundary conditions for aggregation functions we can deduce that aggregation functions are monotone non-decreasing. Note that throughout the paper, all functions which are monotone non-decreasing (monotone non-increasing) on their domains, will be called briefly increasing (decreasing).

Further, if a binary aggregation function $A \colon [0, 1]^2 \to [0, 1]$ has an annihilator $a = 0$, i.e., $A(0, x) = A(x, 0) = 0$ for all $x \in [0, 1]$, then the function $I_A \colon [0, 1]^2 \to [0, 1]$ given by $I_A(x, y) = 1 - A(x, 1 - y)$ is a fuzzy implication function. Similarly, using a fuzzy implication function I, one can introduce a binary aggregation function A_I with annihilator $a = 0$, namely, $A_I(x, y) = 1 - I(x, 1 - y)$. Observe that for each aggregation function A with annihilator $a = 0$ and for each fuzzy implication function I we have $A_{I_A} = A$ and $I_{A_I} = I$.

Note that a similar duality connects binary aggregation functions with annihilator $a = 1$ and fuzzy implication functions, namely, for each such aggregation function $A \in \mathcal{A}_2$ and each $I \in \mathcal{I}$, we have $I^A(x, y) = A(1 - x, y)$, $A^I(x, y) = I(1 - x, y)$, and $I^{A^I} = I$, $A^{I^A} = A$.

As we have already mentioned in Introduction, there exist different construction methods of new fuzzy implication functions from given ones. Recall that for any n-ary aggregation function $A\colon [0,1]^n \to [0,1]$ and binary aggregation functions $A_1, \ldots, A_n\colon [0,1]^2 \to [0,1]$ with annihilator $a = 0$, also the composite $B = A(A_1, \ldots, A_n)\colon [0,1]^2 \to [0,1]$ given by

$$B(x,y) = A(A_1(x,y), \ldots, A_n(x,y))$$

is a binary aggregation function with annihilator 0. Therefore, a similar claim also holds for fuzzy implication functions. Namely, for any n-ary aggregation function A and any fuzzy implication functions $I_1, \ldots, I_n$, also the composite $I = A(I_1, \ldots, I_n)\colon [0,1]^2 \to [0,1]$ given by

$$I(x,y) = A(I_1(x,y), \ldots, I_n(x,y))$$

is a fuzzy implication function. In particular, when A is a weighted arithmetic mean, $A(x_1, \ldots, x_n) = \sum_{i=1}^{n} w_i x_i$, we see that $I = \sum_{i=1}^{n} w_i I_i$, i.e., a convex sum of fuzzy implication functions $I_1, \ldots, I_n$ is a fuzzy implication function.

Similarly, for any binary aggregation function $A\colon [0,1]^2 \to [0,1]$ with annihilator $a = 0$, and any $A_1, A_2, A_3 \in \mathcal{A}_1$, the composite $B = A_1(A(A_2, A_3))\colon [0,1]^2 \to [0,1]$ given by

$$B(x,y) = A_1(A(A_2(x), A_3(y)))$$

is a binary aggregation function with annihilator $a = 0$. Therefore, for any fuzzy implication function I and any $A_1, A_2, A_3 \in \mathcal{A}_1$, the composite J,

$$J = A_1(I(A_2, A_3))\colon [0,1]^2 \to [0,1], \quad \text{given by } J(x,y) = A_1(I(A_2(x), A_3(y))),$$

is also a fuzzy implication function. In particular, if $A_2 = A_3 = \varphi\colon [0,1] \to [0,1]$ is a bijective mapping and $A_1 = \varphi^{-1}$, then J, given by

$$J(x,y) = \varphi^{-1}(I(\varphi(x), \varphi(y))),$$

is an isomorphic transform of I.

3   F-chains Based Construction of Fuzzy Implication Functions

F-chains were introduced in [11] as a background for a particular construction method for aggregation functions.

Definition 3.1. *Let $F\colon [0,1]^n \to [0,1]$ be an aggregation function, and let $\mathbf{c}\colon [0,1] \to [0,1]^n$ be an increasing mapping such that $\mathbf{c}(0) = (0, \ldots, 0)$, $\mathbf{c}(1) = (1, \ldots, 1)$, and satisfying $F(\mathbf{c}(t)) = t$ for all $t \in [0,1]$. Then $\mathbf{c}$ is called an F-chain.*

Each F-chain $\mathbf{c}$ can be written in the form $\mathbf{c} = (c_1, \ldots, c_n)$, where $c_1, \ldots, c_n\colon [0,1] \to [0,1]$ are increasing functions. Note that for any continuous n-ary aggregation function F there exists an F-chain $\mathbf{c}$. Clearly, a necessary condition for the existence of an F-chain is the full range of the considered n-ary aggregation function F, i.e., $\mathrm{Ran}(F) = [0,1]$.

Theorem 3.1. *Let F be an n-ary aggregation function and let $\mathbf{c}$ be an F-chain. Then for each fuzzy implication function $I \in \mathcal{I}$, the function $I_{F,\mathbf{c}} \colon [0,1]^2 \to [0,1]$ given by*

$$I_{F,\mathbf{c}}(x,y) = F(I(c_1(x), c_1(y)), \ldots, I(c_n(x), c_n(y))) \tag{1}$$

is a fuzzy implication function.

Proof: The introduced function $I_{F,\mathbf{c}}$ satisfies the boundary conditions of fuzzy implication functions recalled in Definition 2.1, item (I3). Indeed,

$$
\begin{aligned}
I_{F,\mathbf{c}}(0,0) &= F(I(c_1(0), c_1(0)), \ldots, I(c_n(0), c_n(0))) = F(I(0,0), \ldots, I(0,0)) \\
&= F(1, \ldots, 1) = 1,
\end{aligned}
$$

and similarly, it can be shown that $I_{F,\mathbf{c}}(1,1) = 1$ and $I_{F,\mathbf{c}}(1,0) = 0$.

Further, we have to prove that $I_{F,\mathbf{c}}$ also satisfies the monotonicity conditions (I1) and (I2) from Definition 2.1.

Let us first fix any $y \in [0,1]$ and consider any $x_1, x_2 \in [0,1]$ such that $x_1 \le x_2$. Then $c_i(x_1) \le c_i(x_2)$ for each $i \in \{1, \ldots, n\}$, and thus, due to the property (I1) of I,

$$I(c_i(x_1), c_i(y)) \ge I(c_i(x_2), c_i(y)),$$

and, because of the increasingness of F, we obtain

$$
\begin{aligned}
I_{F,\mathbf{c}}(x_1, y) &= F(I(c_1(x_1), c_1(y)), \ldots, I(c_n(x_1), c_n(y))) \\
&\ge F(I(c_1(x_2), c_1(y)), \ldots, I(c_n(x_2), c_n(y))) = I_{F,\mathbf{c}}(x_2, y),
\end{aligned}
$$

which proves the decreasingness of $I_{F,\mathbf{c}}$ in the first variable.

Similarly, one can show that $I_{F,\mathbf{c}}(x, y_1) \le I_{F,\mathbf{c}}(x, y_2)$ for any fixed $x \in [0,1]$ and arbitrary $y_1, y_2 \in [0,1]$ such that $y_1 \le y_2$, which proves the increasingness of $I_{F,\mathbf{c}}$ in the second variable, and we can conclude that $I_{F,\mathbf{c}}$ is a fuzzy implication function. $\qquad\square$

Construction introduced in Theorem 3.1 preserves some properties of fuzzy implication functions. Before showing a few of them, we recall that an n-ary aggregation function F

– is called self-dual if for each $(x_1, \ldots, x_n) \in [0,1]^n$,

$$F(x_1, \ldots, x_n) = 1 - F(1 - x_1, \ldots, 1 - x_n);$$

– has no unit multipliers if $F(x_1, \ldots, x_n) = 1$ if and only if $x_1 = \cdots = x_n = 1$.

Proposition 3.1. *Let the assumptions of Theorem 3.1 hold. Then we have:*

(i)   *If I satisfies the left neutrality principle* **(NP)**, *then the same holds for $I_{F,\mathbf{c}}$.*

(ii) *If I satisfies the identity principle* **(IP)**, *then the same applies to $I_{F,\mathbf{c}}$.*

Proof:
(i) Let $I \in \mathcal{I}$ satisfy **(NP)**. Then

$$I_{F,\mathbf{c}}(1,y) = F(I(c_1(1), c_1(y)), \ldots, I(c_n(1), c_n(y))) = F(I(1, c_1(y)), \ldots, I(1, c_n(y)))$$
$$= F(c_1(y), \ldots, c_n(y)) = y,$$

which proves that $I_{F,\mathbf{c}}$ also satisfies **(NP)**.
(ii) The proof of this property is similar to the proof of the claim (i), and therefore omitted. □

Under some mild constraints put on F or $\mathbf{c}$, also some other properties of fuzzy implication functions are preserved by construction (1).

Proposition 3.2. *Let the assumptions of Theorem 3.1 hold. Then we have:*

(i) *If the natural negation of I is the standard negation N_s, i.e., $I(x,0) = 1-x$ for all $x \in [0,1]$, and if F is a self-dual aggregation function, then $I_{F,\mathbf{c}}$ also induces the standard fuzzy negation.*
(ii) *If I satisfies the law of contraposition **(CP)** with respect to N_s, and if F is a self-dual aggregation function and the F-chain $\mathbf{c}$ satisfies the symmetry condition*

$$\mathbf{c}(1-t) + \mathbf{c}(t) = (1, \ldots, 1)$$

*for all $t \in [0,1]$, then $I_{F,\mathbf{c}}$ also satisfies **(CP)** with respect to N_s.*
(iii) *If I satisfies the ordering property **(OP)**, and if the F-chain $\mathbf{c}$ is strictly increasing in each component and F has no unit multipliers, then $I_{F,\mathbf{c}}$ also satisfies **(OP)**.*

Proof:
(i) Consider that $I \in \mathcal{I}$ satisfies $N_I = N_s$. Then

$$N_{I_{F,\mathbf{c}}}(x) = I_{F,\mathbf{c}}(x,0) = F(I(c_1(x), c_1(0)), \ldots, I(c_n(x), c_n(0)))$$
$$= F(I(c_1(x),0), \ldots, I(c_n(x),0)) = F(1 - c_1(x), \ldots, 1 - c_n(x))$$
$$= 1 - F(c_1(x), \ldots, c_n(x)) = 1 - x = N_s(x),$$

i.e., the standard negation is also a natural negation for $I_{F,\mathbf{c}}$.
As the proofs of items (ii) and (iii) run in a similar way, we omit the details. □

Example 3.1.
(i) Let $F: [0,1]^2 \to [0,1]$ be the arithmetic mean, $F(x,y) = \frac{x+y}{2}$, and let

$$\mathbf{c}(t) = (t^2, 2t - t^2), \quad t \in [0,1].$$

Then $\mathbf{c}$ is an F-chain and the function $I_{F,\mathbf{c}}: [0,1]^2 \to [0,1]$ given by

$$I_{F,\mathbf{c}}(x,y) = \frac{I(x^2, y^2) + I(2x - x^2, 2y - y^2)}{2} \tag{2}$$

is a fuzzy implication function for each $I \in \mathcal{I}$.

Clearly, if I satisfies (NP), (IP) or (OP), then the same is true for each $I_{F,\mathbf{c}}$ given by (2). Concerning (OP), observe that the given $\mathbf{c}$ is strictly increasing on $[0,1]$, and the arithmetic mean has no unit multipliers.

Note that if $I = I_L$ is the Łukasiewicz implication, where $I_L(x,y) = \min\{1, 1 - x + y\}$ for all $(x,y) \in [0,1]^2$, then I is invariant with respect to construction introduced in (2), i.e., $I_{F,\mathbf{c}} = I$.

Similarly, if $I = I_{GD}$ is the Gödel implication, which is given by

$$I_{GD}(x,y) = \begin{cases} 1 & \text{if } x \leq y, \\ y & \text{otherwise,} \end{cases}$$

then we also have $I_{F,\mathbf{c}} = I$.

However, this is no more true for the Kleene-Dienes implication, i.e., if $I = I_{KD}$, where $I_{KD}(x,y) = \max\{1 - x, y\}$ for all $(x,y) \in [0,1]^2$, because in that case we obtain

$$I_{F,\mathbf{c}}(x,y) = \begin{cases} y & \text{if } x^2 + y^2 \geq 1, \\ 1 - x & \text{if } (1-x)^2 + (1-y)^2 \geq 1, \\ \frac{1 - x^2 + 2y - y^2}{2} & \text{otherwise.} \end{cases} \tag{3}$$

Note that $I_{F,\mathbf{c}}$ given by (3) satisfies (CP), although the F-chain $\mathbf{c}$ does not satisfy the symmetry property, see Proposition 3.2, item (ii). This fact shows that Proposition 3.2 (ii) gives sufficient conditions for preservation of (CP) only, but not necessary.

(ii) Let $F \colon [0,1]^2 \to [0,1]$ be the minimum, $F(x,y) = \min\{x,y\}$, and let

$$\mathbf{c}(t) = (t, 2t - t^2), \ t \in [0,1].$$

Then $\mathbf{c}$ is an F-chain and, for each $I \in \mathcal{I}$, the function $I_{F,\mathbf{c}} \colon [0,1]^2 \to [0,1]$ given by

$$I_{F,\mathbf{c}}(x,y) = \min\{I(x,y), I(2x - x^2, 2y - y^2)\} \tag{4}$$

is also a fuzzy implication function. Observe that if $I = I_L$, we have

$$I_{F,\mathbf{c}}(x,y) = \min\{1, 1 - x + y, 1 + (1-x)^2 \quad (1-y)^2\}, \tag{5}$$

and that $I_{F,\mathbf{c}}$ (see Fig. 1) satisfies (NP), (IP) and (OP), but not (CP). Also observe that a natural negation corresponding to $I_{F,\mathbf{c}}$, which is mentioned in (5), is given by $N_{I_{F,\mathbf{c}}}(x) = I_{F,\mathbf{c}}(x,0) = (1-x)^2$, $x \in [0,1]$.

Example 3.2.
(i) Let $F \colon [0,1]^2 \to [0,1]$ be given by $F(x,y) = \min\{1, x + y\}$ (i.e., let F be the Łukasiewicz t-conorm). Put

$$\mathbf{c}(t) = \begin{cases} \left(\frac{t}{2}, \frac{t}{2}\right) & \text{if } t \in [0,1[, \\ (1,1) & \text{if } t = 1. \end{cases}$$

Then $\mathbf{c}$ is an F-chain, and for each $I \in \mathcal{I}$ satisfying (NP), we obtain the function

$$I_{F,\mathbf{c}}(x,y) = \begin{cases} \min\{1, 2I\left(\frac{x}{2}, \frac{y}{2}\right)\} & \text{if } (x,y) \in [0,1[^2, \\ 1 & \text{if } y = 1, \\ y & \text{if } x = 1, \end{cases}$$

which also belongs to $\mathcal{I}$.

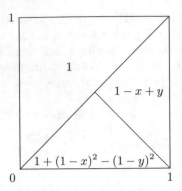

Fig. 1. Fuzzy implication function $I_{F,\mathbf{c}}$ given by (5), satisfying **(NP)**, **(IP)** and **(OP)**, see Example 3.1(ii)

In particular,

– if $I = I_L$, then

$$I_{F,\mathbf{c}}(x,y) = \begin{cases} y & \text{if } x = 1, \\ 1 & \text{otherwise;} \end{cases}$$

– if $I = I_{GG}$, where I_{GG} is the Goguen implication given by

$$I_{GG}(x,y) = \begin{cases} 1 & \text{if } x \le y, \\ \frac{y}{x} & \text{if } x > y, \end{cases}$$

then

$$I_{F,\mathbf{c}}(x,y) = \begin{cases} \min\left\{1, \frac{2y}{x}\right\} & \text{if } x < 1, \\ y & \text{otherwise.} \end{cases}$$

(ii) Consider the product $F \colon [0,1]^n \to [0,1]$, $F(x_1,\ldots,x_n) = \prod_{i=1}^{n} x_i$. For constants $w_1,\ldots,w_n > 0$, such that $\sum_{i=1}^{n} w_i = 1$, put

$$\mathbf{c}(t) = (t^{w_1},\ldots,t^{w_n}), \quad t \in [0,1].$$

Then $\mathbf{c}$ is an F-chain, and for any $I \in \mathcal{I}$ we obtain the function

$$I_{F,\mathbf{c}}(x,y) = \prod_{i=1}^{n} I\left(x^{w_i}, y^{w_i}\right), \tag{6}$$

which also belongs to $\mathcal{I}$.

In particular, if $I = I_L$, then

$$I_{F,\mathbf{c}}(x,y) = \min\left\{1, \prod_{i=1}^{n}(1 - x^{w_i} + y^{w_i})\right\}. \tag{7}$$

Note that if $w_1 = \cdots = w_n = \frac{1}{n}$, then $I_{F,\mathbf{c}}$, given by (7), is a residual implication related to the Schweizer-Sklar t-norm $T_{1/n}^{SS}$ [12], given for all $(x,y) \in [0,1]^2$ by

$$T_{1/n}^{SS}(x,y) = \left(\max\left\{ 0, x^{1/n} + y^{1/n} - 1 \right\} \right)^n.$$

Remind that when applying formula (7) for $w_1 = \cdots = w_n = \frac{1}{n}$, then by letting $n \to \infty$ we obtain the limit

$$\lim_{n\to\infty} \min\left\{ 1, \left(1 - x^{\frac{1}{n}} + y^{\frac{1}{n}} \right)^n \right\} = \begin{cases} 1 & \text{if } x \le y, \\ \frac{y}{x} & \text{otherwise,} \end{cases}$$

i.e., we recover the Goguen fuzzy implication function I_{GG}.

Also observe that if $I = I_{GD}$ or $I = I_{GG}$, then $I_{F,\mathbf{c}} = I$, i.e., these fuzzy implications are invariant with respect to construction described by (6).

4 F-chain-based Ordinal Sums of Fuzzy Implication Functions

In this section, we will consider a fixed weighting vector $\mathbf{w} = (w_1, \ldots, w_n) \in$ $]0,1]^n$ satisfying the property $\sum_{i=1}^{n} w_i = 1$. For each $i = 1, \ldots, n$ we will write $v_i = w_1 + \cdots + w_i$, and we denote by $W_{\mathbf{w}}$ the related weighted arithmetic mean, i.e., $W_{\mathbf{w}} : [0,1]^n \to [0,1]$ is an aggregation function given by

$$W_{\mathbf{w}}(\mathbf{x}) - \sum_{i=1}^{n} w_i x_i.$$

Further, we denote by $\mathbf{c_w}$ a $W_{\mathbf{w}}$-chain $\mathbf{c_w} : [0,1] \to [0,1]^n$, introduced in [11], and given by

$$\mathbf{c_w}(t) = \begin{cases} \left(\frac{t}{w_1}, 0, \ldots, 0 \right) & \text{if } t \in [0, w_1], \\ \left(1, \ldots, 1, \frac{t - v_{i-1}}{w_i}, 0, \ldots, 0 \right) & \text{if } t \in]v_{i-1}, v_i], \end{cases}$$

where in the latter case, 1 is taken $(i-1)$-times and 0 is taken $(n-i)$-times.

Finally, for each $I \in \mathcal{I}$ we introduce a fuzzy implication function $I_{\mathbf{w}} : [0,1]^2 \to [0,1]$, given as proposed in (1) when $F = W_{\mathbf{w}}$ and $\mathbf{c} = \mathbf{c_w}$ (i.e., we have abbreviated $I_{W_{\mathbf{w}},\mathbf{c_w}}$ to $I_{\mathbf{w}}$). Thus, $I_{\mathbf{w}}$ is given as follows:

– If $(x,y) \in [v_{i-1}, v_i]^2$ (with convention $v_0 = 0$), then

$$I_{\mathbf{w}}(x,y) = 1 - w_i + w_i \cdot I\left(\frac{x - v_{i-1}}{w_i}, \frac{y - v_{i-1}}{w_i} \right);$$

– If $(x,y) \in [v_{k-1}, v_k] \times [v_{j-1}, v_j]$, $k \ne j$, then

$$I_{\mathbf{w}}(x,y) = \sum_{i=1}^{n} (1 - w_i + w_i \cdot I(s_i, t_i)) - (n-1),$$

where $s_i = \max\{0, \min\{1, \frac{x - v_{i-1}}{w_i}\}\}$, $t_i = \max\{0, \min\{1, \frac{y - v_{i-1}}{w_i}\}\}$, for each $i = 1, \ldots, n$.

Observe that a similar approach was considered by De Baets and Mesiar in [9] when introducing ordinal sums of aggregation functions.

Our approach for introducing an ordinal sum of fuzzy implication functions mentioned above is based on a single fuzzy implication function I. It is only a matter of verification of the axioms for fuzzy implication functions to prove the next result based on n fuzzy implication functions $I_1, \ldots, I_n$.

Theorem 4.1. *Let* $\mathbf{w} \in]0,1]^n$ *be a fixed weighting vector and let* $I_1, \ldots, I_n \colon [0,1]^2 \to [0,1]$ *be fuzzy implication functions. Then the ordinal sum* $I_{\mathbf{w},I_1,\ldots,I_n} \colon [0,1]^2 \to [0,1]$ *given as follows:* *if* $(x,y) \in [v_{i-1}, v_i]^2$ *(with convention* $v_0 = 0$*),*

$$I_{\mathbf{w},I_1,\ldots,I_n}(x,y) = 1 - w_i + w_i \cdot I_i \left(\frac{x - v_{i-1}}{w_i}, \frac{y - v_{i-1}}{w_i} \right),$$

and if $(x,y) \in [v_{k-1}, v_k] \times [v_{j-1}, v_j]$*,* $k \neq j$*,*

$$I_{\mathbf{w},I_1,\ldots,I_n}(x,y) = \sum_{i=1}^{n} (1 - w_i + w_i \cdot I_i(s_i, t_i)) - (n - 1),$$

where $s_i = \max\{0, \min\{1, \frac{x - v_{i-1}}{w_i}\}\}$*,* $t_i = \max\{0, \min\{1, \frac{y - v_{i-1}}{w_i}\}\}$*,* $i = 1, \ldots, n$*,* *is also a fuzzy implication function.*

The ordinal sum construction method given in Theorem 4.1 preserves several properties of fuzzy implication functions, for example:

Proposition 4.1. *Let* $\mathbf{w} \in]0,1]^n$ *be a fixed weighting vector and let* $I_1, \ldots, I_n$ *be fuzzy implication functions, all of them satisfying the properties* **(NP)**, **(IP)**, **(OP)** *and* $N_{I_i} = N_s$*, respectively. Then the ordinal sum* $I_{\mathbf{w},I_1,\ldots,I_n}$*, introduced in Theorem 4.1, also satisfies the properties* **(NP)**, **(IP)**, **(OP)** *and* $N_{I_{\mathbf{w},I_1,\ldots,I_n}} = N_s$*.*

5   Conclusion

The construction method for fuzzy implication functions introduced in this paper enables one to construct a (possibly new) fuzzy implication function from any given one. In addition to introducing the construction method, we have also studied preserving some of the properties of an original fuzzy implication function by this construction. We have shown that the proposed construction always preserves the left neutrality principle and identity principle, and we have provided sufficient conditions for preserving the ordering property and the law of contraposition. Finally, we have introduced and discussed ordinal sums for fuzzy implication functions. Note that for a fixed aggregation function F and a related F-chain $\mathbf{c}$, the mapping $\xi \colon \mathcal{I} \to \mathcal{I}$, given by $\xi(I) = I_{F,\mathbf{c}}$, can be seen as a transformation of fuzzy implication functions. Thus, in further research, some other problems can be solved, for example, characterization of invariant fuzzy implication functions, i.e., those satisfying $\xi(I) = I$, or investigation of properties of composites $(\xi)^n = \xi \circ \cdots \circ \xi$, etc.

Acknowledgement. R. Mesiar kindly acknowledges the support of the grant VEGA 1/0006/19 and A. Kolesárová is grateful for the support of the grant VEGA 1/0614/18.

References

1. Aguiló, I., Suñer, J., Torrens, J.: New types of contrapositivisation of fuzzy implications with respect to fuzzy negations. Inform. Sci. **322**, 223–236 (2015)
2. Aguiló, I., Suñer, J., Torrens, J.: A new look on fuzzy implication functions: FNI-implications. In: Carvalho, J.P., et al. (eds.) Proceedings of the 16th International Conference IPMU 2016, Part I, Part of the Communications in Computer and Information Science, vol. 610, pp. 375–386. Springer (2016)
3. Baczyński, M., Beliakov, G., Bustince, H., Pradera, A.: Advances in Fuzzy Implication Functions. Sudies in Fuzziness and Soft Computing Series, vol. 300. Springer, Heidelberg (2013)
4. Baczyński, M., Jayaram, B.: Fuzzy Implications. Sudies in Fuzziness and Soft Computing Series, vol. 231. Springer, Heidelberg (2008)
5. Baczyński, M., Jayaram, B., Massanet, S., Torrens, J.: Fuzzy implications: past, present, and future. In: Kacprzyk, J., Pedrycz, W. (eds.) Springer Handbook of Computational Intelligence, pp. 183–202. Springer, Heidelberg (2015)
6. Beliakov, G., Pradera, A., Calvo, T.: Aggregation Functions: A Guide for Practitioners. Springer, Heidelberg (2007)
7. Beliakov, G., Bustince, H., Calvo, T.: A Practical Guide to Averaging Functions. Springer, Heidelberg (2016)
8. Calvo, T., Kolesárová, A., Komorníková, M., Mesiar, R.: Aggregation operators: properties, classes and construction methods. In: Calvo, T., Mayor, G., Mesiar, R. (eds.) Aggregation Operators. New Trends and Applications, pp. 3–107. Physica-Verlag, Heidelberg (2002)
9. De Baets, B., Mesiar, R.: Ordinal sums of aggregation operators. In: Bouchon-Meunier, B., et al. (eds.) Technologies for Constructing Intelligent Systems 2, STUDFUZZ, vol. 90, pp. 137–147 (2002)
10. Grabisch, M., Marichal, J.-L., Mesiar, R., Pap, E.: Aggregation Functions. Cambridge University Press, Cambridge (2009)
11. Jin, L., Mesiar, R., Kalina, M., Špirková, J., Borkotokey, S.: Generalized phi-transformation of aggregation funtions. Fuzzy Sets and Systems. https://doi.org/10.1016/j.fss.2018.09.016
12. Klement, E.P., Mesiar, R., Pap, E.: Triangular Norms. Kluwer Academic Publishers, Dordrecht (2000)
13. Kolesárová, A., Mesiar, R.: On linear and quadratic constructions of aggregation functions. Fuzzy Sets Syst. **268**, 1–14 (2015)
14. Kolesárová, A., Mayor, G., Mesiar, R.: Quadratic construction of copulas. Inform. Sci. **310**, 69–76 (2015)
15. Kolesárová, A., Massanet, S., Mesiar, R., Riera, J.V., Torrens, J.: Polynomial constructions of fuzzy implication functions: the quadratic case. Inform. Sci. (2018, submitted)
16. Mas, M., Monserrat, M., Torrens, J., Trillas, E.: A survey on fuzzy implication functions. IEEE Trans. Fuzzy Syst. **15**(6), 1107–1121 (2007)
17. Massanet, S., Torrens, J.: On some properties of threshold generated implications. Fuzzy Sets Syst. **205**, 30–49 (2012)
18. Massanet, S., Torrens, J.: Threshold generation method of construction of a new implication from two given ones. Fuzzy Sets Syst. **205**, 50–75 (2012)

19. Massanet, S., Torrens, J.: On the vertical threshold method of fuzzy implication and its properties. Fuzzy Sets Syst. **226**, 32–52 (2013)
20. Massanet, S., Torrens, J.: An overview of construction methods of fuzzy implications. In: Baczyński, M., Beliakov, G., Bustince, H., Pradera, A. (eds.) Advances in Fuzzy Implication Functions. Sudies in Fuzziness and Soft Computing Series, vol. 300, pp. 1–30. Springer, Heidelberg (2013)
21. Massanet, S., Riera, J.V., Torrens, J.: On linear and quadratic constructions of fuzzy implication functions. In: Medina, J., et al. (eds.) IPMU 2018. CCIS, vol. 853, pp. 623–635. Springer International Publishing AG (2018)
22. Mesiar, R., Kolesárová, A., Bustince, H., Fernandez, J.: Dualities in the class of extended Boolean functions. Fuzzy Sets Syst. **332**, 78–92 (2018)
23. Pradera, A., Beliakov, G., Bustince, H., De Baets, B.: A review of the relationships between implication, negation and aggregation functions from the point of view of material implication. Inform. Sci. **329**, 357–380 (2016)
24. Shi, Y., Gasse, B., Ruan, D., Kerre, E.: On dependencies and independencies of fuzzy implications axioms. Fuzzy Sets Syst. **161**(10), 1388–1405 (2010)
25. Vemuri, N.R., Jayaram, B.: Representation through a monoid on the set of fuzzy implications. Fuzzy Sets Syst. **247**, 51–67 (2014)
26. Vemuri, N.R., Jayaram, B.: The ⊛-composition of fuzzy implications: closures with respect to properties, powers and families. Fuzzy Sets Syst. **275**, 58–87 (2015)

Modalities Based on Double Negation

József Dombi$^{(\boxtimes)}$

Department of Informatics, University of Szeged, Szeged, Hungary
dombi@inf.u-szeged.hu

Abstract. Modal operators play an important role in fuzzy theory, and in recent years researchers have devoted more effort on this topic. Here we concentrate on continuous strictly monotonously increasing Archimedian t-norms. In our study, we will construct modal operators related to negation operators and we introduce graded modal operators.

Keywords: Negation · Modalities · Pliant logic ·
Necessity and possibility operators

1 Introduction

In logic, modal operators have a variety of applications and even from a theoretical perspective it is interesting to study the continuous extension of these operators. Our approach is different from other authors, because we would like to find proper algebraic expressions for these operators based on some basic considerations. On the one hand, continuous-valued logic can be studied from a logical point of view (axiomatization, completeness, possible extensions, predicat calculi, etc.). On the other hand it can be studied from algebraic point of view to find the proper operator, as we do it in conjunctive and disjunctive operators and now we have Frank, Hamacher, Einstein mean operators etc. With the latter, he have to solve functional equations. If we have different continuous-valued logical system (i.e. operators), we have to build different modal operators.

Our objective is to find these operators. If a logical operator is given we construct its unary operator. Different unary operators are studied in continuous valued (fuzzy) logic as modal operators (necessity and possibility, hedges (strengthened and weakened operators, truth-value modifiers, truth-stresser, truth-depresser), etc). Here, we will present approaches for obtaining the concrete form of the necessity and possibility operators. These may be expressed in a simple parametrical form. By modifying the parameter value, we get different unary operators, namely modality, hedge and negation operators.

In this paper we deal with continuous valued, Archimedian t-norm based logic. It is Hájek BL with the exception of Gödel logic, since in the latter $x \wedge x = x$ holds. If we use BL formalism (i.e. strong negation to define modalities), then the most general approach to deal with involution in t-norm based logic is the paper of Flaminio, Marchioni [10]. Here the authors set of logical frame to Esteva and Godo monodial logic MTL, which contains BL. So we can say that this is the most general logic from this point of view. Cintulas paper only deals with involutive expansions of the logic SBL, which includes Gödel logic. We have to mention that Esteva, Godo and Noguera [6] study the probably most general logic for truth-hedges. It is the closest to the system which is given in this paper. Cintula et al. [1] carried out a study on fuzzy logic with an additional involutive negation operator. This was a survey paper and they presented a propositional logic extended with an involutive negation. With this concept, Cintula improved the expressive potential of mathematical logic.

In Hájek's paper [8], a system called basic logic (BL) was defined. Not long ago, a survey paper was published [7] that discussed the state-of-art development of BL. The problem with this logic is that the implication is defined by the residual of the t-norm, the negation operator is defined by $\sim x = x \to 0$ and in the strict operator case, the negation operator is not involutive. In fact it is a drastic negation operator. Neither the implication operator nor the negation operator is continuous. From an application point of view, the continuity property is always indispensable.

In Esteva et al. [5], logics with involutive negation were introduced. This negation is different from implication based negation "$\sim$" and it functions as a basic negation ($\text{not}(x) = 1 - x$). But this negation operator is not related to the residual implication in strict monotonous operator case.

Modal logic has been used in rough sets as well, where the sets are approximated by elements of a partition induced by an equivalence relation. A natural choice for rough set logic is S5 (Orlowska [11]). Here, the possibility and necessity modalities express outer and inner approximation operators.

In our previous article [3], we looked for strictly monotonously increasing Archimedian t-norms and t-conorms (called conjunctive and disjunctive operators) for which the De Morgan identity is valid with infinitely many negation operators. In this article, we will denote these operators by $c(x, y)$ and $d(x, y)$, respectively.

2   Basic Considerations of Negation

Here, we will interpret 1 as the true value and 0 as the false value. Now we will state definitions and properties of negation operator.

Definition 1. *We say that $\eta(x)$ is a strong negation if $\eta \colon [0, 1] \to [0, 1]$ satisfies the following conditions:*
    C1: $\eta \colon [0, 1] \to [0, 1]$ is bijective and continuous (Continuity and Bijectivity)
    C2: $\eta(0) = 1$, $\eta(1) = 0$ *(Boundary conditions)*
    C3: $\eta(x) < \eta(y)$ for $x > y$ *(Monotonicity)*
    C4: $\eta(\eta(x)) = x$ *(Involution)*

Remark. *The boundary condition C2 can be inferred by using C1 and C3.*

From *C1*, *C2* and *C3*, it follows that there exists a fixed point (or neutral value) $\nu \in [0, 1]$ of the negation where

$$\eta(\nu) = \nu \tag{1}$$

Later on we will characterise the negation operator in terms of the ν parameter.

Definition 2. *We will say that a negation $\eta_{\nu_1}(x)$ is stricter than $\eta_{\nu_2}(x)$ if $\nu_1 < \nu_2$.*

For the strong negation, two representation theorems are known. Trillas [13] showed that every involutive negation operator has the following form, and here we denoted this negation by $n(x)$.

$$n(x) = g^{-1}(1 - g(x)) \tag{2}$$

where $g : [0, 1] \rightarrow [0, 1]$ is a continuous strictly increasing (or decreasing) function. This generator function corresponds to nilpotent operators (nilpotent t-norms [9,12,14]). Examples for the negation: $n_\alpha(x) = (1 - x^\alpha)^{\frac{1}{\alpha}}$ (Yager negation), $n_a(x) = \frac{1-x}{1+ax}$ (Hamacher and Sugeno negation). We can express the parameter of the negation operator in terms of its fixed point (or neutral value). The Yager negation operator has the form

$$n_\nu(x) = \left(1 - x^{-\frac{\ln\nu}{\ln 2}}\right)^{-\frac{\ln 2}{\ln\nu}}$$

In a similar way, we get the new form of the Hamacher negation operator:

$$n_\nu(x) = \frac{1}{1 + (\frac{1-\nu}{\nu})^2 \frac{x}{1-x}}.$$

This form of the negation operator can be found in [2].

For the strictly monotonously increasing t-norms, another form of negation operator is given in [3,4]. It is

$$\eta_\nu(x) = f^{-1}\left(\frac{f^2(\nu)}{f(x)}\right) \tag{3}$$

where $f : [0, 1] \rightarrow [0, \infty]$ is a continuous, increasing (or decreasing) function and f is the generator function of a strict monotone t-norm, or t-conorm. This negation operator is an element of the Pliant system [3,4].

Here we show that (2) and (3) are equivalent, when $f(\nu) = 1$.

Proposition 1. *Let $n(x)$ and $\eta(x)$ be defined by (2) and (3). If $f(\nu) = 1$ and*

$$f(x) = \frac{1 - g(x)}{g(x)}, \qquad g(x) = \frac{1}{1 + f(x)}$$

then

$$n(x) = \eta(x)$$

Proof. The following expression is valid

$$f^{-1}(x) = g^{-1}\left(\frac{1}{x+1}\right).$$

So we get

$$f^{-1}\left(\frac{1}{f(x)}\right) = g^{-1}\left(\frac{1}{1+\frac{1}{f(x)}}\right) =$$

$$= g^{-1}\left(\frac{1}{1+\frac{g(x)}{1-g(x)}}\right) = g^{-1}(1-g(x)).$$

The properties of the functions (f,g) can be easily verified.   □

Next, we will use (3) to represent the negation operator because here we are just considering strict monotone operators.

In Fig. 1, we sketch the shape of the negation function and we demonstrate the meaning of the ν value. We can introduce a non-continuous negation [3,4].

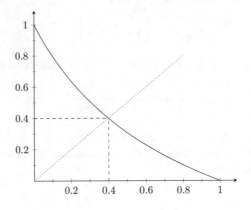

Fig. 1. The shape of the negation function when $f(x) = \frac{1-x}{x}$ and $\nu = 0.4$

Definition 3 (Drastic negation). *We call $\eta_0(x)$ and $\eta_1(x)$ drastic negations when*

$$\eta_0(x) = \begin{cases} 1 \ if \ x = 0 \\ 0 \ if \ x \neq 0 \end{cases} \quad \eta_1(x) = \begin{cases} 1 \ if \ x \neq 1 \\ 0 \ if \ x = 1 \end{cases}$$

Here, $\eta_0(x)$ is the strictest negation, while $\eta_1(x)$ is the least strict negation. They are non-continuous negation operators, so they are not negation operators in the original sense (see Fig. 2).

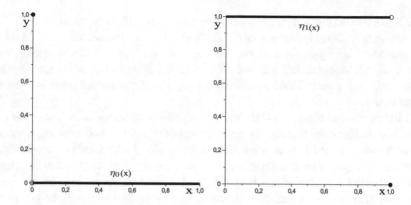

Fig. 2. $\eta_0(x)$ and $\eta_1(x)$ are drastic negation operators

Let $g(x) = \dfrac{1}{1 + \frac{1-\nu}{\nu} \frac{x}{1-x}}$. Then

$$g^{-1}(1 - g(x)) = \dfrac{1}{1 + \left(\frac{1-\nu}{\nu}\right)^2 \frac{x}{1-x}} = \dfrac{1-x}{1+ax},$$

where $a + 1 = \left(\frac{1-\nu}{\nu}\right)^2$ and $\nu \in (0,1)$.
Let $f(x) = \frac{1-x}{x}$. Then

$$f^{-1}\left(\frac{f^2(\nu)}{f(x)}\right) = \dfrac{1}{1 + \left(\frac{1-\nu}{\nu}\right)^2 \cdot \frac{x}{1-x}} \tag{4}$$

Remark. *There are strictly monotone operators (t-norm, t-conorm) that build a DeMorgan system with infinitely many negations. This operator is called Pliant operator* [3].

This system is useful for building modal operators. In the Pliant system the negation operator closely related to the t-norm and t-conorm. Based on different negations, in the next we deal with modalities.

3   Modalities Induced by Two Different Negation Operators

To obtain this structure, we equip it with another type of negation operator. In modal logic, it is called an intuitionistic negation operator. In our system, the modalities induced by a suitable composition of the two negation operators generate a modal system with the full distributivity property of the modal operators. The necessity operator is simultaneously distributive over the conjunctive and disjunctive operators and the possibility operator is also simultaneously distributive over the conjunctive and disjunctive operators.

With this starting point, the necessity and possibility operators used in fuzzy logic are based on an extension of modal logic to the continuous case. We begin with the negation operator and we make use of two types of this operator, one is strict, and one is less strict. We will show that with these two negation operators we can define the modal hedges. Next, we use the classical notation for the sake of convention.

In intuitionistic logic, another kind of negation operator also has to be taken into account. Here $\sim x$ means the negated value of x. $\sim_1 x$ and $\sim_2 x$ are two negation operators. We will construct linguistic modal hedges called necessity and possibility hedges. The construction is based on the fact that modal operators can be realized by combining two kinds of negation operators. In our modal logic, $\sim_1 x$ means x is impossible. In other words, $\sim_1$ is a stronger negation than $not(x)$, i.e. $\sim_2 x$.

We can write

$$\sim_1 x := impossible(x)$$

$$\sim_2 x := not(x)$$

As we mentioned above, in modal logic we have two more operators than the classical logic case, namely necessity and possibility; and in modal logic there are two basic identities. These are:

$$\sim_1 x = impossible(x) = necessity(not(x)) = \Box \sim_2 x \tag{5}$$

$$\Diamond x = possible(x) = not(impossible(x)) = \sim_2(\sim_1 x) \tag{6}$$

In our context, we model $impossible(x)$ with a stricter negation operator than $not(x)$. Equation (6) also serves as a definition for the possibility operator.

If in Eq. (5) we replace x by $\sim_2 x$ and using the fact that $\sim_2 x$ is involutive, we get

$$\Box x = \sim_1(\sim_2 x), \tag{7}$$

and with Eq. (6), we have

$$\Diamond x = \sim_2(\sim_1 x). \tag{8}$$

The necessity and possibility operators have a common form, i.e. they can be expressed by double negation equipped by different neutral values. Here, "not" and "impossible" are two different negations.

If ν is small, we can say that the negation operator is strict; otherwise it is not strict. "Impossible" is a stricter negation compared with "not".

Based on the above considerations, we can formally define the necessity and possibility modifiers.

Definition 4. *The general form of the modal operator is*

$$\tau_{\nu_1,\nu_2}(x) = \eta_{\nu_1}\left(\eta_{\nu_2}(x)\right) \qquad or \qquad \tau_{\nu_1,\nu_2}(x) = n_{\nu_1}\left(n_{\nu_2}(x)\right) \tag{9}$$

and ν_1, ν_2 are neutral values of the negation operator. If $\nu_1 < \nu_2$, then $\tau_\nu(x)$ is a necessity operator, and if $\nu_2 < \nu_1$, then $\tau_\nu(x)$ is a possibility operator. If $\nu_1 = \nu_2$ then $\tau_\nu(x)$ is the identity operator and $\tau_\nu(x) = x$.

With this notion, we can make use of Eqs. (7) and (8).

The necessity and possibility operators using the representation of the negation operator (τ) have a common form

$$\tau_{\nu_1,\nu_2}(x) = \eta_{\nu_1}(\eta_{\nu_2}(x)) = f^{-1}\left(\frac{f^2(\nu_1)}{f^2(\nu_2)}f(x)\right)$$

$$\tau_{\nu_1,\nu_2}(x) = \eta_{\nu_1}(\eta_{\nu_2}(x)) = f_1^{-1}\left(2\nu_1 - f_1\left(f_2^{-1}\left(2\nu_2 - f_2(x)\right)\right)\right)$$

We can define the dual possibility and necessiy operators like so:

Definition 5. *A necessity operator and a possibility operator are dual if*

$$\nu_1 = \eta(\nu_2) \qquad\qquad \nu_1 = f^{-1}\left(\frac{1}{f(\nu_2)}\right)$$

We will show that both modal operators belong to the same class of unary operators, and also show that because they have a common form, we can denote both of them by $\tau_\nu(x)$. Depending on the ν value, we get the necessity operator or the possibility operator.

Definition 6. *The dual possibility and necessity operators are*

$$\square_\nu(x) = \tau_\nu^N(x) = f^{-1}\left(\frac{f(x)}{f^2(\nu)}\right) \qquad\qquad and \qquad\qquad (10)$$

$$\lozenge_\nu(x) = \tau_\nu^P(x) = f^{-1}\left(f^2(\nu)f(x)\right) \qquad\qquad (11)$$

when $f(\nu) < 1$

Previously we defined the drastic negation operator. Here we will define the drastic necessity and possibility operators by using the drastic negation operators:

Definition 7. *Drastic model operators are the following:*

Drastic necessity $\qquad\qquad \square_1(x) = \tau_1^N(x) = \begin{cases} 1 & if\, x = 1 \\ 0 & if\, x \neq 1 \end{cases} \qquad (12)$

Drastic possibility $\qquad\qquad \lozenge_0(x) = \tau_0^P(x) = \begin{cases} 0 & if\, x = 0 \\ 1 & if\, x \neq 0 \end{cases} \qquad (13)$

See Fig. 3 below

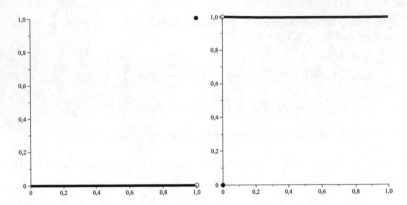

Fig. 3. The drastic necessity $\tau_1(x)$ and the drastic possibility $\tau_0(x)$

Remark. *Drastic necessity and possibility operators can be obtained by using drastic negations (See Definition 3).*

(12) and (13) are known as Baas-Monteiro Δ operator and its dual ∇, respectively. Δ and ∇ are definable by an involution and strict negation [7].

The necessity operator ($\Box$) will be denoted by $\tau^N(x)$, and the possibility operator ($\Diamond$) will be denoted by $\tau^P(x)$. Because both operators can be deduced from each other, we handled them together.

Now let $\tau^N[0,1] \to [0,1]$ and $\tau^P[0,1] \to [0,1]$ be two unary operators that satisfy the following conditions for the necessity operator and the possibility operator:

$N1.$ $\tau^N(1) = 1$ $P1.$ $\tau^P(0) = 0$

$N2.$ $\tau^N(x) \leq x$ $P2.$ $x \leq \tau^P(x)$

$N3.$ $x \leq y$ implies $\tau^N(x) \leq \tau^N(y)$ $P3.$ $x \leq y$ implies $\tau^P(x) \leq \tau^P(y)$

$N4.$ $\tau^P(x) = \eta\left(\tau^N\left(\eta(x)\right)\right)$ $P4.$ $\tau^N(x) = \eta\left(\tau^P(\eta(x))\right)$

$[N5.$ $\tau^P(x) = \tau^N\left(\tau^P(x)\right)$ $P5.$ $\tau^N(x) = \tau^P\left(\tau^N(x)\right)]$

Remark. *In our system, $(N5)$ and $(P5)$ are not required. Only a special parametrical form of τ^P and τ^N satisfies $(N5)$ and $(P5)$.*

Instead of $(N5)$ and $(P5)$, our demand is the so-called neutrality principle, i.e.

$$N'(5) \quad \tau^N\left(\tau^P(x)\right) = x \qquad P'(5) \quad \tau^P\left(\tau^N(x)\right) = x$$

Next, we will show that the basic properties are fulfilled.

Proposition 2. *$\tau_\nu^N(x)$ and $\tau_\nu^P(x)$ satisfy the basic properties of modalities: $\{N$ principle, T principle, K principle, $DF\Diamond$ principle, N^* principle, P principle, T principle, K principle, $DF\Box$ principle, P^* principle$\}$*

Proof. We prove only the Necessity case. The possibility case can be proven in a similar way. We will assume that f is strictly decreasing and that $f(\nu) < 1$

N1. $\tau_\nu^N(1) = f^{-1}\left(\dfrac{f(1)}{f^2(\nu)}\right) = f^{-1}(0) = 1$

N2. $\tau_\nu^N(x) < x \qquad x \in (0,1) \qquad$ So:

$$f^{-1}\left(\frac{f(x)}{f^2(\nu)}\right) < x$$

$$1 < f^2(\nu)$$

N3.   if $x < y$ then

$$f^{-1}\left(\frac{f(x)}{f^2(\nu)}\right) < f^{-1}\left(\frac{f(y)}{f^2(\nu)}\right),$$

so $f(x) > f(y)$

N4. $\tau_\nu^N(x) = f^{-1}\left(\dfrac{f(x)}{f^2(\nu)}\right)$, $\tau_\nu^N(\eta(x)) = f^{-1}\left(\dfrac{1}{f(x)f^2(\nu)}\right)$

$\eta(\tau_\nu^N(\eta(x))) = f^{-1}\left(f^2(\nu)f(x)\right) = \tau_\nu^P(x)$

N'5 $\tau_\nu^N(\tau_\nu^P(x)) = x,\ f^{-1}\left(\dfrac{f^2(\nu)f(x)}{f^2(\nu)}\right) = x$

$\square$

Proposition 3. *For the composition of the drastic modal operator the following are valid:*

(A)   i $\square_1(\Diamond_\nu(x)) = \tau_1^N(\tau_\nu^P(x)) = \tau_1^N(x) = \sqcap_1(x)$
      ii $\square_\nu(\Diamond_0(x)) = \tau_\nu^N(\tau_0^P(x)) = \tau_0^P(x) = \Diamond_0(x)$
(B)   i $\Diamond_0(\square_\nu(x)) = \tau_0^P(\tau_\nu^N(x)) = \tau_0^P(x) = \Diamond_0(x)$
      ii $\Diamond_\nu(\square_1(x)) = \tau_\nu^P(\tau_1^N(x)) = \tau_1^N(x) = \square_1(x)$
(C)   i $\square_1(\Diamond_0(x)) = \tau_1^N(\tau_0^P(x)) = \tau_0^P(x) = \Diamond_0(x)$
      ii $\Diamond_0(\square_1(x)) = \tau_0^P(\tau_1^N(x)) = \tau_1^N(x) = \square_1(x)$

The proofs are based on the definition of drastic modal operators stated above.

Remark. *N5 and P5 are also valid when the possibility operator is drastic.*

For N5 see: A/ii, and C/i
For P5 see: B/ii, and C/ii

By making use of (2) and (3), we can define the concrete forms of the necessity and possibility operators.

Example 1.
Here, we use the Yager operator and the representation theorem of Trillas [13]

$$\tau(x) = n_{\nu_1}(n_{\nu_2}(x)) = \left(1 - \left(1 - x^{-\frac{\ln 2}{\ln \nu_2}}\right)^{\frac{\ln \nu_2}{\ln \nu_1}}\right)^{-\frac{\ln \nu_1}{\ln 2}}$$

Example 2.

$$\tau_\nu^P(x) = \frac{1}{1 + \left(\frac{1-\nu}{\nu}\right)^2 \frac{1-x}{x}} \quad \text{or} \quad \tau_\nu^N(x) = \frac{1}{1 + \left(\frac{\nu}{1-\nu}\right)^2 \frac{1-x}{x}} \quad \text{when } \nu > \frac{1}{2} \quad (14)$$

See both plots below (Fig. 4),

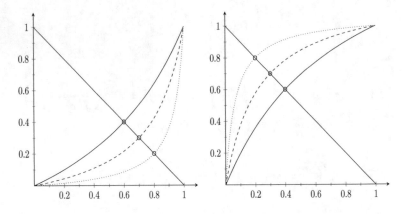

Fig. 4. The *necessity* and *possibility* operators with different ν values ($\nu = 0.2, 0.3, 0.4$)

If we use the definition of $\tau(x) = \Box x$ or $\Diamond x$ (i.e. necessity, or possibility x), then we can introduce different necessity and possibility operators.

$$\Box_\nu^{(2)} x = \Box_\nu\left(\Box_\nu(x)\right) = \tau_\nu^N\left(\tau_\nu^N(x)\right)$$
$$\Diamond_\nu^{(2)} x = \Diamond_\nu\left(\Diamond_\nu(x)\right) = \tau_\nu^P\left(\tau_\nu^P(x)\right)$$

Definition 8. *We call graded modalities a k composition of the modalities.*

$$\Box_\nu^{(k)}(x) = \Box_\nu(\underbrace{\Box_\nu(\ldots \quad \Box_\nu}_{k}(x))\ldots) = \tau_\nu^N\left(\tau_\nu^N(\ldots \quad \tau_\nu^N(x))\right) \quad (15)$$

$$\Diamond_\nu^{(k)}(x) = \Diamond_\nu(\underbrace{\Diamond_\nu(\ldots \quad \Diamond_\nu}_{k})(x))\ldots) = \tau_\nu^P\left(\tau_\nu^P(\ldots \quad \tau_\nu^P(x))\right) \quad (16)$$

Proposition 4. *The composition of a modal operator is a closed operation.*

$$\Box_\nu^{(k)}(x) = \Box_{\nu^*}(x) \qquad\qquad \Diamond_\nu^{(k)}(x) = \Diamond_{\nu^*}(x)$$

where

$$\nu^* = f^{-1}\left(f^{2k}(\nu)\right) \tag{17}$$

Proof. In the possible modal operator case:

$$f^{-1}\left(f^2(\nu)\left(f^2(\nu)\ldots\left(f^2(\nu)f(x)\right)\ldots\right)\right) = f^{-1}\left(f^{2k}(\nu)f(x)\right)$$

$\Box$

Proposition 5. *The following properties hold for the composition of modal operators*

Properties Using Classical Notations

1. $\Box_\nu^{(n)}(\Box_\nu^{(m)}(x)) = (\tau_\nu^N(\tau_\nu^N(x))^{(m)})^{(n)} = \left(\tau_\nu^N(x)\right)^{(n+m)} = \Box_\nu^{(n+m)}(x)$
2. $\Diamond_\nu^{(n)}(\Diamond_\nu^{(m)}(x)) = (\tau_\nu^P(\tau_\nu^P(x))^{(m)})^{(n)} = \left(\tau_\nu^P(x)\right)^{(n+m)} = \Diamond_\nu^{(n+m)}(x)$
3. $\Diamond_\nu^{(n)}(\Box_\nu^{(m)}(x)) = (\tau_\nu^P(\tau_\nu^N(x))^{(m)})^{(n)} = \begin{cases} \Diamond_\nu^{(n-m)}(x) & \text{if} \quad n-m>0 \\ x & \text{if} \quad n=m=0 \\ \Box_\nu^{(m-n)}(x) & \text{if} \quad n-m<0 \end{cases}$
4. $\Box_\nu^{(n)}(\Diamond_\nu^{(m)}(x)) = (\tau_\nu^N(\tau_\nu^P(x))^{(m)})^{(n)} = \begin{cases} \Box_\nu^{(n-m)}(x) & \text{if} \quad n-m>0 \\ x & \text{if} \quad n=m-0 \\ \Diamond_\nu^{(m-n)}(x) & \text{if} \quad n-m<0 \end{cases}$
5. $\lim\limits_{K\to\infty}(\tau_\nu^N(x))^{(K)} = \tau_1(x)$
6. $\lim\limits_{K\to\infty}(\tau_\nu^P(x))^{(K)} = \tau_0(x)$

Proof: They follow from direct calculation.

Using the Dombi operator when $\nu > \frac{1}{2}$, we have

$$\tau_\nu^{(N)}(x) = \cfrac{1}{1+\left(\frac{1}{\nu}\frac{\nu}{\nu}\right)^{2k}\frac{1-x}{x}} \qquad\qquad \tau_\nu^{(P)}(x) = \cfrac{1}{1+\left(\frac{\nu}{1-\nu}\right)^{2k}\frac{1-x}{x}}$$

4 Conclusions

We defined necessity and possibility operators using double negation where the fixed points are different. The composition of the modal operators is closed. So we have series of modal operators with different degrees. We defined these modal operators using a generator function of the operator system and these functions are the generator functions of the operators of the Pliant logical system. There are several open questions as how can we apply this approach to the nilpotent operator class, or how the logic behind this structure can be characterized, etc. It seems that the Pliant structure has an outstanding position. Particularly its special case namely the Dombi operator plays an important role in the practical applications.

References

1. Cintula, P., Klement, E.P., Mesiar, R., Navara, M.: Fuzzy logics with an additional involutive negation. Fuzzy Sets Syst. **161**, 390–411 (2010)
2. Dombi, J.: Towards a general class of operators for fuzzy systems. IEEE Trans. Fuzzy Syst. **16**, 477–484 (2008)
3. Dombi, J.: De Morgan systems with an infinitely many negations in the strict monotone operator case. Inf. Sci. **181**, 1440–1453 (2011)
4. Dombi, J.: On a certain class of aggregative operators. Inf. Sci. **245**(1), 313–328 (2013)
5. Esteva, F., Godo, L., Hájek, P., Navara, M.: Residuated fuzzy logics with an involutive negation. Arch. Math. Log. **39**(2), 103–124 (2000)
6. Esteva, F., Godo, L., Noguera, C.: A logical approach to fuzzy truth hedges. Inf. Sci. **232**, 366–385 (2013)
7. Gottwald, S., Hájek, P.: Triangular norm-based mathematical fuzzy logics. Logical Algebraic, Analytic and Probabilistic Aspects of Triangular Norms, pp. 275–299 (2005)
8. Hájek, P.: Metamathematics of Fuzzy Logic. Kluwer Academic Publishers, Dordrecht (1998)
9. Lukasiewicz, J.: Two-valued logic. Przeglad Filozoficzny **13**, 189–205 (1921)
10. Flaminio, T., Marchioni, E.: T-norm-based logics with an independent involutive negation. Fuzzy Sets Syst. **157**(24), 3125–3144 (2006)
11. Orlowska, E.: A logic of indiscernibility relations. LNCS, vol. 208, pp. 177–186. Springer, Berlin (1985)
12. Pavelka, J.: On fuzzy logic I. Many-valued rules of inference. Zeitschrift fur mathematische Logik und Grundlagen der Mathematik **25**, 45–52 (1979)
13. Trillas, E.: Sobre functiones de negacion en la teoria de conjuntas difusos. Stochastica **3**, 47–60 (1979)
14. Zadeh, L.A.: The concept of a linguistic variable and its application to approximate reasoning, part 1. Inf. Sci. **8**, 199–249 (1975)

Condorcet Winners on Bounded and Distributive Lattices

Marta Cardin[(⊠)]

Department of Economics, Ca' Foscari University of Venice,
Sestiere Cannaregio 873, Venezia, Italy
mcardin@unive.it

Abstract. Aggregating preferences for finding a consensus between several agents is an important topic in social choice theory. We obtain several axiomatic characterizations of some significant subclasses of voting rules defined on bounded and distributive lattices.

Keywords: Lattice · Preference · Voting rule

1 Introduction

Social choice theory concerns the analysis of methods for aggregating the preferences of individual agents over some abstract set of alternatives into one collective preference relation or, more generally, to associate to a profile of preferences a set of alternatives. In fact in many practical settings one is interested in a set of acceptable alternatives rather than a collective preference relation.

Examples of such methods include voting procedures, which are used to aggregate the preferences of voters over a set of candidates standing for election to determine which candidate should win the election.

A group decision method or a voting rule associates with each profile of preferences the "winning" alternatives or the alternatives representing the "social choice". This means that group decision methods are represented by functions from set of preferences to sets of possible outcomes.

It is worth noting that we consider only preferences compatible with a lattice structure and that when the lattice is finite every compatible preference is completely determined by its top element. Hence in our approach and in the finite case voting rules are aggregation functions defined on a lattice.

The present work considers different classes of group decision methods where the set of alternatives is a bounded and distributive lattice and the set of admissible preferences is the set of preferences compatible with the lattice structure.

The remainder of this paper is organized as follows. The following Sect. 2 defines compatible preferences on a lattice, Sect. 3 introduces our framework while Sect. 4 propose some results about characterization of voting rules. We end by pointing out some further lines of research.

© Springer Nature Switzerland AG 2019
R. Halaš et al. (Eds.): AGOP 2019, AISC 981, pp. 339–346, 2019.
https://doi.org/10.1007/978-3-030-19494-9_31

2   Compatible Preference on a Lattice

First we recall some basic notions in lattice and ordered set theory.

An element z of a lattice L is called *join irreducible* if $z = x \vee y$ for $x, y \in L$ implies that $z = x$ or $z = y$. The notion of meet-irreducible element is defined dually. The set of join-irreducible elements of a lattice L is denoted by $J(L)$.

An element z of a lattice L is called *join prime* i if $z \leq x \vee y$ for $x, y \in L$ implies that $z \leq x$ or $z \leq y$. The notion of meet-prime element is defined dually. By The set of join-prime elements of a lattice L is denoted by $JI(L)$. It can be proved that a join-prime element is join-irreducible and if L is a finite distributive lattice we have that $J(L) = JI(L)$.

A *filter* of a lattice L is a nonempty subset F such that

(i) if $x \in F$ and $x \leq y$ then $y \in F$,
(ii) $x, y \in F$ then $x \wedge y \in F$.

Sets satisfying Condition (i) of a filter are called upsets. The dual notation is that of an *ideal*. If $x \in L$ we define the *principal filter* generated by x as $\uparrow x = \{y \in L : y \geq x\}$. It is easy to prove that $\uparrow x$ is a filter for every $x \in L$. It can be proved that in a finite lattice each filter and each ideal are principal.

A *proper filter* is a filter that is neither empty nor the whole lattice while a *prime filter* is a proper filter P such that if $x \vee y \in P$ then $x \in P$ or $y \in P$. An element x of a lattice L is join-prime if and only if $\uparrow x$ is prime. A filter F is prime if and only if $L \setminus F$ is an ideal, which is then a prime ideal.

Throughout this paper lattice means bounded and distributive lattice. We note that if L is a bounded and distributive lattice every element is characterized by the set of prime filters which contain the given element since a duality between the lattice and the power set of the set of prime filters of L ordered by inclusion as is proved in [9].

In this section we introduce relations that respect the lattice operations. A binary relation $R \subseteq L \times L$ in a lattice L is said to be *compatible* whenever it preserves the join and the meet i.e.

(i) if $(x, y) \in R$ then $(x \wedge z, y \wedge z) \in R$ for each $z \in L$;
(ii) if $(x, y) \in R$ then $(x \vee z, y \vee z) \in R$ for each $z \in L$.

A binary relation $R \subseteq L \times L$ in a lattice L is said to be *monotone* if when $x \geq y$ then $(x, y) \in R$. It is straightforward to prove that a compatible binary relation on a lattice L is monotone if and only if for every $x \in L$, $(x, 0) \in R$.

A *preference* on a lattice L is a monotone, compatible and transitive transitive relation on L. If a preference $R \subseteq L \times L$ in a lattice L, is such that for every $x, y \in L$, $(x, y) \in R$ or $(y, x) \in R$ the binary relation is a complete preference.

Our aim is to characterize preference on a lattice. If $B \subseteq L$ define, for every $x \in L$ $B(x) = \{b \in B : b \leq x\}$. The following proposition proved in [3] characterizes compatible and monotone preorders.

Proposition 1. *If L is a finite lattice, a relation P is a preference relation on L if and only if there exists a set $B \subseteq J(L)$, such that*

$$xPy \quad \text{if and only if} \quad B(x) \supseteq B(y) \tag{1}$$

The following result characterizes complete preferences on a finite lattice.

Proposition 2. *If $B \subseteq J(L)$ and P is a preference on L defined with respect to the set B, P is a complete if and only if the set B is a chain.*

The results above depend on the existence of many join-irreducible elements in a finite distributive lattice. In an infinite distributive lattice there may be no join-irreducible element. In the infinite case, the role of join-irreducible elements is taken by prime filters.

Let $\mathcal{F}$ the set of prime filters of a lattice L. If $\mathcal{B} \subseteq \mathcal{F}$ and $x \in L$ we define $\mathcal{B}(x) = \{F \in \mathcal{B} : x \in F\}$.

The following proposition generalizes Proposition 1 to the case of a lattice L not necessarily finite.

Proposition 3. *If L is a lattice a relation P is a preference relation on L if and only if there exists a set $\mathcal{B} \subseteq \mathcal{F}$ and P is defined by*

$$xPy \quad \text{if and only if} \quad \mathcal{B}(x) \supseteq \mathcal{B}(y). \tag{2}$$

Also in this case we can characterize complete preferences.

Proposition 4. *If $\mathcal{B} \subseteq \mathcal{F}$ and P is a preference on L defined with respect to the set $\mathcal{B}$, P is a complete preference if and only if the set $\mathcal{B}$ is a chain.*

3   Voting Rules

We begin by introducing the key definitions of the social choice model (see [2,6,10] and [11]). Let $N = \{1, \ldots, n\}$ with $n \geq 2$ be a group of two or more individuals that seeks to make collective judgments or a set of voters and X be a (possibly infinite)set of alternatives. We suppose that X is equipped with a lattice structure and we consider the set $\mathcal{P}$ of monotone, compatible and complete preferences on X.

A *profile of preferences* or simply a *profile* is a list of preferences across individuals or voters. A profile is an element of $\mathcal{P}^N$ and will be represented by $\mathbf{P} = (P_1 \ldots P_n \ldots)$. If we consider an element $\mathbf{P} = (P_1 \ldots P_n \ldots)$ of $\mathcal{P}^N$, we denote by (P', P_{-i}) the element of $\mathcal{P}^N$ whose i- component is the element $P' \in \mathcal{P}$ and whose j-component is P_j for every $j \neq i$. If $P \in \mathcal{P}$ let S the strict relation induced by P.

A *voting rule* is a function $V \colon \mathcal{P}^N \to X$ that assigns to each profile a collective choice element. Our framework is very general, we do not assume that the sets X is finite or that the map $V \colon \mathcal{P}^N \to X$ is surjective. We are going to consider some properties of a voting rule. We focus on voting rules that induce truth-telling as a (weakly) dominant strategy. So if the voting rule is non manipulable or strategy-proof it is a dominant strategy for a voter to honestly assign to an element the evaluation that he believes is the correct one and there is no incentive to misreport the evaluations (see [10] and [12]). Note that manipulative behaviour may produce undesirable results especially if there are coalitions of several voters with the same behaviour. We can also note that in many situations there is the possibility that voters might not only be led by their true preferences, but also by other calculations, when filling their ballots. Anyone who has sometimes rooted for a loser must have debated whether to actually vote for the hopeless candidate, or else select the least undesirable candidate among the potential winners, and cast a "useful" vote.

Definition 1. *A voting rule* $V : \mathcal{P}^N \to X$, *is strategy-proof if for every* $\mathbf{P} \in \mathcal{P}^N$ *and every* $P' \in \mathcal{P}$, *we have that* $V(\mathbf{P})P_i V(P', P_{-i})$.

Note that the in this case the voter with preference P_i prefers the outcome $\mathbf{P}$ with respect to the outcome $V(P', P_{-i})$.

If P_i is an element of $\mathcal{P}$ then $\mathcal{B}_i \subseteq \mathcal{F}$ is the subset of elements of $\mathcal{F}$ such that (2) is satisfied. If $\mathbf{P} \in \mathcal{P}^N$ and $F \in \mathcal{B}$ we define

$$N(\mathbf{P}, F) = \{i \in N : F \in \mathcal{B}_i\}.$$

It is important to note that $\mathcal{B}_i$ is a is a chain for every $\mathbf{P} \in \mathcal{P}^N$ and every $i, 1 \leq i \leq n$ since P_i is a complete preference for every $i, 1 \leq i \leq n$.

Now we consider also a monotonicity property of voting rules that in [10] is named monotonicity in properties.

Definition 2. *A voting rule* $V : \mathcal{P}^N \to X$ *is monotone if when* $\mathbf{P}, \mathbf{P}' \in \mathcal{P}^N$, $F \in \mathcal{B}$ *and* $N(\mathbf{P}', F) \subseteq N(\mathbf{P}, F)$, *if* $V(\mathbf{P}') \in F$ *then* $V(\mathbf{P}) \in F$.

A voting rule $V : \mathcal{P}^N \to 2^X$ *is strongly monotone if when* $\mathbf{P}, \mathbf{P}' \in \mathcal{P}^N$, $F, F' \in \mathcal{B}$ *and* $N(\mathbf{P}', F) \subseteq N(\mathbf{P}, F')$ *if* $V(\mathbf{P}') \in F$, *then* $V(\mathbf{P}) \in F'$.

Anonimity requires that the candidates are treated equally and is a fundamental requirement in most elections.

Definition 3. *A voting rule* $V : \mathbf{P}^N \to 2^X$ *is anonymous if for every permutation* σ *in* N *and* $P \in \mathcal{P}^N$, $V(\mathbf{P}) = V(P_{\sigma(1)} \dots P_{\sigma(n)})$.

4   Characterization of Voting Rules

We introduce in our framework the well known method of voting by committees (see also [1]). A committee is a group of voters that can made final decisions regarding various issues. So we can consider a committee as a group of experts or a group of influential people. Who have some special right, reward, skill etc that other people do not have. Committees exist in any organization that delegates power to a small group of people. Note that in many situations some power must be delegated to a group of expert insiders so that an organization can function.

A non empty collection of non empty sets $\mathcal{A} \subseteq 2^N$ is said to be a *committee* in N if $A \in \mathcal{A}$ and $A \subset B$ implies that $B \in \mathcal{A}$(see [12] for a similar definition).

Definition 4. *A voting rule* $V : \mathbf{P}^N \to X$ *is a generalized committee voting rule if for every* $F \in \mathcal{F}$ *there exists a committee* $\mathcal{A}_F$ *in* N *such that*

$$V(\mathbf{P}) \subseteq F \text{ if and only if } N(\mathbf{P}, F) \in \mathcal{A}_F.$$

A voting rule is a committee voting rule if it is a generalized committee voting rule such that $\mathcal{A}_F$ *does not depend on* F.

If X is a finite lattice for every $P \in \mathcal{P}(X)$ there exists x_P such that

$$x_P = \bigwedge \{y \in X : yPx \text{ for every } x \in X\}$$

or such that $x_P = \bigwedge \{y \in X : yI1\}$. Moreover if X is a finite lattice and $F \in \mathcal{F}$ then there exists $z \in J(L)$ such that $F = \uparrow z$.

Then a voting rule $V : \mathbf{P}^N \to 2^X$ in a finite lattice X is a generalized committee voting rule if and only if for every $z \in J(L)$ there exists $\mathcal{A}_z$ in N such that

$$V(\mathbf{P}) = \bigvee \{z \in J(L) \text{ such that } \{x_{P_i} \geq z\} \in \mathcal{A}_z\}$$

As we said before it is important to note that when the lattice L is finite a generalized committee voting rule depends only on the elements $x_{P_i}, 1, \ldots, n$.

The following proposition characterizes a class of voting rules.

Proposition 5. *If* $V : \mathbf{P}^N \to X$ *is a voting rule the following conditions are equivalent:*

(i) V *is strategy-proof*
(ii) V *is monotone*
(iii) V *is a generalized committee voting rule*

Proof. Let V a non monotone voting rule. Then there exist two elements $\mathbf{P}, \mathbf{P}'$ in $\mathcal{P}^N$ such that $\mathbf{P}' = (P_i', P_{-i})$, $P_i' \in \mathcal{P}$ and $F \in \mathcal{F}$ such that $F \in \mathcal{B}'$ but $F \notin \mathcal{B}$ where $\mathcal{B}$ and $\mathcal{B}'$ are the subsets of elements of $\mathcal{F}$ associated to P_i and P_i'.

Then we get that $\mathbf{P}')S_i\mathbf{P}$ and so V is not strategy-proof.

Conversely suppose that V is a monotone voting rule. If $\mathbf{P} \in \mathcal{P}^N$ and $P_i' \in \mathcal{P}$ we consider the element of $\mathcal{P}^N$, $\mathbf{P}' = (P_i', P_{-i})$. It is easy to prove that for every $F \in \mathcal{F}$, $N(\mathbf{P}', F) \subseteq N(\mathbf{P}, F)$ so we can conclude that if $V(\mathbf{P}') \in F$ then $V(P) \in F$ and then $V(\mathbf{P})P_iV(P', P_{-i})$.

Since a committee is a family of sets closed under taking supersets a generalized committee voting rule is monotone.

Now, if $V : \mathbf{P}^N \to 2^X$ is is a monotone voting rule we define for every $F \in \mathcal{F}$ the family of subset of N,

$$\mathcal{A}_F = \{A \subseteq N : \text{if } N(\mathbf{P}, F) = A \text{ then } V(\mathbf{P}) \in F\}.$$

By monotonicity of V the definition of $\mathcal{A}_F$ does not depend on the choice of the element $\mathbf{P} \in \mathcal{P}^N$ and it can be easily proved that $\mathcal{A}_F$ is a committee for every $F \in \mathcal{F}$.

As a consequence of the previous results we obtain the following proposition.

Proposition 6. *A voting rule* $V \colon \mathbf{P}^N \to X$ *is strongly monotone if and only if* V *is a committee voting rule.*

Proof. It is easy to prove that a committee voting rule is strongly monotone. In order to prove the converse we need only to note that if $V \colon \mathbf{P}^N \to 2^X$ is a strongly monotone voting rule then the set $A_F = \{A \subseteq N : \text{if } N(\mathbf{P}, F) = A \text{ then } V(\mathbf{P}) \in F\}$ does not depend on F.

Now we consider the class of quota rules. Here a characteristic is collectively accepted if and only if the number of individuals accepting it is greater than or equal to some threshold, which may depend on the characteristic in question. Majority voting is a special quota rule with a simple majority threshold for every characteristic.

Definition 5. *A voting rule* $V \colon \mathbf{P}^N \to \mathcal{P}(X)$ *is a generalized quota rule if for every* $F \in \mathcal{F}$ *there exists* $n_F \in \{1, \ldots, n\}$ *and*

$$V(\mathbf{P}) \in F \text{ if and only if } |N(\mathbf{P}, F)| \geq n_F.$$

A generalized quota rules is a quota rules if n_F *does not depend on* F.

The following proposition characterizes a class of anonymous voting rules.

Proposition 7. *The voting rule* $V \colon \mathbf{P}^N \to X$ *is anonymous and monotone if and only if is a generalized quota rule. The voting rule* $V \colon \mathbf{P}^N \to X$ *is anonymous and strongly monotone if and only if is a quota rule.*

5   Condorcet Winners

As it is well known an alternative defeating every other alternative in pairwise majority comparisons is called a Condorcet winner.

Definition 6. *A voting rule* $V \colon \mathbf{P}^N \to 2^X$ *is a generalized Condorcet rule if there exists a committee* $\mathcal{A}$ *in* N *such that* $A \in \mathcal{A}$ *if and only if* $A^c \notin \mathcal{A}$ *for every* $A \in \mathcal{A}$ *and* $x \in V(\mathbf{P})$ *for* $\mathbf{P} \in \mathbf{P}^N$ *if and only if for every* $y \in X$

$$\{i \in N : x P_i y\} \in \mathcal{A}.$$

Note that the existence of a Condorcet winner cannot be guaranteed and so in this case $F(\mathbf{P})$ could be a subset of alternatives or even the empty set.

Note that if we consider the committee of elements of cardinality greater or equal to $\frac{n}{2}$ we obtain the usual Condorcet rule.

Proposition 8. *If* $V \colon \mathbf{P}^N \to 2^X$ *is a generalized Condorcet rule with respect to the committee* $\mathcal{A}$ *and* $\overline{x} \in V(\mathbf{P})$ *then* $\overline{x} \in V'(\mathbf{P})$ *where* $V' \colon \mathbf{P}^N \to 2^X$ *is a generalized committee voting rule with respect to the committee* $\mathcal{A}$.

Proof. Let $\overline{x}$ is an element of X that is a generalized Condorcet winner with respect to the committee $\mathcal{A}$ and $\overline{x} \in V(\mathbf{P})$ for $\mathbf{P} \in \mathcal{P}^N$.

If we suppose $\overline{x} \notin V'(\mathbf{P})$ there exists $\overline{F} \in$ if and only if such that $N(\mathbf{P}, \overline{F}) \in \mathcal{A}$ and $\overline{x} \notin \overline{F}$.

If we consider an element $y \in \overline{F}$ and such that

$$x \in F \text{ if and only if } y \in F \text{ for every } F \in \mathcal{A}, F \neq \overline{F}$$

we get that yS_ix for every $i \in N(\mathbf{P}, \overline{F}) \in \mathcal{A}$. Then since $A \in \mathcal{A}$ if and only if $A^c \notin \mathcal{A}$ we have that $\{i \in N : xP_iy\} \notin \mathcal{A}$ that is a contradiction.

6 Concluding Remarks

We have characterized aggregation of compatible preferences on bounded and distributive lattices that are monotone with respect to properties. We have shown that monotonicity with respect to properties is equivalent to strategy-proofness and we have considered also the case of symmetric aggregation functions and a class of function that generalizes the well known Condorcet rule.

We plan to consider aggregation of preferences in more general domains, and to find more applications of our results in future work.

References

1. Barberá, S., Sonnenschein, H., Zhou, L.: Voting by committees. Econometrica **59**, 595–609 (1991)
2. Buechel, B.: Condorcet winners on median spaces. Soc. Choice Welf. **42**, 735–750 (2014)
3. Cardin, M.: Benchmarking over Distributive Lattices. Communications in Computer and Information Science, vol. 610, pp. 117–125. Springer (2016)
4. Cardin, M.: Aggregation over property-based preference domains. In: Torra, V., Mesiar, R., Baets, B. (eds.) Aggregation Functions in Theory and in Practice, AGOP 2017. Advances in Intelligent Systems and Computing, vol. 581, pp. 400–407. Springer (2018)
5. Cardin, M.: Sugeno integral on property-based preference domains. In: Kacprzyk, J., Szmidt, E., Zadrożny, S., Atanassov, K., Krawczak, M. (eds.) Advances in Fuzzy Logic and Technology 2017, EUSFLAT 2017. Advances in Intelligent Systems and Computing, vol. 641, pp. 400–407 (2017)
6. Gordon, S.: Unanimity in attribute-based preference domains. Soc. Choice Welf. **44**, 13–29 (2015)
7. Leclerc, B., Monjardet, B.: Aggregation and Residuation. Order **30**, 261–268 (2013)
8. Monjardet, B.: Arrowian characterization of latticial federation consensus functions. Math. Soc. Sci. **20**, 51–71 (1990)
9. Morandi, P.: Dualities in Lattice Theory, Mathematical Notes. http://sierra.nmsu.edu/morandi/
10. Nehring, K., Puppe, C.: The structure of strategy-proof social choice - part I: general characterization and possibility results on median spaces. J. Econ. Theory **135**(1), 269–305 (2007)

11. Nehring, K., Puppe, C.: Abstract Arrowian aggregation. J. Econ. Theory **145**, 467–494 (2010)
12. Savaglio, E., Vannucci, S.: Strategy-proofness and single peakedness in bounded distributive lattices. Soc. Choice Welf. **52**(2), 295–327 (2019)
13. van de Vel, M.L.J.: Theory of Convex Structures. North-Holland Mathematical Library, vol. 50. Elsevier, Amsterdam (1993)
14. Vannucci, S.: Weakly unimodal domains, anti-exchange properties, and coalitional strategy-proofness of aggregation rules. Math. Soc. Sci. **84**, 50–67 (2016)

Author Index

R. Halaš et al. (Eds.): AGOP 2019, AISC 981, pp. 347–348, 2019.
https://doi.org/10.1007/978-3-030-19494-9